Grammars and Automata for String Processing

From Mathematics and Computer Science to Biology, and Back

TOPICS IN COMPUTER MATHEMATICS

A series edited by David J. Evans, Loughborough University, UK

Grammars and Automata for String Processing

From Mathematics and Computer Science to Biology, and Back

Essays in Honour of Gheorghe Păun

Edited by

Carlos Martín-Vide
Rovira i Virgili University, Tarragona, Spain

and

Victor Mitrana
University of Bucharest, Romania

Taylor & Francis
Taylor & Francis Group

LONDON AND NEW YORK

First published 2003 by Taylor & Francis
11 New Fetter Lane, London EC4P 4EE

Simultaneously published in the USA and Canada
by Taylor & Francis Inc,
29 West 35th Street, New York, NY 10001

Taylor & Francis is an imprint of the Taylor & Francis Group

Printer's note:
This book was prepared from camera-ready copy supplied by the authors.
Printed and bound in Great Britain by TJ International Ltd, Padstow, Cornwall.

Every effort has been made to ensure that the advice and information in this book is true
and accurate at the time of going to press. However, neither the publisher nor the authors
can accept any legal responsibility or liability for any errors or omissions that may be
made. In the case of drug administration, any medical procedure or the use of technical
equipment mentioned within this book, you are strongly advised to consult the
manufacturer's guidelines.

British Library Cataloguing in Publication Data
A catalogue record for this book is available from the British Library

Library of Congress Cataloging in Publication Data
A catalog record for this book is available from the Library of Congress

ISBN 0-415-29885-7

CONTENTS

II. AUTOMATA

III. LOGICS, LANGUAGES AND COMBINATORICS

Contents

CONTRIBUTING AUTHORS

Agapie, Alexandru
Alfonseca, Manuel
Beek, Maurice H. ter
Boon-van der Nat, Marloes
Bordihn, Henning
Braitenberg, Valentino
Buszkowski, Wojciech
Buzoianu, Manuela
Căzănescu, Virgil E.
Chiurtu, Ruxandra
Costagliola, Gennaro
Coven, Ethan M.
Crespi Reghizzi, Stefano
Currie, James D.
Deufemia, Vincenzo
Dima, Cătălin
Ewert, Sigrid
Fernau, Henning
Ferrucci, Filomena
Freund, Franziska
Freund, Rudolf
García, Pedro
Giuclea, Marius
Gravino, Carmine
Harbusch, Karin
Hirose, Sadaki
Holcombe, Mike
Holzer, Markus
Honkala, Juha
Ilie, Lucian
Imreh, Balázs
Ipate, Florentin
Ito, Masami
Jiménez-López, M. Dolores
Jirásková, Galina

Jonoska, Nataša
Kappes, Martin
Kolář, Dušan
Král, Jaroslav
Kudlek, Manfred
Martín-Vide, Carlos
Maurel, Denis
Meduna, Alexander
Mereghetti, Carlo
Mitrana, Victor
Morita, Kenichi
Niemann, Gundula
Niemi, Valtteri
Okawa, Satoshi
Oltean, Florin
Ortega, Alfonso
Otto, Friedrich
Petre, Ion
Pighizzini, Giovanni
Polkowski, Lech
Rahonis, George
Rodríguez-Patón, Alfonso
Rozenberg, Grzegorz
Ruiz, José
Salurso, Marianna
Soufi, Loutfi
Steinby, Magnus
Suárez, Alberto
Trakhtman, Avraam N.
Vasilco, Roxana
Vaszil, György
Walt, Andries van der
Watson, Bruce W.
Žemlička, Michal

PREFACE

The present book contains a collection of articles, clustered in sections, that are directly or indirectly related to areas where Gheorghe Păun has made major contributions.

The volume opens with an essay discussing a conjecture about windmill movements formulated by Gheorghe during one of his visits to Leiden, The Netherlands. This is a good example, among many others, of his curiosity for all kinds of games and problems, not just mathematical ones.

The first section, *Grammars and Grammar Systems*, includes a number of papers related to an important concept of the theory of formal languages: that of grammar. Some results in "classical" areas of grammar theory are presented: computational power of grammar systems with Lindenmayer components – Gheorghe Păun is one of the inventors of that framework – as well as a few practical applications of the theory (a grammar system approach to the study of natural language, and how an eco-grammar system might model George Orwell's 'Animal Farm'), new variants of contextual grammars, descriptional complexity (a survey of all the results regarding the number of nonterminals in multi-parallel grammars), parsability approaches for contextual grammars and context-free grammars with left recursive symbols, the power and limitation of regulated rewriting in image generation.

The other classical part of formal language theory, that of *Automata*, is the subject of the second section. Several types of automata (cellular automata, directable automata, X-machines, finite automata, real-time automata) are investigated in search of new theoretical properties and some potential applications in software engineering, linguistics and ecology.

A large number of contributions are grouped in the third section, *Logics, Languages and Combinatorics*. Homomorphical characterizations of the language classes in the Chomsky hierarchy, languages for picture descriptions, results regarding semilinear power series and D0L power series, unary languages, relationships between different classes of languages and the languages associated with rewriting systems are presented. Other contributions consider logical aspects: for instance, a simple logic element, called "rotary memory", is shown to be reversible and logically universal, while some topological structures of spatial reasoning are considered in the framework of rough mereology.

The last section, *Models of Molecular Computing*, is dedicated to a very hot and exciting topic in computer science: computing with molecules. Both experiments and theoretical models are presented. Amongst other things, it is shown how a mathematical model is able to cope with metabolic reactions in bacteria, and also some relationships between the entropies of DNA based computing models (Adleman's model, splicing systems) and tiling shifts. Furthermore, operations inspired by gene recombination and DNA strand assembly are considered as formal operations on strings and languages.

All the papers are contributed by Gheorghe Păun's collaborators, colleagues, friends and students from five continents, who wanted to show their recognition to him for his tremendous intellectual work on the occasion of his 50th birthday in this way. We have collected 40 papers by 69 authors here. (Another set of 38 papers by 75 authors was recently published: C. Martín-Vide & V. Mitrana, eds. (2001), *Where Mathematics, Computer Science, Linguistics and Biology Meet*. Kluwer, Dordrecht.) The subtitle of the present volume intends to reflect the sequence of Gheorghe Păun's scientific interests over a period of time.

Summing up, this book makes an interdisciplinary journey from classical formal grammars and automata topics, which still constitute the core of mathematical linguistics, to some of their most recent applications, particularly in the field of molecular biological processing.

The editors would like to emphasize Loli Jiménez's technical help in the preparation of the volume, Gemma Bel's contribution to it in its final stage, as well as the publisher's warm receptiveness to the proposal from the beginning. We hope this book will be a further step in the reviving of formal language theory as a highly interdisciplinary field, and will be understood as an act of scientific justice and gratitude towards Gheorghe Păun.

Tarragona, November 2000

Carlos Martín-Vide
Victor Mitrana

Gheorghe Păun and the Windmill Curiosity

Marloes Boon-van der Nat

Leiden Institute of Advanced Computer Science (LIACS)
Leiden University
The Netherlands
marloes@liacs.nl

Grzegorz Rozenberg

Leiden Institute of Advanced Computer Science (LIACS)
Leiden University
The Netherlands
and
Department of Computer Science
University of Colorado at Boulder
U.S.A.
rozenber@liacs.nl

Gheorghe is a frequent visitor in Leiden - by now we are really very good friends, and we know him pretty well. The most characteristic feature of Gheorghe is his enormous (and very contagious) enthusiasm for research which comes from his natural curiosity for everything around. This curiosity is best illustrated by the following.

We have had certainly hundreds of visitors to our institute in the past, but none of them has ever asked any question about the most symbolic feature of the Dutch landscape: windmills. One day, after a standard ride from Bilthoven to Leiden (passing several windmills on the way), Gheorghe commented on something unusual about the windmills: their wings always turn *counterclockwise*. He immediately wanted an explanation (because once he gets an explanation of anything, he can formulate a theory of it!).

This has turned out to be not an easy question to answer. We have passed it to many of our colleagues in Holland, but *nobody* knew the answer. In this way, an interesting question by Gheorghe has led us to some interesting research. Through this, our knowledge of something that is as Dutch as possible, has increased considerably. We have understood something very Dutch (something that, in the first place we should have already known) only because Gheorghe has asked a question. It is this posing of questions and searching for answers that makes Gheorghe such a good scientist.

We feel that explaining to Gheorghe possible reasons for the counterclockwise turning of the wings of Dutch windmills will be very much appreciated by him (we are just curious how many interesting questions Gheorghe has posed during the 50 years of his life).

First of all, in order to turn this question into a truly scientific problem, we gave it a name: the *windmill wings invariant problem*, abbreviated as the *WWI problem*. It has turned out to be a genuinely interdisciplinary problem, and our solution involves a combination of historical, ergonomic, and engineering arguments.

From the *historical* point of view, one has to realize that before the windmill was invented, the hand-operated stone-grinding mill was used. This is shown schematically in Fig.1. Now comes the very important *ergonomic* argument: about 90 % of people are right-handed (this must have also been true at the time that hand operated stone-grinding mills were used), and the natural direction for the right hand to turn the handle during long periods of grinding is anticlockwise, as indicated in Fig. 1.

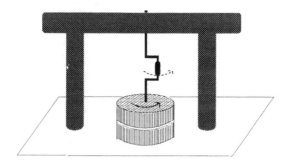

Figure 1: A handmill.

Since turning the grinding stones in this way for long periods of time must have been really exhausting, people sought help from the forces of nature. A scheme for converting the force of wind to turn the grinding stones, illustrated in Fig. 2, was thus invented – the *engineering instinct* of mankind had manifested itself once again! Because of the original (hand-operated) scheme, it was natural to use the new scheme in such a way that the grinding stone still turned in the same direction, hence anticlockwise. This meant that the vertical "gear" in the "gear transmission" would turn clockwise if observed from the "inside" (observer a in Fig. 2). Consequently, the same gear observed from the outside (observer b in Fig. 2) turns anticlockwise, as will the wings when observed from outside (observer c in Fig. 2). Thus, here we are (if we consider the construction from Fig. 2 as the precursor of the current windmills):

"THE WINGS OF THE WINDMILL TURN ANTICLOCKWISE"

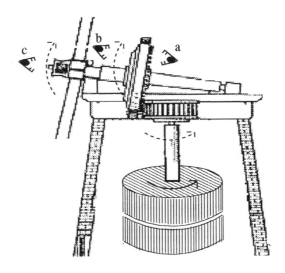

Figure 2: From handmill to windmill.

Soon afterwards, someone must have observed that the force used to turn the grinding stone can be used to turn *several* grinding stones (it is important to note here that efficiency arguments were known long before computer science became interested in the efficiency of algorithms!). By using just one more gear wheel, one could now power several grinding stones, each of which having its own ("small") gear wheel powered by (turned by) the additional ("big") gear wheel. In this way we get the standard construction of windmills illustrated in Fig. 3. Note that now, although nothing has changed from the outside (the wings still turn anticlockwise), the situation inside the mill has changed quite dramatically ... each individual grinding stone now turns in the clockwise direction! Hence the main trace of the original motivation has been wiped out! This made our research even more challenging.

We are sure that Gheorghe, after reading the above solution to the WWI problem, will right away formulate a language theoretic model of the situation. But then ... this is the effect that we have expected from our paper

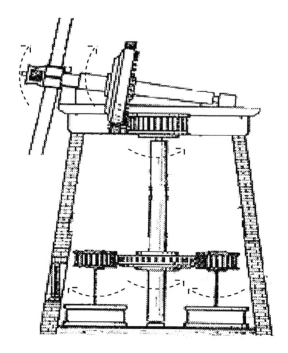

Figure 3: The inside of a modern windmill.

anyhow, and in this way our contribution becomes a gift for Gheorghe of the sort that he likes the most.

Let us then conclude this article with (anticlockwise!) birthday wishes from the two of us and the rest of the Leiden group:

P.S. George: please notice that two *membranes* suffice to express the *best* birthday wishes from *all* of us. But to this aim one has to use *words* rather than arbitrary objects.

Acknowledgements

The authors are indebted to Maurice ter Beek, our local graphics wizzard, for helping with the illustrations, and to "The Dutch Windmill" Society for providing a lot of valuable information.

I

GRAMMARS AND GRAMMAR SYSTEMS

Animal Farm: An Eco-Grammar System

Maurice H. ter Beek

Leiden Institute for Advanced Computer Science
Leiden University
The Netherlands
`mtbeek@liacs.nl`

Abstract. An eco-grammar system is used to model George Orwell's
Animal Farm: A Fairy Story.

1 Introduction

Eco-grammar systems were originally introduced in [3] as a framework motivated by Artificial Life (cf. [8]) and able to model life-like interactions. Subsequently, many variants were introduced and studied (for an extensive survey cf. [2]), mostly with a strong focus on their generative power. The articles in [15] give a nice overview.

This paper provides a glimpse of the modelling power of eco-grammar systems through a rather enhanced eco-grammar system that models George Orwell's acclaimed *Animal Farm: A Fairy Story.* I decided to write the paper for many reasons.

To begin with, I hope to inspire those working on eco-grammar systems to pursue further research with a perspective other than generative power in mind. A return to the original motivation of eco-grammar systems calls for them to be used to model issues stemming from Artificial Life. Even though I merely provide a humorous example in the style of Jürgen Dassow's eco-grammar system modelling MIT's Herbert as a can collecting robot ([6]); it is my belief that eco-grammar systems can also model more scientifically challenging issues from Artificial Life. Valeria Mihalache set a good example in this direction in [10] by using eco-grammar systems to simulate games.

This article also hints that eco-grammar systems can be used to generate stories. One of the postulates of the multidisciplinary research field Narratology — the study of narrative structure (cf. [1]) — is that stories within the same literary genre follow a common pattern. In [16], Vladimir Propp interpreted a hundred fairy tales in terms of their smallest narrative units, so-called "narratemes", and found that they all displayed the same narrative structure. This led to the field Semiotic Narratology, at the crossroads of Narratology and Semiotics — the science of signs (cf. [11]). This field focuses heavily on minimal narrative units which constitute the so-called "grammar of the plot" or "story grammars".

Closer to home, and more recently, Solomon Marcus and others associated formal languages to many Romanian fairy tales (cf. [9]). Hierarchies known from Formal Language Theory consequently enabled certain fairy tales to be classified as more sophisticated than others (cf. [7]). The above considerations show that it would be interesting to model more stories by eco-grammar systems and to search for structural equivalences between them. The same naturally holds for games.

Finally, I celebrate Gheorghe Păun's 50th birthday by bringing together two of his "hobbies". For it is Gheorghe who is the (co-)author of many of the papers on eco-grammar systems — including the one that introduced the framework — and it is Gheorghe who wrote a sequel ([14]) to George Orwell's other classic novel: *Nineteen Eighty-Four* ([13]).

2 Eco-Grammar Systems

I assume the reader to be familiar with Formal Language Theory (otherwise cf. [17]) — in particular with eco-grammar systems (otherwise cf. [2]) — and to have read George Orwell's *Animal Farm: A Fairy Story* (otherwise read [12]).

Since the specific eco-grammar system used here is based on variants that are well known from the literature – e.g. simple eco-grammar systems ([5]) and reproductive eco-grammar systems ([4]) –, the definition is given with little intuitive explanation.

An *eco-grammar system* (of degree n, $n \geq 0$) is a construct $\Sigma = (E, \mathcal{A}_1, \mathcal{A}_2, \ldots, \mathcal{A}_n)$, where:

- $E = (V_E, P_E)$, where:

  - V_E is a finite alphabet, the *environmental alphabet*, and
  - P_E is a finite and complete set of P0L rewriting rules of the form $x \rightarrow y$ with $x \in V_E$ and $y \in V_E^+$, the *environmental evolution rules*,

and for $1 \leq i \leq n$:

- $\mathcal{A}_i$ is a multiset $\langle A_{i_1}, A_{i_2}, \ldots, A_{i_{k_i}} \rangle$ of *animals* $A_{i_j} = (V_{i_j} \cup \{\sqcup, \dagger\}, P_{i_j}, R_{i_j})$ *of the i-th type*, $1 \leq j \leq k_i$, *where*:

  - V_{i_j} are finite alphabets, the *alphabets of the animals of the i-th type*, $\sqcup \notin V_{i_j}$ is the *reproduction symbol*, and $\dagger \notin V_{i_j}$ is the *death symbol*,

  - P_{i_j} are finite and complete sets of P0L rewriting rules of the form $x \rightarrow y$ with $x \in V_{i_j} \cup \{\dagger\}$ and $y \in V_{i_j} \cup \{\dagger\}$, united with pure context-free productions of the form $\alpha \rightarrow \beta_1 \sqcup \beta_2 \sqcup \cdots \sqcup \beta_p$ with $\alpha, \beta_f \in V_{i_j}$, $p \geq 2$, and $1 \leq f \leq p$, the *evolution rules of the animals of the i-th type*, and

  - R_{i_j} is a finite set of pure context-sensitive productions of the form $\alpha \rightarrow \beta$ for $\alpha, \beta \in V_E^+$, the *action rules of the animals of the i-th type*.

A *state* of Σ is a construct $\sigma = (w_E, W_1, W_2, \ldots, W_n)$, where $w_E \in V_E^+$ and each W_i, $1 \leq i \leq n$, is a multiset $\langle w_{i_1}, w_{i_2}, \ldots, w_{i_{k_i}} \rangle$ of symbols $w_{i_j} \in V_{i_j} \cup \{\dagger\}$, $1 \leq j \leq k_i$. This w_E is the *environmental state* and these w_{i_1} to $w_{i_{k_i}}$ are the *states of the currently existing animals of the i-th type*.

The state of Σ changes when the state of every animal and the environment evolves at each position, except for those positions where animals perform actions. Note that the application of a production containing the reproduction symbol results in an increase in the number of symbols in the multisets of animals.

A state $\sigma = (w_E, W_1, W_2, \ldots, W_n)$ of Σ derives a state $\sigma' = (w'_E, W'_1, W'_2, \ldots, W'_n)$ in one step, written as $\sigma \Longrightarrow_\Sigma \sigma'$, iff:

- $w_E = x_1 \alpha_{r_1} x_2 \alpha_{r_2} \cdots x_s \alpha_{r_s} x_{s+1}$ and $w'_E = y_1 \beta_{r_1} y_2 \beta_{r_2} \cdots y_s \beta_{r_s} y_{s+1}$, where:

  - $\{r_1, r_2, \ldots, r_s\} \subseteq \{i_j \mid 1 \leq j \leq k_i, \ 1 \leq i \leq n\}$,
  - $\alpha_{r_l} \rightarrow \beta_{r_l} \in R_{r_l}$, for $1 \leq l \leq s$, and
  - $y_1 y_2 \cdots y_s y_{s+1}$ is the result of applying rules from P_E to $x_1 x_2 \cdots x_s x_{s+1}$,

and for $1 \leq i \leq n$:

- W'_i is the multiset obtained from $W_i = \langle w_{i_1}, w_{i_2}, \ldots, w_{i_{k_i}} \rangle$ by putting in W'_i for each w_{i_j}, $1 \leq j \leq k_i$, either:

  - w'_{i_j} if $w_{i_j} \rightarrow w'_{i_j} \in P_{i_j}$ with $w'_{i_j} \in V_{i_j} \cup \{\dagger\}$, or

- $w_{i_j}^{[1]}, w_{i_j}^{[2]}, \ldots,$ and $w_{i_j}^{[p]}$ if $w_{i_j} \to w_{i_j}^{[1]} \sqcup w_{i_j}^{[2]} \sqcup \cdots \sqcup w_{i_j}^{[p]} \in P_{i_j}$ with $w_{i_j}^{[f]} \in V_{i_j}$, for $1 \le f \le p$.

Given Σ and an initial state σ_0, the set of state sequences of Σ is defined by $Seq(\Sigma, \sigma_0) = \{\{\sigma_i\}_{i=0}^\infty \mid \sigma_0 \Longrightarrow_\Sigma \sigma_1 \Longrightarrow_\Sigma \sigma_2 \Longrightarrow_\Sigma \cdots\}$. This set thus contains the *evolution stages* (a.k.a. the *developmental behaviour*) of both the environment and the animals.

3   Animal Farm

In this section, I present an eco-grammar system A modelling *Animal Farm: A Fairy Story*.

The environment E of A consists of a house — originally named **Manor** Farm — and plenty of **corn**, **hay**, and **straw**. Hence $V_E = \{M, c, h, s\}$ can serve as the environmental alphabet. Naturally the quantity of corn, hay and straw continuously grows. The evolution of the environment can thus be captured well by the environmental evolution rules $P_E = \{c \to c, c \to cc, h \to h, h \to hh, s \to s, s \to ss\}$.

Next I describe the animals. They all have essentially the same basic structure, i.e. the same alphabet, the same evolution rules, and the same action rules. However, some of the leading animals — i.e. animals with a name — undertake more actions throughout the book and thus have more action rules.

Let L be an animal. This animal is either alive on the farm (modelled by L), alive outside the farm (modelled by ?), or dead (modelled by †). The alphabet of L is thus $V_L = \{L, ?\}$, and its mortality and mobility are guaranteed by its evolution rules $P_L = \{L \to L, L \to ?, L \to †, ? \to ?, ? \to L, ? \to †, † \to †\}$. In the course of the book, all animals participate in harvesting corn and hay, and they use straw. I thus choose the actions of L to be $R_L = \{ccc \to c, hhh \to h, ss \to s\}$. Hence $L = (V_L \cup \{\sqcup, †\}, P_L, R_L)$. Consequently I build the multisets of animals of A by replacing L and L by symbols modelling the animals of the book.

Consider, for example, the most featured animals in the book: pigs. The vast majority of them are not leading animals and the multiset Pig thus contains quite a number of animals $P = (V_P \cup \{\sqcup, †\}, P_P, R_P)$, where $V_P = \{P, ?\}$, $P_P = \{P \to P, P \to ?, P \to †, ? \to ?, ? \to P, ? \to †, † \to †\}$, and $R_P = \{ccc \to c, hhh \to h, ss \to s\}$. Moreover the pigs Minimus, Pinkeye, and Squealer are also well described by Minimus = Pinkeye = Squealer = $(V_P \cup \{\sqcup, †\}, P_P, R_P)$. However, the main leading animals — the pigs Snowball and Napoleon — are not. For it is Snowball who, in Chapter 2, paints out Manor Farm and in its place paints Animal Farm, and it is Napoleon who makes it Manor Farm again in Chapter 10. Hence, they are well described by Snowball $= (V_P \cup \{\sqcup, \dagger\}, P_P, R_P \cup \{M \to A\})$ and Napoleon $= (V_P \cup \{\sqcup, \dagger\}, P_P, R_P \cup \{A \to M\})$.

To keep A as simple as possible I do not model birth even though I model death. The only exception is the birth of nine puppies between the dogs Bluebell and Jessie in Chapter 3, as this is an important event in the book. Therefore, the animals Bluebell and Jessie are added to the multiset Dog. Compared to other dogs D, they have an augmented set of evolution rules, viz. Bluebell $= (V_D \cup \{\sqcup, \dagger\}, P_D \cup \{D \to D \sqcup D \sqcup D \sqcup D\}, R_D)$ and Jessie $= (V_D \cup \{\sqcup, \dagger\}, P_D \cup \{D \to D \sqcup D \sqcup D \sqcup D \sqcup D\}, R_D)$.

Following the basic structure of animals L sketched above, the nearly complete list of animals featuring in *Animal Farm: A Fairy Story* — the capitalized letter in the sort name indicates the symbol that replaces L and L, and the name of the leading animals of that sort are added between brackets — becomes Chicken, Dog (Bluebell, Jessie, Pincher), goosE, Goat (Muriel), Horse (Mollie, Boxer, Clover), Man (Mr. Jones, Mrs. Jones, Mr. Frederick, Mr. Whymper, Mr. Pilkington), Pig (Old Major, Snowball, Napoleon, Minimus, Pinkeye, Squealer), Raven (Moses), Sheep, coW, and donkeY (Benjamin). For reasons of space, I only mention the generic names of the animals, i.e. I do not use the specific names for the female and for the male.

Finally, note that Bluebell, Jessie, and Pincher are the only dogs, Muriel is the only goat, Moses is the only raven, and Benjamin is the only donkey. Then the eco-grammar system A modelling *Animal Farm: A Fairy Story* is:

$$A = ((V_E, P_E), \text{Chicken}, \text{Dog}, \text{goosE}, \langle \text{Muriel} \rangle, \text{Horse}, \text{Man}, \text{Pig}, \langle \text{Moses} \rangle,$$
$$\text{Sheep}, \text{coW}, \langle \text{Benjamin} \rangle),$$

where:

Chicken $= \langle C, C, \ldots, C \rangle$,
Dog $= \langle \text{Bluebell}, \text{Jessie}, \text{Pincher} \rangle$,
goosE $= \langle E, E, \ldots, E \rangle$,
Horse $= \langle \text{Mollie}, \text{Boxer}, \text{Clover}, H, H, \ldots, H \rangle$,
Man $= \langle \text{Mr. Jones}, \text{Mrs. Jones}, \text{Mr. Frederick}, \text{Mr. Whymper},$
Mr. Pilkington$, M, \ldots, M, M \rangle$,
Pig $= \langle \text{Old Major}, \text{Snowball}, \text{Napoleon}, \text{Minimus}, \text{Pinkeye},$
Squealer$, P, P, \ldots, P \rangle$,
Sheep $= \langle S, S, \ldots, S \rangle$, and
coW $= \langle W, W, \ldots, W \rangle$.

4   A Fairy Story

In this section I show how A can generate the fairy story of the book.

In the beginning there is Manor Farm and plenty — where plenty is modelled by the presence of 9 symbols — of corn, hay, and straw. Of the humans, only Mr. Jones, Mrs. Jones and their four men are present. Hence:

$$\sigma_0 = (Mc^9h^9s^9, \langle C, \ldots, C \rangle, \langle D, D, D \rangle, \langle E, \ldots, E \rangle, \langle G \rangle,$$
$$\langle H, \ldots, H \rangle, \langle M, M, M, M, M, M, ?, \ldots, ? \rangle, \langle P, \ldots, P \rangle, \langle R \rangle,$$
$$\langle S, \ldots, S \rangle, \langle W, \ldots, W \rangle, \langle Y \rangle).$$

Next, I summarize Chapter 1 to 10 of the book and for some chapters I display the state modelling the situation after that chapter. I leave it to the reader to display the other states and to spell out precisely which rules must be applied to obtain them.

In Chapter 1 — as in all chapters — the corn, hay, and straw naturally grow a little, while at the same time some straw is used. Hence σ_1 is the same as σ_0 except that the environmental state has become $Mc^{13}h^{13}s^9$. From now on I will no longer mention the growth of corn, hay, and straw, nor the decrease in straw.

In Chapter 2, Old Major dies. The rebellion then causes Mr. Jones, Mrs. Jones, and their four men to flee from the farm, and causes Snowball to change the name of the farm to Animal Farm. Moreover, a hay harvest takes place.

In Chapter 3, Bluebell and Jessie whelp and there is a corn harvest. Hence:

$$\sigma_3 = (Ac^6h^{10}s^9, \langle C, \ldots, C \rangle, \langle D, D, D, D, D, D, D, D, D, D, D, D \rangle,$$
$$\langle E, \ldots, E \rangle, \langle G \rangle, \langle H, \ldots, H \rangle, \langle ?, \ldots, ? \rangle, \langle \dagger, P, \ldots, P \rangle, \langle R \rangle, \langle S, \ldots, S \rangle,$$
$$\langle W, \ldots, W \rangle, \langle Y \rangle).$$

In Chapter 4, Mr. Jones and his four men return to the farm only to be expelled again by the animals — at the cost of only one sheep — during the Battle of the Cowshed. Around this time, Moses flies off.

In Chapter 5, Mollie disappears, and at the height of Animalism Snowball is chased off the farm by Bluebell's and Jessie's nine puppies, which have grown tremendously under Napoleon's control. Hence:

$$\sigma_5 = (Ac^{20}h^{20}s^9, \langle C, \ldots, C \rangle, \langle D, D, D, D, D, D, D, D, D, D, D, D \rangle,$$
$$\langle E, \ldots, E \rangle, \langle G \rangle, \langle ?, H, \ldots, H \rangle, \langle ?, \ldots, ? \rangle, \langle \dagger, ?, P, \ldots, P \rangle, \langle ? \rangle,$$
$$\langle S, \ldots, S, \dagger \rangle, \langle W, \ldots, W \rangle, \langle Y \rangle).$$

In Chapter 6, nothing much happens.

In Chapter 7, nine hens die after Comrade Napoleon orders them to starve themselves. He also orders his dogs to kill the four pigs, three hens, one goose, and three sheep that confess to have rebelled against him. Hence:

$$\sigma_7 = (Ac^{30}h^{28}s^9, \langle \dagger, \dagger, \dagger, C, \ldots, C \rangle, \langle D, D, D, D, D, D, D, D, D, D, D, D \rangle,$$
$$\langle E, \ldots, E, \dagger \rangle, \langle G \rangle, \langle ?, H, \ldots, H \rangle, \langle ?, \ldots, ? \rangle, \langle \dagger, ?, P, \ldots, P, \dagger, \dagger, \dagger, \dagger \rangle,$$
$$\langle ? \rangle, \langle S, \ldots, S, \dagger, \dagger, \dagger, \dagger \rangle, \langle W, \ldots, W \rangle, \langle Y \rangle).$$

In Chapter 8, there are more harvests and another battle — at the cost of one cow, three sheep, and two geese this time — after Mr. Frederick and his men attack the farm.

In Chapter 9, Moses reappears and Boxer dies. Hence:

$$\sigma_9 = (Ac^{10}h^{11}s^9, \langle\dagger, \dagger, \dagger, C, \ldots, C\rangle, \langle D, D, D, D, D, D, D, D, D, D, D, D\rangle,$$
$$\langle E, \ldots, E, \dagger, \dagger, \dagger\rangle, \langle G\rangle, \langle?, \dagger, H, \ldots, H\rangle, \langle?, \ldots, ?\rangle,$$
$$\langle\dagger, ?, P, \ldots, P, \dagger, \dagger, \dagger, \dagger\rangle, \langle R\rangle, \langle S, \ldots, S, \dagger, \ldots, \dagger\rangle, \langle W, \ldots, W, \dagger\rangle, \langle Y\rangle).$$

In Chapter 10, Bluebell, Jessie, Pincher, Muriel, three horses, and Mr. Jones die. Furthermore, Mr. Pilkington is now an appreciated neighbour, in whose presence Napoleon changes the name of the farm back to Manor Farm. Hence:

$$\sigma_{10} = (Mc^{20}h^{22}s^9, \langle\dagger, \dagger, \dagger, C, \ldots, C\rangle, \langle\dagger, \dagger, \dagger, D, D, D, D, D, D, D, D, D\rangle,$$
$$\langle E, \ldots, E, \dagger, \dagger, \dagger\rangle, \langle\dagger\rangle, \langle?, \dagger, H, \ldots, H, \dagger, \dagger, \dagger\rangle, \langle\dagger, M, ?, \ldots, ?\rangle,$$
$$\langle\dagger, ?, P, \ldots, P, \dagger, \dagger, \dagger, \dagger\rangle, \langle R\rangle, \langle S, \ldots, S, \dagger, \ldots, \dagger\rangle, \langle W, \ldots, W, \dagger\rangle, \langle Y\rangle).$$

The story of the book naturally is only one of the possible stories that A can generate. I leave it to the reader to play with A and to enjoy other outcomes of this fairy story.

Acknowledgements

I wish to thank Erzsébet Csuhaj-Varjú, Judit Csima, Nikè van Vugt, and Nadia Pisanti for useful comments and suggestions on a preliminary version of this paper.

References

[1] M. Bal, *Narratology: Introduction to the Theory of Narrative.* Toronto University Press, Toronto, 1985.

[2] E. Csuhaj-Varjú, Eco-grammar systems: recent results and perspectives. In Gh. Păun (ed.), *Artificial Life: Grammatical Models.* Black Sea University Press, Bucharest, 1995, 79–103.

[3] E. Csuhaj-Varjú, J. Kelemen, A. Kelemenová and Gh. Păun, Eco(grammar)systems: a generative model of artificial life, 1993, manuscript.

[4] E. Csuhaj-Varjú, J. Kelemen, A. Kelemenová and Gh. Păun, Eco(grammar)systems: a preview. In R. Trappl (ed.), *Cybernetics and Systems'94.* World Scientific, Singapore, 1994, vol. 1, 941–948.

[5] E. Csuhaj-Varjú, J. Kelemen, A. Kelemenová and Gh. Păun, Eco-grammar systems: a grammatical framework for studying life-like interactions, *Artificial Life*, 3.1 (1997), 1–28.

[6] J. Dassow, An example of an eco-grammar system: a can collecting robot. In Gh. Păun (ed.), *Artificial Life: Grammatical Models*. Black Sea University Press, Bucharest, 1995, 240–244.

[7] J. Dassow and Gh. Păun, *Regulated Rewriting in Formal Language Theory*. Springer, Berlin, 1989.

[8] Ch.G. Langton (ed.), *Artificial Life: An Overview*. MIT Press, Cambridge, Mass., 1995.

[9] S. Marcus, *La sémiotique formelle du folklore*. Klincksieck, Paris, 1978.

[10] V. Mihalache, General artificial intelligence systems as eco-grammar systems. In Gh. Păun (ed.), *Artificial Life: Grammatical Models*. Black Sea University Press, Bucharest, 1995, 245–259.

[11] W. Nöth, *Handbook of Semiotics*. Indiana University Press, Bloomington, In., 1990.

[12] G. Orwell, *Animal Farm: A Fairy Story*. Martin Secker & Warburg, London, 1945.

[13] G. Orwell, *Nineteen Eighty-Four*. Martin Secker & Warburg, London, 1949.

[14] Gh. Păun, *O mie nouă sute nouăzeci şi patru*. Ecce Homo, Bucureşti, 1993. English translation published as *Nineteen Ninety-Four, or The Changeless Change*. Minerva, London, 1997.

[15] Gh. Păun (ed.), *Artificial Life: Grammatical Models*. Black Sea University Press, Bucharest, 1995.

[16] V.Ia. Propp, *Morfologiia skazki*. Academia, Leningrad, 1928. English translation published as *Morphology of the Folktale*. Indiana University and The University of Texas Press, Austin, Tx., 1968.

[17] G. Rozenberg and A. Salomaa (eds.), *Handbook of Formal Languages*. Springer, Berlin, 1997.

Towards a Brain Compatible Theory of Syntax Based on Local Testability[1]

Stefano Crespi Reghizzi

Department of Electronics and Information
Polytechnical University of Milan
Italy
crespi@elet.polimi.it

Valentino Braitenberg

Laboratory of Cognitive Science
University of Trento
Rovereto, Italy
and
Max Planck Institute for Biological Cybernetics
Tübingen, Germany
lsc@inf.unitn.it

Abstract. Chomsky's theory of syntax came after criticism of probabilistic associative models of word order in sentences. Immediate constituent structures are plausible but their description by generative grammars has met with difficulties. Type 2 (context-free) grammars account for constituent structure, but they go beyond the mathematical capacity required by language, because they generate unnatural mathematical sets as a result of being based on recursive function theory.

[1]This work was presented at the *Workshop on Interdisciplinary Approaches to a New Understanding of Cognition and Consciousness*, Villa Vigoni, Menaggio, 1997. We acknowledge the support of Forschung für anwendungsorientierte Wissenverarbeitung, Ulm and of CNR-CESTIA.

Abstract associative models investigated by formal language theoreticians (Schützenberger, McNaughton, Papert, Brzozowski, Simon) are known as locally testable models. We propose a combination of locally testable and constituent structures models under the name of Associative Language Description and we argue that this combination has the same explanatory power as type 2 grammars but is compatible with brain models. We exemplify and discuss two versions of ALD, one of which is based on modulation while the other is based on pattern rules. In conclusion, we provide an outline of brain organization in terms of cell assemblies and synfire chains.

1   Introduction

Chomsky's theory of syntax came after his criticism of probabilistic associative models of word order in sentences, but the inadequacy of probabilistic left-to-right models (Markov process) had already been noticed by Lashley [21], who anticipated Chomsky's arguments [7] by observing that probabilities between adjacent words in a sentence have little relation to the grammaticality of the string. Although associative models provide an intuitively appealing explanation of many linguistic regularities, they are also aligned with current views on information processing in the cortex. A classical argument pro syntax is that the choice of an element is determined by a much earlier element to be remembered across gaps filled by intervening clauses (constituents). Ambiguity too provides a strong indication that sentences carry a structure. The established model for immediate constituent analysis relies on context-free (CF) grammars. Their rules assign names (nonterminal symbols) to different kinds of constituents (syntax classes). Such grammars cannot handle many non-elementary aspects of language, but go beyond the mathematical capacity required by language, because they generate unnatural mathematical sets. In our opinion this is a clear indication that this model is misdirected. Related, more complex models, such as context sensitive grammars, are even more subject to the same criticism: for instance, they generate such mathematical languages as the set of strings whose length is a prime number [25], a consequence of being based on recursive function theory.

On this basis and motivated by the search for a linguistic theory that is more consistent with the findings of brain science, we propose a new model, called associative language description (ALD). This grafts the immediate constituent structure onto the old associative models. The associative theories we build upon were investigated in the 60's by mathematicians (notably Brzozowski, McNaughton, Papert, Schützenberger, Simon, Zalcstein) and are known as locally testable (LT) models. In sect. 2 we recall the LT definitions and we present the ALD models. The plural indicates that ALD is a sort of general approach that can be realised in different ways. The original ALD model exploits LT to specify both the structure of constituents and their permitted contexts; the bounded ALD model of [11] is a mathematically simpler version that has been used to prove formal properties and also in technical applications for programming languages [14]. Then we briefly compare ALD and CF models, and we show that ALD are easier to infer from examples of structured sentences. In sect. 3 we sketch a brain organization for ALD processing in terms of neural mechanisms. In the conclusion we refer to early related research and discuss possible avenues for future work

2   Associative Language Descriptions

2.1   Local Testability

Certain frequent patterns of language can be described by considering pairs of items that occur next to each other in a sentence. The precise nature of the items depends on the level of language description. The items are the phonemes at the phonological level, but lexemes or word categories at the syntactical level. In our discussion we do not specify the level of language description, since what we are proposing is an abstract model of grammar that in future development will have to be instantiated to account for various linguistic phenomena. The items will accordingly be represented as characters from a terminal alphabet $\Sigma = \{a, b, \ldots\}$.

Looking at a string such as $x = abccbcccc$ we note that x contains the following substrings of length 2: ab, bc, cc, cb. These are called *digrams* or *2-grams*. We notice that ab, the prefix of the string, is the initial digram; similarly cc is the final digram or suffix. The study of the digrams (or more generally the k-grams, $k \geq 2$) that appear in phrases has been suggested many times as a technique for characterizing to some extent grammatically valid strings. In particular, the limits of a Markovian model based on the relative frequency of k-grams occurring in the language corpus have been examined by Chomsky and Miller [8]. Checking for the presence of certain k-grams in a given string to be analysed is a simple operation for a computer, and one that could very easily be performed by cortical structures, as already noticed by Wickelgren [26], who shows that serial order can be enforced by small associative memories. The recognition algorithm needs a finite memory to be used as a sliding window that is capable of storing k characters. The window is initially positioned the left to of the string to be analyzed. Its content is checked against the set of permitted k-grams. If it matches, the window advances by one position along the string, and the same check is performed, until the window reaches the right edge of the string. If the check is positive in all window positions, the string is accepted; otherwise it is rejected. Because the sliding window essentially performs a series of local inspections, the languages thus defined have been aptly called *locally testable* (LT). As this algorithm uses a finite memory, independently of the length of the string, its discriminatory power cannot exceed that of a finite-state automaton. Not all finite-state languages are LT, but this loss of generative capacity primarily concerns certain periodic patterns, which are irrelevant for modeling human languages. One example would be a string over the alphabet a, b containing any number of a's and a number of b's that is a multiple of 3. Such a property cannot be checked by local inspections over the string. The formal properties of LT languages have been investigated by mathematicians using algebraic and automata-theoretical approaches. An early comprehensive reference is the book on non-counting languages by McNaughton and Papert [22]. Several variations to the notion of LT have been proposed by theoreticians (e.g. see [23]), but we stick to the simplest definition, since nothing is gained by considering more refined models at this early stage of our investigation.

We use a special character $\perp$, not present in the terminal alphabet, called the *terminator*, which encloses the sentences of the language.

For $k \geq 1$, a k-gram x is either a string containing exactly k characters of Σ or a possibly shorter string starting or ending by the terminator. Formally:
$x \in \Sigma^k \cup \perp(\Sigma \cup \varepsilon)^k \cup (\Sigma \cup \varepsilon)^k \perp$

Definition 1 *For a string x of length greater than k, we consider three sets: $\alpha_k(x)$ the* initial k-gram *of x; $\gamma_k(x)$ the* final k-gram *of x; and, if x is longer than $k+1$, $\beta_k(x)$, the set of* internal k-grams *(those that occur in any position other than on the left and right edges). (Notice that internal k-grams may not contain terminators.) A locally testable description (LTD) of order $k \geq 1$ consists of three sets of k-grams: A (initial), B (internal), C (final). A LTD $D = (A, B, C)$ defines a formal language, denoted $L(D)$, by the next condition. A string x is in $L(D)$ if, and only if, its initial, internal, and final k-grams are resp. included in A, B, and C. More precisely, the condition is: $x \in L$ iff, $\alpha_k(x) \in A \land \beta_k(x) \in B \land \gamma_k(x) \subseteq C$.*

As a consequence, if two strings x and y have the same sets α, β, and γ either both are valid sentences of $L(D)$ or neither one is. It is often convenient to avoid specifying the exact width of the k-grams. A language L is called *locally testable* if there exists a finite integer k such that L admits a LTD of order k.

2.2   From Local Testability to Structure Definition

It is obvious that an LTD falls short of the capacity required by natural language, as it enables only some finite-state languages to be defined. Its major weakness is its inability to define constituent structures, a necessary feature of any syntax model. A simple way to introduce constituent structures into an LTD is now proposed. A constituent has to be considered as a single item, encoded by a new terminal symbol Δ called a *place holder* (PH). By this expedient a constituent containing other constituents gives rise to k-grams containing the PH.

Next, we define the tree structures that are relevant for ALD. A tree (see Fig. 1) whose internal nodes are labeled by Δ and whose leaves are labeled by characters of Σ is called a *stencil tree*.

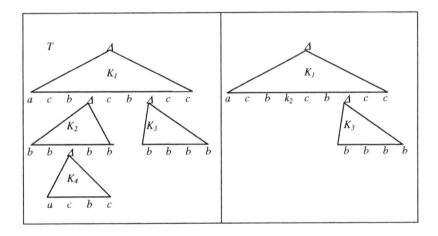

Figure 1: A stencil tree T with four constituents schematised as triangles. To the right, a condensed tree.

A tree is composed of juxtaposed subtrees of height one and leaves with labels in $\Sigma \cup \{\Delta\}$, called *constituents*. The frontier of a stencil tree T is denoted by $\tau(T)$, while $\tau(K_i)$ denotes the frontier of the constituent K_i.

Definition 2 *For an internal node i of a stencil tree T, labeled by Δ, let K_i and T_i be resp. the constituent and the maximal subtree of T having root i. Introduce a new terminal symbol k_i, and consider the 'condensed' tree $T_{K'i}$ obtained by replacing the subtree T_i in T with k_i. Consider the frontier $\tau(T_{K'i})$ of the condensed tree, which can be written as the concatenation* *of three parts:* $\tau(T_K'i) = sk_it$, *where s and t are possibly empty terminal strings. The strings $\perp s$ and $t \perp$ are called, resp., the* left/right context *of the constituent K_i (or of the subtree T_i) in T: $left(K_i, T) = left(T_i, T) = \perp s$ $right(K_i, T) = right(T_i, T) = t \perp$.*

For instance, in Fig. 1 the left context of K_3 is $acbbbacbcbbcb$ and the right context of K_1 is $\perp$: notice that terminators are automatically pre-fixed/appended.

2.3 ALD Based on Local Testability

The original idea of Associative Language Descriptions (ALD) was entirely founded on the concepts of LT. Each class of constituent is associated to an LT description, i.e. a triple $\langle A, B, C \rangle$, with the provision that k-grams contain place holders, if they contain nested constituents. For graduality, we start with a simpler model, where any constituent can freely occur anywhere an PH occurs.

Definition 3 *A* free ALD *F consists of a collection $\{D_1, D_2, \ldots, D_m\}$ of locally testable descriptions of order k, each one of the form $D_i = \langle A_i, B_i, C_i \rangle$, where each one of the three symbols represents a set of k-grams possibly containing PHs. Each triple D_i is called a* syntax class.

A free ALD $F = \{D_1, D_2, \ldots, D_m\}$ defines a formal language, $L(F)$, consisting of a set of stencil trees, by the following condition. A tree T is in $L(F)$ iff for each constituent K of T the frontier $\tau(T_K)$ belongs to the language defined by some syntax class D_i of F. Formally:

$$T \in L(F) \Longleftrightarrow \forall K \in T : (\exists D_i \in F : (\tau(T_K) \in L(D_i))).$$

Example 1 *Take the alphabet $\Sigma = \{a, +, \times\}$ of certain arithmetic expressions, where a stands for some numeric value. Sums, e.g. $a + a + a$, are defined by the LTD $\langle A = \{a+\}, B = \{a+, +a\}, C = \{+a\} \rangle$ and products are similarly defined by $\langle A = \{a\times\}, B = \{a\times, \times a\}, C = \{\times a\} \rangle$.*

Now we extend the language by allowing any variable to be replaced with any expression as constituent, thus giving rise to a tree language, which is defined by the free ALD $F = \langle D_1, D_2 \rangle$, where:
$D_1 = \langle A_1 = \{a+, \Delta+\}, B_1 = \{a+, \Delta+, +a, +\Delta\}, C_1 = \{+a, +\Delta\} \rangle$,
$D_2 = \langle A_2 = \{a\times, \Delta\times\}, B_2 = \{a\times, \Delta \times \times a, \times\Delta\}, C_2 = \{\times a, \times\Delta\} \rangle$.
For instance, the stencil tree $\underline{a + a \times a} + \underline{a} + \underline{a \times a}$, where the constituents are underscored, is valid since the triples α, β, γ exhibited by each constituent are included in D_1 or in D_2. On the other hand, the tree $a + \underline{a \times a} + a$ is not valid since the underscored constituent yields $\langle \{a\times\}, \{\times a, a+\}, \{+a\} \rangle$, not included either in D_1 or in D_2.

The license to replace any constituent by another one would cause obvious over generalisation in a grammar, say, of English. For instance, if a preposition clause (*in the garden*) occurs in a certain position (*the man in the garden*), then any other clause (e.g. the noun clause *the sharp knife*) would be permitted in the same position, causing acceptance of the ungrammatical string *the man the sharp knife*. In other words, all syntax classes of a free ALD are equivalent with respect to their context of occurrence. This insufficient discriminatory capacity is remedied by the next development. The improvement consists of adding, to each syntax class, the indication of the valid contexts of occurrence. Two different manners of so doing have been considered. In the original proposal, the contexts are specified by listing the k-grams that may occur at the left and right edge of a constituent. The second manner, introduced in [11], uses patterns to specify at once a syntax class and its permitted contexts.

Definition 4 *An ALD G with modulation[2] consists of a free ALD:*

$$F = \{D_1, D_2, \ldots, D_m\}, m \geq 1$$

with the following additions. For each LT description D_i, two non-empty sets of k-grams without PHs[3] are given: the opening (or left) passage L_i and the closing (or right) passage R_i. Therefore, G is defined by a finite collection of components:

$$G = \{E_1, E_2, \ldots, E_m\},$$

where $E_i = \langle D_i, L_i, R_i \rangle$ and $D_i = \langle A_i, B_i, C_i \rangle$. As before, each component E_i defines a syntax class.

Before presenting the acceptance condition, we need to formalise the k-grams occurring on the borders of a constituent.

Definition 5 *The left border $\lambda(K, T)$ of a constituent K of stencil tree T is the set of k-grams, with no PH, such that they occur in the frontier $\tau(T)$ of T, and they partially overlap the left edge of $\tau(K)$. Reformulating more precisely the second condition, after segmenting the string $\tau(T)$ as $\tau(T) = u\tau(K)v$, we have:*
$\lambda(K, T) = \{x | (x$ is a k-gram over $\Sigma \cup \bot) \wedge (x = x'x''$, where x' is a suffix of $\bot u) \wedge (x''$ is a prefix of $\tau(K)v\bot)\}$. The right border $\rho(T, K)$ of a constituent is symmetrically defined.

---

[2]In music, 'modulation' indicates a passage leading from one section of a piece to another.

[3]The reason for not allowing PHs is procedural. In order to parse a string, passages must be recognised, which would be inconvenient if the PH, itself the result of a parse, had to be detected in advance. This choice is, however, one of several possible variations.

As an example, from Fig. 1 we have the following borders for $k = 2$:

$$\lambda(K_2, T) = \{bb\}, \rho(K_2, T) = \{bc\}, \lambda(K_1, T) = \{\perp a\}, \rho(K_1, T) = \{c\perp\}.$$

For $k = 4$: $\lambda(K_3, T) = \{bcbb, cbbb, bbbb\}; \rho(K_3, T) = \{ bbbc, bbcc, bcc\perp\}.$

Intuitively for a stencil tree to be valid each constituent must be preannounced by the occurrence of some k-grams belonging to the opening passage of its syntax class; a symmetrical condition applies to the closing passage.

Example 2 *To illustrate the discriminative capacity of modulation, we change Ex. 2, by permitting one containment only: multiplicative expression inside additive expressions. By inspecting a few instances of stencil trees one easily comes out with the ALD:*

*$A_1 = \{a+, \Delta+\}, B_1 = \{a+, \Delta+, +a, +\Delta\}, C_1 = \{+a, +\Delta\}, L_1 = \{\perp a\},$
$R_1 = \{a\perp\}.$*

$A_2 = \{a\times\}, B_2 = \{a\times, \times a\}, C_2 = \{\times a\}, L_2 = \{a + a, +a\times, \perp a\times\}, R_2 = \{\times a+, a + a, \times a\perp\}.$

Notice that at least 3-grams are required to make modulation effective for L_2 and R_2, but for brevity 2-grams have been used in all other sets.

As a test, one can verify that the tree $a \times \underline{a + a}$ is rejected because $\times a+$ is not present in the left passage L_1. On the other hand, the tree $a + \underline{a \times a}$ is accepted since the left border $\{a + a, +a\times\}$ of the constituent is included in L_2 and its right border $\{\times a\perp\}$ is included in R_2.

2.4   Bounded ALD

Recently the original definition has been reshaped into a mathematically simpler model [11], which is more directly comparable with CF grammars, yet it does not decrease capacity. The model will here be called *bounded ALD* to distinguish it from ALD with modulation.

In a sense, bounded ALD stay to ALD with modulation in a similar relation as CF grammars stay to CF grammars with regular expressions: stencil trees of bounded ALD are bounded in degree, because recursion is used instead of iteration to produce repetitive structures. The other difference between the two models has to do with the manner in which permitted contexts are specified.

Definition 6 *A bounded ALD A consists of a finite collection of rules of the form $\langle x, y, z \rangle$, usually written $x[y]z$, where $x \in (\perp \Sigma*) \cup \Sigma^+$, $y \in (\Sigma * \perp) \cup \Sigma^+$, and $z \in (\Sigma \cup \{\Delta\})*$. For a rule, the string z is called the* pattern *and the strings x and y are called the permissible left/right contexts. If a left/right permissible context is irrelevant, it can be replaced the "don't care" symbol '_'.*

A tree T is valid *for a bounded ALD A iff for each constituent K_i of T there exists a rule $u[\tau(K_i)]v$ where u is a suffix of the left context of K_i in T and v is a prefix of the right context of K_i in T.*

The language $L(A)$ defined by A is the set of stencil trees valid for A.

Sometimes we also consider the set of strings corresponding to the frontier of a tree language. This allows us to talk of the *string language* defined by an ALD.

Example 3 *The string language $\{a^n c b^n | n \geq 1\}$ is defined by the rules: $\perp[a \Delta b] \perp, a[a \Delta b]b, a[c]b$. Both contexts can be dropped from the first two rules, and the right (or left, but not both) context can be omitted from the last rule to give the equivalent ALD: $-[a\Delta b]-, a[c]-$.*

The following remarks are rigorously justified for bounded ALD [11], but we expect them to hold for ALD with modulation too, with little change.

Ambiguity: The phenomenon of ambiguity occurs in ALD much as in CF grammars. For example, the following ALD A ambiguously defines the Dyck language over the alphabet $\{b, e\}$: $-[\Delta\Delta]-, -[b\Delta e]-, -[\varepsilon]-$, because a sentence like *bebe* has two distinct trees in $L(A)$.

Other formal properties [11]: Bounded ALD languages (both tree and string) form a strict subfamily of CF. Actually an algorithm enables a CF grammar to be constructed that is structurally equivalent to a bounded ALD. More precisely, the ALD tree languages enjoy the *Non-Counting* property of CF languages [12]. This property is believed to be a linguistic universal of all natural and artificial languages intended for human communication.

The family of ALD string languages is not closed with respect to the basic operations of concatenation, star, union, and complementation: a lack of nice mathematical properties that does not affect the potential uses of ALD for real languages, as witnessed by the successful definition of Pascal [14]. Indeed it is a common misconception that any language family is useless unless it is closed with respect to union and concatenation. Practical examples abound to the contrary, since very rarely is it possible to unite two languages without modification.

ALD models differ from Chomsky's grammars in another aspect: they are not generative models, because the notion of deriving a sentence by successive application of rules is not present. On the other hand, an ALD can be used to check the validity of a string by a parsing process. Preliminary analysis of the problem indicates that the classical parsing methods for CF grammars can be adapted to our case.

2.5   Grammar by Example

One of the appealing features of ALD is that they are suitable for *grammar inference*, the learning process that enables a language on the limit [16] to be identified from a series of positive and negative examples. We already observed that it is straightforward to construct a LTD by inspecting a given sample of strings and extracting the k-grams. If the value of k is unknown, the learner can start with $k = 2$, and gradually increase the length if the inferred LTD is too general. Over generalisation means that some strings are accepted, and these are tagged or negative by the informant. In the basic model of language learnability theory, the sentences presented to the learner are flat strings of words, but other studies have considered another model, in which sentences are presented in the form of structures (such as stencil trees in [9] or functor-argument structures in [20]). The availability of structure in the information has been defended on several grounds such as the presence of semantic tagging. In practice, learning from structures is considerably simpler than learning from strings, because the problem space is more constrained. Constructing the ALD from a sample of stencil trees is a simple, determinate task that consists of extracting and collecting the relevant sets of k-grams, much as in the LTD case, since for a given integer k the ALD which is compatible with a given sample of trees is essentially unique. In the case of programming languages such as Pascal or C, small values of k have proved sufficient, so this grammar inference approach can be applied without combinatorial explosion. As a consequence, ALD can be specified by examples, rather than by rules, because of the direct bijective relation between two sets of positive and negative stencil trees and the ALD. In synthesis, the k-gram extraction method is a good procedure for extrapolating an unbounded set of valid structures from a given sample. In contrast, for CF grammars there may exist many structurally equivalent grammars which are compatible with a sample of stencil trees, and the inference process is more complex and undetermined.

3   Mapping ALD on Brain Structures

From the point of view of neuronal modelling, associative language description has the great advantage of naturally blending into some of the most accredited theories of brain function. This is not the place to give a full account of the experimental evidence [5] on which these theories rest, and we will only sketch some of the main results. We start by recalling the basic mechanisms assumed.

Cell assemblies (CA) [18]: It seems that the things and events of our experience are represented by ensembles of neurons (= nerve cells) which are strongly connected to each other and therefore (a) tend to become active all together ('ignite') even if only some of them are activated, and (b) stay active even after the external excitation ceases.

Sequences of cell assemblies: Although the individual CA, when activated, may follow a certain temporal order in the activation of its component neurons, it is not a sufficient physiological basis for the sequential order which characterizes much of behaviour: sequences of items in some animal (bird song) and human (singing, speech) vocalisations or skilled behaviour such as occurs in crafts or musical performance. There must be control mechanisms in the brain which extinguish an active CA and ignite the following one in well defined sequences, although possibly with varying rhythm. These may be genetically determined or acquired by learning.

Synfire chains [1] [2]: There is evidence of very precisely timed sequences of neural activations in chains of neurons, or groups of neurons, which conduct neural activity but cannot arrest it or store it anywhere along the way. These so-called synfire chains are probably responsible for the timing of events within a time span of a few tenths of a second (whereas sequences of cell assemblies may have a duration of several seconds). These, too, are the result of learning processes.

We may ask some questions in connection with associative grammar. First, in what way are the sequences which define the local rules of grammar (k-grams) learned and stored. The items which occur in ordered groups of k elements are words (though in the previous definitions they are denoted by single letters), composed of one or a few syllables, and therefore with a duration of between 0.2 and 1 (or at most 2) seconds (the duration of a syllable being about 0.2 seconds). A trigram composed of such elements would span a time of several seconds, too long for synfire chains but quite in the order of magnitude of various kinds of skilled behaviour (such as sports, musical performance, etc.). As in these and other performances, the ability of the brain to learn sequences of events is evident and in some cases can even be related to some detailed neurophysiology. It is not impossible to imagine neuronal networks containing representatives of such learned sequences which are only activated when the correct sequence is presented in the input. They are stored in parallel and may be prevented from being activated more than one at a time by some mechanism of reciprocal inhibition. In analysing a sequence of input events, they may also be activated in partially overlapping temporal episodes as required by the 'sliding' window idea. Of course the number of 3-grams is enormous in language, but the number of neurons involved in language processing in the brain is also very large, perhaps of the order of 10^8 and the number of the useful combinations of these neurons is possibly greater. Moreover, in reality we do not imagine that the system requires all k-grams to have the same value k (as in the formal definition of ALD), but rather we suggest that the value will vary in different contexts, in order to minimize the memory requirements. It is conceivable that the minimal values required for k are discovered in the learning phase, assuming that the grammar by example algorithms outlined in Sect. 2 is deployed . This mechanism for matching k-grams is absolutely essential for ALD as it is needed not only for recognizing constituents, but also for detecting the k-grams passage that announces the modulation between two constituents.

We notice that some k-grams play a special role when they act as the initial (or final) k-gram of a constituent. The requirement of specially marking some k-grams is consistent with the so called 'X bar' theory of syntax [19], which affirms that each syntax class requires the presence of a specific lexeme. It is not difficult to imagine how the proposed neuronal scheme would detect such marked or compulsory k-grams.

Another question is how sequences of items are embodied in synaptic networks and how they are learned. Connections between neurons or groups of neurons are statistically symmetrical, but again in the cortical 'wiring', which is mostly stochastic, there is ample opportunity for asymmetrical connection that may embody the relation of temporal order or sequence. Moreover it is clear that synaptic relations between neurons in the cortex are to a large extent determined by learning (or experience), and it is certain that much of what is learnt in the way of succession of events (causal relations etc.) will determine the unidirectional influence of one group of neurons on another. There is a technical difficulty, however, in the physiology of learning when the sequence which is learned involves considerable delays, such as in trigrams (sequences of three words) which span several seconds. We are forced to assume that there are different delays between the input and internal representatives of trigrams. The synfire chains already mentioned may provide an adequate mechanism for delays with the possibility of representing asynchronous input in a synchronous way to the internal representative.

Finally the nesting rules (e.g. in Dyck's language of parentheses) have always been a problem when translated in neurological terms. A model based on the concepts of decaying activity in CA that serve as quasi-bistable memory elements has been shown to incorporate the virtues required for 'push-down' memories. The problem of inserting phrases into phrases to a certain extent involves recognizing legal or unusual k-grams on the borders; this requires a brain mechanism which can interrupt the embedding phrase throughout the duration of the embedded phrase. Once the embedded phrase has finished, the embedding phrase resumes. For this there are plausible neurological models, if the elements of the phrase are represented in the brain by concatenated CAs. Due to the nature of the CA, which is stable in both its active and inactive state, the information for continuing the embedding phrase may be preserved in the activity of the last active cell assembly before the interruption. When there are several embedded constituents, it is well known that the order of closure of the various phrases is inverse to the order of opening. This is fairly easily explained if we assume that the activity of the CA, which indicates an interrupted phrase, slowly decays in time and that completion starts with the most active CA, the most recently activated one [24] [3].

Sequences of cell assemblies at a rate of $4 - 5$ a second have been postulated in many contexts of behaviour (e.g. vision) and are probably related to the periodic action of a mechanism controlling the state of activity of the cortex. This mechanism also involves inhibitory links, which not only isolate individual cell assemblies from the background activity (by the principle of 'winner takes all') but also see to it that in the sequencing of cell assemblies one is just extinguishing when the next one is activated. This explanation essentially answers the question of how the 'place holders' of the theory are represented in the brain. But details are, of course, not known.

It is certainly true that not all the elements that spell out the grammaticality of a sentence can be identified with the elements at the surface level of language. Traditional grammar rules involve such things as grammatical (or lexical) categories. In theory we may think of tags that are attached to words by means of the all pervading associative mechanism typical of the cortex. But it is not at all clear how these categories are extracted from language in the early phase of learning.

A final note on the non-counting property that all human languages seem to have. Modulo-counting is important in various forms of behaviour, particularly in music: complex periodic structures occur in many compositions. The brain is quite effective at processing such periodic vocalisations or motor sequences, as in percussion playing. We are therefore inclined to think that the non-counting property of language has to do with other constraints external to the brain, such as the excessive noise-sensitivity of such languages: a cough could easily transform an odd string into an even one, thus subverting the grammaticality and meaning of an utterance.

4   Conclusion

We believe that the associative language description model explains fundamental syntactic phenomena more convincingly, in terms of brain behaviour, than earlier attempts with context-free grammars. In spite of obvious limitations and drastic simplification, the model should provide a good basis for extension and refinement. We summarise our findings and add a few remarks. The ALD model is based on constituent structures and associative memory. Whereas the first concept provides the basis of the classical Chomskian grammars, the latter has been rejected by linguists as inadequate for discriminating sentences from non-sentences. Yet associative processing is a fundamental mechanism of the brain and plays an important role in speech understanding and language production. The new model is based on the mathematical theory of local testability and counter-free languages, research into which began in the 1960s, but until now it has not been combined with constituent structures. Loosely related ideas have been studied in the area of programming languages, where the concept of precedence of operators was introduced by Floyd [15], to make parsing deterministic. He proposes that digrams (called precedence relations) be used to detect the left and right border of a constituent. The fact that non-counting operator precedence grammars can be easily inferred from examples was noticed in [9] and the formal model was studied by [13].

Several mathematical aspects could be investigated in the vein of formal language theory. One such aspect concerns weak generative capacity: can any regular language be generated by an ALD? (the same question for CF languages had a negative answer in [11]). But a more relevant research project would be to assess the linguistic adequacy of the model, preferably by implementing the learning procedures that would allow ALD to be automatically produced from a linguistic body. Such a research programme would enable the local testability model to be tuned so that, for example, compulsory k-grams and long-distance relations could be specified and Boolean operations could be applied to sets. The possibility of validating the brain model depends on timing analysis and using psycholinguistic findings for purposes of comparison. Likewise, the computer simulation of the proposed cell assemblies should also be feasible.

Acknowledgment

We would like to thank Alessandra Cherubini, Pierluigi San Pietro, and Friedemann Pulvermüller.

References

[1] M. Abeles, *Local Cortical Circuits: An Electrophysiological Study.* Springer, New York, 1982.

[2] M. Abeles, *Corticonics: Neural Circuits of the Cerebral Cortex.* Cambridge University Press, Cambridge, 1991.

[3] V. Braitenberg, *Il Gusto della Lingua: Meccanismi Cerebrali e Strutture Grammaticali.* Alpha Beta, Merano, 1996.

[4] V. Braitenberg and F. Pulvermüller, Entwurf einer neurologischen Theorie der Sprache, *Naturwissenschaften,* 79 (1992), 103–117.

[5] V. Braitenberg and A. Schüz, *Anatomy of the Cortex: Statistics and Geometry.* Springer, New York, 1991.

[6] J.A. Brzozowski, Hierarchies of aperiodic languages, *RAIRO Informatique Théorique,* 10 (1976), 33–49.

[7] N. Chomsky, *Syntactic Structures.* Mouton, The Hague, 1957.

[8] N. Chomsky and G. Miller, Finitary models of language users. In R. Luce, R. Bush and E. Galanter (eds.), *Handbook of Mathematical Psychology.* John Wiley, New York, 1963, 112–136.

[9] S. Crespi Reghizzi, Reduction of enumeration in grammar acquisition. In *Proceedings of the Second International Conference on Artificial Intelligence*, London, 1971, 546–552.

[10] S. Crespi Reghizzi, An effective model for grammar inference. In *Proceedings of Information Processing 71*, Ljubliana, 1972, 524–529.

[11] S. Crespi Reghizzi, A. Cherubini and P.L. San Pietro, Languages based on structural local testability. In C.S. Calude and M.J. Dinneen (eds.), *Combinatorics, Computation and Logic*. Springer, Berlin, 1999, 159–174.

[12] S. Crespi Reghizzi, G. Guida and D. Mandrioli, Non-counting context-free languages, *Journal of the ACM*, 25 (1978), 571–580.

[13] S. Crespi Reghizzi, G. Guida and D. Mandrioli, Operators precedence grammars and the non-counting property, *SIAM Journal of Computing*, 10 (1981), 174–191.

[14] S. Crespi Reghizzi, M. Pradella and P.L. San Pietro, Conciseness of associative language descriptions. In J. Dassow and D. Wotschke (eds.), *Proceedings of Descriptional Complexity of Automata, Grammars and Related Structures*, Universität Magdeburg, 1999, 99–108.

[15] R.W. Floyd, Syntactic analysis and operator precedence, *Journal of the ACM*, 10 (1963), 316–333.

[16] E.M. Gold, Language identification in the limit, *Information and Control*, 10 (1967), 447-474.

[17] S.A. Greibach, The hardest context-free language, *SIAM Journal of Computing*, 2 (1973), 304–310.

[18] D.O. Hebb, *The Organization of Behaviour: A Neuropsychological Theory*. John Wiley, New York, 1949.

[19] R. Jackendoff, *X' Syntax: A Study of Phrase Structure*. MIT Press, Cambridge, Mass., 1977.

[20] M. Kanazawa, *Learnable Classes of Categorial Grammars*. CSLI, Stanford, Ca., 1998.

[21] K.S. Lashley, The problem of serial order in behavior. In L.A. Jeffress (ed.), *Cerebral Mechanisms in Behavior*. John Wiley, New York, 1951, 112–136.

[22] R. McNaughton and S. Papert, *Counter-Free Automata*. MIT Press, Cambridge, Mass., 1971.

[23] J.E. Pin, *Variétés de Langages Formels*. Masson, Paris, 1984.

[24] F. Pulvermüller, Syntax und Hirnmechanismen: Perspektive einer multidisziplinären Sprachwissenschaft, *Kognitionswissenschaft*, 4 (1994), 17–31.

[25] A. Salomaa, *Formal Languages*. Academic Press, New York, 1973.

[26] A. Wickelgren, Context-sensitive coding, associative memory, and serial order in (speech) behavior, *Psychological Review*, 76 (1969).

The Power and Limitations of Random Context

Sigrid Ewert[1]

Department of Computer Science
University of Bremen
Germany
sewert@informatik.uni-bremen.de

Andries van der Walt

Department of Computer Science
University of Stellenbosch
South Africa
apjw@cs.sun.ac.za

Abstract. We use random context picture grammars to generate pictures through successive refinement. The productions of such grammars are context-free, but their application is regulated—'permitted' or 'forbidden'—by contexts that are randomly distributed in the developing picture. Grammars using this relatively weak context often succeed where context-free grammars fail, eg., in generating the typical iteration sequence of the Sierpiński carpet. We were also able to develop iteration theorems for three subclasses of these grammars; finding necessary conditions is problematic for most models of context-free picture grammars with context-sensing ability, since they consider a variable and its context as a connected unit.

We give two detailed examples of picture sets generated with random context picture grammars. Then we show how to construct a picture set that cannot be generated using random context only.

---

[1]Postdoctoral fellow with a scholarship from the National Research Foundation, South Africa.

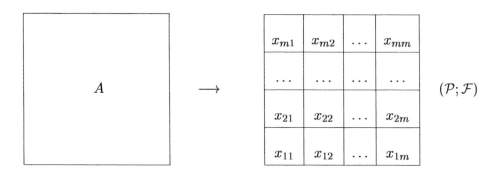

x_{m1}	x_{m2}	$\cdots$	x_{mm}
$\cdots$	$\cdots$	$\cdots$	$\cdots$
x_{21}	x_{22}	$\cdots$	x_{2m}
x_{11}	x_{12}	$\cdots$	x_{1m}

Figure 1: Production of a random context picture grammar

1 Introduction

Random context picture grammars (RCPGs), a method of syntactic picture generation, have been described and studied elsewhere [1], [3], [2], [4]. The model was generalized in [5].

We formally introduce RCPGs in Section 2. In Section 3, we give two detailed examples of picture sets generated with these grammars. Finally, in Section 4, we show how to construct a picture set that cannot be generated using random context only.

2 Definitions

We generate pictures using productions such as those in Figure 1, where A is a variable, $m \in \{1, 2, 3, \ldots\}$, $x_{11}, x_{12}, \ldots, x_{mm}$ are variables or terminals, and $\mathcal{P}$ and $\mathcal{F}$ are sets of variables. The interpretation is as follows: if a developing picture contains a square labeled A and if all variables of $\mathcal{P}$ and none of $\mathcal{F}$ label the squares in the picture, then the square labeled A may be divided into equal squares with labels $x_{11}, x_{12}, \ldots, x_{mm}$.

In order to cast the formulation into a more linear form, we denote the square with sides parallel to the axes, lower lefthand vertex at (s, t) and upper righthand vertex at (u, v) by $((s, t), (u, v))$. We use lowercase Greek letters for such constructs. Thus, (A, α) denotes a square α labeled A. Furthermore, if α is such a square, $\alpha_{11}, \alpha_{12}, \ldots, \alpha_{mm}$ denote the equal squares into which α can be divided, with, eg., α_{11} denoting the bottom left one.

A *random context picture grammar* (RCPG) $G = (V_N, V_T, P, (S, \sigma))$ has a finite alphabet V of *labels*, consisting of disjoint subsets V_N of *variables* and V_T of *terminals*. P is a finite set of *productions* of the form $A \to [x_{11}, x_{12}, \ldots, x_{mm}]\ (\mathcal{P}; \mathcal{F})$, where $m \in \{1, 2, 3, \ldots\}$, $A \in V_N$, $x_{11}, x_{12}, \ldots, x_{mm} \in V$ and $\mathcal{P}, \mathcal{F} \subseteq V_N$. Finally, there is an *initial labeled square*

(S, σ) with $S \in V_{\mathrm{N}}$.

A *pictorial form* is any finite set of nonoverlapping labeled squares in the plane. If Π is a pictorial form, we denote by $l(\Pi)$ the set of labels used in Π. The *size* of a pictorial form Π is the number of squares contained in it, i.e. $|\Pi|$.

For an RCPG G and pictorial forms Π and Γ, we write $\Pi \Longrightarrow_{\mathrm{G}} \Gamma$ if there is a production $A \to [x_{11}, x_{12}, \ldots, x_{mm}]$ $(\mathcal{P}; \mathcal{F})$ in G, Π contains a labeled square (A, α), $l(\Pi \setminus \{(A, \alpha)\}) \supseteq \mathcal{P}$ and $l(\Pi \setminus \{(A, \alpha)\}) \cap \mathcal{F} = \emptyset$, and $\Gamma = (\Pi \setminus \{(A, \alpha)\}) \cup \{(x_1, \alpha_{11}), (x_2, \alpha_{12}), \ldots, (x_{mm}, \alpha_{mm})\}$. As usual, $\Longrightarrow_{\mathrm{G}}^*$ denotes the reflexive transitive closure of $\Longrightarrow_{\mathrm{G}}$.

A *picture* is a pictorial form Π with $l(\Pi) \subseteq V_{\mathrm{T}}$. The *gallery* $\mathcal{G}(G)$ *generated by a grammar* $G = (V_{\mathrm{N}}, V_{\mathrm{T}}, P, (S, \sigma))$ is the set of pictures Π such that $\{(S, \sigma)\} \Longrightarrow_{\mathrm{G}}^* \Pi$.

Note. To improve legibility, we write productions of the type $A \to [x_{11}]$ $(\mathcal{P}; \mathcal{F})$ as $A \to x_{11}$ $(\mathcal{P}; \mathcal{F})$.

3 The Power of Random Context

We give an indication of the power of random context picture grammars by showing two examples of galleries that can be generated using this type of context.

Example 1 (Sierpiński carpet)

Consider the iteration sequence of the Sierpiński carpet, members of which are shown in Figures 2 and 3. A context-free grammar cannot generate such a sequence, because it cannot ensure that all regions of a given picture have the same degree of refinement.

This gallery can be created with the RCPG $G_{\mathrm{carpet}} = (\{S, T, U, F\}, \{w, b\}, P, (S, ((0,0), (1,1))))$, where P is the set:

$$S \quad \to \quad [T, T, T, T, w, T, T, T, T] \; (\{\} ; \{U\}) \tag{1}$$

$$T \quad \to \quad U \; (\{\} ; \{S, F\}) \mid \tag{2}$$
$$F \; (\{\} ; \{S, U, F\}) \mid \tag{3}$$
$$b \; (\{F\} ; \{\}) \tag{4}$$

$$U \quad \to \quad S \; (\{\} ; \{T\}) \tag{5}$$

$$F \quad \to \quad b \; (\{\} ; \{T\}) \tag{6}$$

We associate the colours 'light' and 'dark' with the terminals 'w' and 'b', respectively.

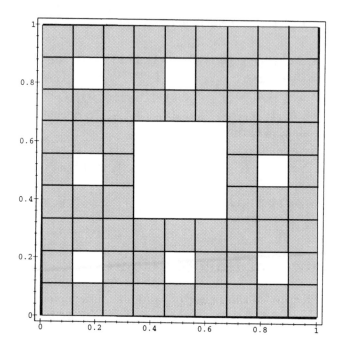

Figure 2: Sierpiński carpet: first refinement

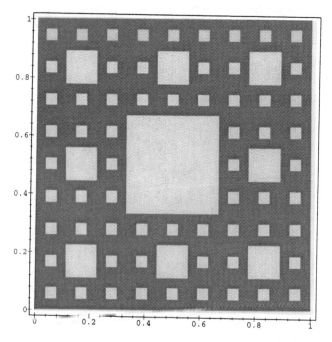

Figure 3: Sierpiński carpet: second refinement

The initial square labeled S is divided into nine equally big squares, of which the middle square is labeled w and the others T (1). This pictorial form can now derive a picture or a more refined pictorial form. The decision is made by any T.

If T decides to terminate, it produces an F (3). The other T's, on sensing the F, each produce a b (4). Once there are no T's left in the pictorial form, F also produces a b (6).

Alternatively, T produces a U and all other T's follow suit (2). Once this has been done, each U is replaced by an S (5) and the process is repeated.

Example 2 *Consider the following gallery. A picture consists of 2^i, $i \geq 0$, identical lanes. The upper half of a lane consists of 2^i identical isosceles triangles which are next to each other, and which are grey on a light background. The lower half of a lane is again divided in 2^j, $j \geq 1$, sub-lanes, the upper half of which is dark and the lower half light. Examples are given in Figures 4 and 5.*

This gallery can be created with the RCPG $G_{\text{triangles+stripes}} = (\{S, T, U\} \cup \{I_{r-\text{tri}}, I_{l-\text{tri}}, I_{\text{stripe}}\} \cup \{T_{r-\text{tri}}, T_{l-\text{tri}}, T_{\text{stripe}}\} \cup \{U_{r-\text{tri}}, U_{l-\text{tri}}, U_{\text{stripe}}\} \cup \{F_{\text{tri}}, F_{\text{stripe}}\}, \{w, b, f\}, P, (S, ((0,0), (1,1))))$, where P is the set:

$$S \quad \rightarrow \quad [T, T, T, T] \; (\{\} \, ; \{U, I_{\text{stripe}}\}) \; | \tag{7}$$

$$[I_{\text{stripe}}, I_{\text{stripe}}, I_{l-\text{tri}}, I_{r-\text{tri}}] \; (\{\} \, ; \{U, T\}) \tag{8}$$

$$T \quad \rightarrow \quad U \; (\{\} \, ; \{S\}) \tag{9}$$

$$U \quad \rightarrow \quad S \; (\{\} \, ; \{T\}) \tag{10}$$

$$I_{r-\text{tri}} \quad \rightarrow \quad [b, T_{r-\text{tri}}, T_{r-\text{tri}}, w] \; (\{\} \, ; \{S, U_{l-\text{tri}}, U_{r-\text{tri}}, F_{\text{tri}}\}) \; | \tag{11}$$

$$[b, b, F_{\text{tri}}, w] \; (\{\} \, ; \{S, U_{l-\text{tri}}, U_{r-\text{tri}}, T_{r-\text{tri}}, F_{\text{tri}}\}) \; | \tag{12}$$

$$[b, b, b, w] \; (\{F_{\text{tri}}\} \, ; \{\}) \tag{13}$$

$$T_{r-\text{tri}} \quad \rightarrow \quad U_{r-\text{tri}} \; (\{\} \, ; \{I_{r-\text{tri}}, I_{l-\text{tri}}\}) \tag{14}$$

$$U_{r-\text{tri}} \quad \rightarrow \quad I_{r-\text{tri}} \; (\{\} \, ; \{T_{r-\text{tri}}, T_{l-\text{tri}}\}) \tag{15}$$

$$F_{\text{tri}} \quad \rightarrow \quad b \; (\{\} \, ; \{I_{r-\text{tri}}, I_{l-\text{tri}}\}) \tag{16}$$

$$I_{l-\text{tri}} \quad \rightarrow \quad [T_{l-\text{tri}}, b, w, T_{l-\text{tri}}] \; (\{T_{r-\text{tri}}\} \, ; \{\}) \; | \tag{17}$$

$$[b, b, w, b] \; (\{F_{\text{tri}}\} \, ; \{\}) \tag{18}$$

$$T_{l-\text{tri}} \quad \rightarrow \quad U_{l-\text{tri}} \; (\{\} \, ; \{I_{l-\text{tri}}, I_{r-\text{tri}}\}) \tag{19}$$

$$U_{l-\text{tri}} \quad \rightarrow \quad I_{l-\text{tri}} \; (\{\} \, ; \{T_{l-\text{tri}}, T_{r-\text{tri}}\}) \tag{20}$$

$$I_{\text{stripe}} \quad \rightarrow \quad [T_{\text{stripe}}, T_{\text{stripe}}, T_{\text{stripe}}, T_{\text{stripe}}]$$

$$(\{\} \, ; \{S, U_{\text{stripe}}, F_{\text{stripe}}\}) \; | \tag{21}$$

$$[w, w, F_{\text{stripe}}, f] \; (\{\} \, ; \{S, U_{\text{stripe}}, T_{\text{stripe}}, F_{\text{stripe}}\}) \; | \tag{22}$$

$$[w, w, f, f] \, (\{F_{\text{stripe}}\} \,; \{\}) \tag{23}$$

$$T_{\text{stripe}} \;\rightarrow\; U_{\text{stripe}} \, (\{\} \,; \{I_{\text{stripe}}\}) \tag{24}$$

$$U_{\text{stripe}} \;\rightarrow\; I_{\text{stripe}} \, (\{\} \,; \{T_{\text{stripe}}\}) \tag{25}$$

$$F_{\text{stripe}} \;\rightarrow\; f \, (\{\} \,; \{I_{\text{stripe}}\}) \tag{26}$$

We associate the colours 'light', 'dark' and 'grey' with the terminals 'w', 'b' and 'f', respectively.

A picture in this gallery is generated in two phases. In the first phase, the canvas is divided into lanes (7, 9, 10). In the second phase, a row of triangles is generated in the upper half of each lane (8, 11–20), while the lower half is divided into sub-lanes (8, 21–26). The triangles and sub-lanes are generated independently of each other.

The first phase proceeds as follows. At the beginning of the ith, $i \geq 1$, iteration of the loop (7, 9, 10), the pictorial form Π_i consists of 4^{i-1} equally big squares, all labeled S. This pictorial form can derive the pictorial form Π_{i+1} or start filling in the existing squares with triangles and stripes. The decision is made by any S.

If S decides to generate more squares, the square containing it is divided in four and each quarter labeled T; all other S's follow suit (7). Once this has been done, each T is stored in U (9), the S's are restored (10) and an S must again make the decision described above.

If, on the other hand, S decides to start decorating the existing lanes with triangles and stripes, the square containing it is divided in four and the lower two quarters labeled I_{stripe}, while the upper left and right quarters are labeled $I_{\text{l-tri}}$ and $I_{\text{r-tri}}$, respectively (8).

Each $I_{\text{r-tri}}$ generates an isosceles right triangle with the right angle on the left, while each $I_{\text{l-tri}}$ generates an isosceles right triangle with the right angle on the right. The smoothness of the hypotenuse is determined by any $I_{\text{r-tri}}$.

If $I_{\text{r-tri}}$ decides to smoothen the hypotenuse further, the square containing it is divided in four and the lower left quarter labeled b, the upper right quarter w, and the remaining two quarters $T_{\text{r-tri}}$; all other occurrences of $I_{\text{r-tri}}$ follow suit (11). Each $I_{\text{l-tri}}$, on sensing $T_{\text{r-tri}}$ in the pictorial form, smoothens the hypotenuse it is part of: the square containing it is divided in four and the lower right quarter labeled b, the upper left w, and the remaining two quarters $T_{\text{l-tri}}$ (17).

Once this has been done, each $T_{\text{r-tri}}$ is stored in $U_{\text{r-tri}}$ (14) and each $T_{\text{l-tri}}$ in $U_{\text{l-tri}}$ (19). Then $I_{\text{r-tri}}$ and $I_{\text{l-tri}}$ are restored (15 and 20, respectively) and the aforementioned decision by an $I_{\text{r-tri}}$ must be made again.

If I_{r-tri} decides to complete the existing triangles, the square containing it is divided in four and the two lower quarters labeled b, the upper left F_{tri} and the upper right w (12).

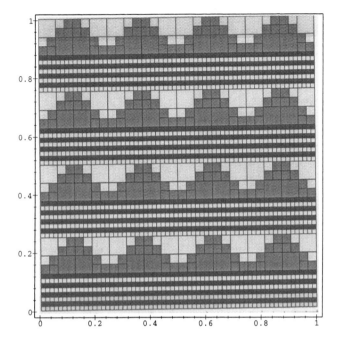

Figure 4: Triangles and stripes: medium refinement

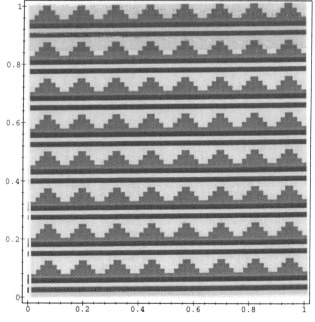

Figure 5: Triangles and stripes: high refinement

All other I_{r-tri}'s, on sensing F_{tri} in the pictorial form, generate the final edge, which consists of a w in the upper right quarter and b's in the rest (13). Similarly, each I_{l-tri}, on sensing F_{tri}, generates a final edge consisting of a w in the upper left quarter and b's in the rest (18).

Each I_{stripe} generates 2^i, $i \geq 0$, sub-lanes. With each execution of the loop (21, 24 and 25), the number of sub-lanes is doubled. Whether or not the loop is executed is decided by any I_{stripe}.

If I_{stripe} decides to repeat the loop, it creates four quarters all labeled T_{stripe}; all I_{stripe}'s follow suit (21). Once this has been done, each T_{stripe} is stored in U_{stripe} (24), the I_{stripe}'s are restored (25) and the decision by an I_{stripe} must be made again.

If I_{stripe} decides to colour the existing sub-lanes, the square containing it is divided in four and the two lower quarters are labeled w, while the upper left is labeled F_{stripe} and the upper right f (22). All other occurrences of I_{stripe}, on sensing F_{stripe} in the pictorial form, generate four quarters, of which the two lower ones are labeled w and the upper ones f (23). Once this has been done, F_{stripe} produces an f (26).

4 The Limitations of Random Context

We now turn to the limitations of RCPGs. We prove that every gallery generated by an RCPG of a certain type has a property which we call *commutativity at level m*. This enables us to construct a picture set that cannot be generated using random context only.

To facilitate the description, we consider only pictures occupying the unit square $((0,0),(1,1))$. A picture is called *n-divided*, for $n \geq 1$, if it consists of 4^n equal subsquares, each labeled with a terminal. A *level-m subsquare* of an n-divided picture, with $1 \leq m \leq n$, is a square:

$$\left((x2^{-m}, y2^{-m}), ((x+1)2^{-m}, (y+1)2^{-m}) \right),$$

where x and y are integers and $0 \leq x, y < 2^m$. Note that, for $m < n$, a level-m subsquare consists of all 4^{n-m} labeled subsquares contained in it.

Two n-divided pictures Φ_1 and Φ_2 are said to *commute at level m* if Φ_1 contains two different level-m subsquares α and β such that Φ_2 can be obtained by simply interchanging the labeling of α and β. This implies that Φ_1 is obtained when the similarly situated subsquares of Φ_2 are interchanged. Clearly, if Φ_1 and Φ_2 commute at level m, then they also commute at level t for $m \leq t \leq n$. For example, the two 2-divided pictures in Figure 6 commute at level 1.

A picture Φ_1 is called *self-commutative at level m* if Φ_1 and Φ_1 commute at level m.

Figure 6: Two 2-divided pictures that commute at level 1

For example, the 2-divided picture in Figure 7 is self-commutative at level 1.

Theorem 1 *Let $G = (V_N, V_T, P, (S, \sigma))$ be an RCPG that generates an infinite gallery of n-divided pictures. Then there exist an m and a c such that each picture that is c-divided is either self-commutative at level m or commutes with another picture in the gallery at level m.*

Proof. We may assume, without loss of generality, that every production in P that effects a subdivision produces exactly 4 subsquares. Let $k = |V_N|$ and consider an n-divided picture Φ_1 with $n > k$. In any given derivation of Φ_1, let Π be the first pictorial form that contains a level-k subsquare. Then Π will contain at least $3k + 1$ subsquares, all labeled by variables. This means that there will be a variable, say A, which appears at least twice in Π, say as the label of squares α and β. Then α and β are of equal size, else Π could derive a picture that is not n-divided. So α and β are both level-h subsquares for some $h \leq k$. Π, then, not only derives Φ_1, but also a –not necessarily different– picture Φ_2 with the property that it can be obtained from Φ_1 by simply interchanging the subsquares α and β. Thus we could take $m = k$ and $c = k + 1$. □

Figure 7: A 2-divided picture that is self-commutative at level 1

Suppose $m \geq 1$ is given. Consider the $2m$-divided picture Φ that is constructed as follows. For any level-m subsquare α in Φ, if α is in row i and column j of Φ, then the level-$2m$ subsquare in row i and column j of α is dark. All the other level-$2m$ subsquares are light. Then Φ is not self-commutative at level m. Thus, we have:

Theorem 2 *There exists a set of pictures that cannot be generated by an RCPG.*

Acknowledgement

The first author would like to thank Frank Drewes and Renate Klempien-Hinrichs for stimulating discussions.

References

[1] S. Ewert and A. van der Walt, Generating pictures using random forbidding context, *International Journal of Pattern Recognition and Artificial Intelligence*, 12.7 (1998), 939–950.

[2] S. Ewert and A. van der Walt, A hierarchy result for random forbidding context picture grammars, *International Journal of Pattern Recognition and Artificial Intelligence*, 13.7 (1999).

[3] S. Ewert and A. van der Walt, Generating pictures using random permitting context, *International Journal of Pattern Recognition and Artificial Intelligence*, 13.3 (1999), 339–355.

[4] S. Ewert and A. van der Walt, Random context picture grammars, *Publicationes Mathematicae (Debrecen)*, 54 (Supp) (1999), 763–786.

[5] S. Ewert and A. van der Walt, Shrink indecomposable fractals, *Journal of Universal Computer Science*, 5.9 (1999). http://www.iicm.edu/jucs_5_9.

Parsing Contextual Grammars with Linear, Regular and Context–Free Selectors

Karin Harbusch

Computer Science Department
University of Koblenz–Landau
Germany
harbusch@informatik.uni-koblenz.de

Abstract. *Contextual Grammars (CGs)* provide an appropriate description of natural languages. Unfortunately, no parser which runs in polynomial time is known for some linguistically relevant classes.

This paper presents an intertwined two–level Earley–based parser for CGs with finite, regular and context–free selectors. Both phases define context–free grammars which identify individual selectors and contexts in the input string and which find context–selector pairs. The Earley algorithms provide an efficient data structure so that results can be exchanged between the two phases (i.e. phase one $\longrightarrow$ phase two: contexts and selectors, phase two $\longrightarrow$ phase one: erased contexts according to a production) and intermediate results can be reused effectively when repeating new contexts and selectors are predicted for each removed context in the input string, thus identifying new productions.

While we use a polynomial parser for CGs with context–free selectors, the linguistic relevance of this family of languages (i.e. $CL_\alpha(CF)$) is discussed here. We show that some center–embedded phenomena which cannot be generated by a CG with regular selectors, according to the recursion definition Mg belong to $CL_{Mg}(CF)$.

1   Introduction

Contextual Grammars (CGs) were introduced by Solomon Marcus in 1969
[6] as *"intrinsic grammars"* without auxiliary symbols. They were based
only on the fundamental linguistic operation of inserting words into given
phrases according to certain contextual dependencies. The definition of a
CG is simple and intuitive. Contextual Grammars include *contexts*, i.e. pairs
of words, associated with *selectors* (sets of words). A context can be adjoined
to any associated selector element. In this way, starting from a finite set of
words (*axioms*), the language is generated.

Many of the properties of CGs have been studied. For instance, they
do not fit into the Chomsky hierarchy (cf. [10] for more formal properties).
CGs have also been shown to provide an appropriate description of natural
languages (cf. e.g. [8]). For instance, the language $L_1 = \{a^n c b^m c a^n c b^m \mid$
$n, m \geq 1\}$, which is not a context–free language, circumscribes phenomena
in a dialect of German spoken around Zurich (Switzerland) [14], and allows
constructions of the form $NP_a^n \ NP_d^m \ NP_a^n 1 \ NP_d^m$.[1] However, a surprising
limitation of CGs with maximal global (Mg) use of selectors is discussed in
[8]. There exist center–embedded structures that cannot be generated by
such grammars even if regular selectors are imposed. Specifically, let us consider the language $L_2 = \{a^n c b^m c b^m c a^n \mid n, m \geq 1\}$. Note that this is a linear
language in Chomsky's sense and that it belongs to the families $CL_{in}(FIN)$
and $CL_{Ml}(FIN)$. However, L_2 is not in the family $CL_{Mg}(REG)$. As we use
a parser for CGs with context-free selectors, we discuss the linguistic relevance of CGs with context-free selectors in section 2. We present a grammar
which generates L_2 according to Mg.

In order to prove that CG is an appropriate formalism for natural–
language processing we must have an efficient parser. Unfortunately, for
some linguistically relevant classes there is no known parser which runs in
polynomial time. Here, an intertwined two–level Earley–based parser is presented for Contextual Grammars with finite, regular or context–free selectors. It is based on the three derivation definitions *in*, *Mg* and *Ml*, i.e.
contexts can be unrestrictedly replaced according to a selector, maximal selectors can only be replaced according to all selectors or maximal selectors
can be replaced according to the same selector, respectively. The parser
consists of two passes. Basically, in the first pass all individual contexts
and selectors are identified and stored in items denoting the left and right
boundary of these fragments. This task is performed by an ordinary Earley parser. All these items are used in the second– phase Earley–parser in
order to check the only context–free rule–type ($sel_i \longrightarrow con_{i_k l} \ sel_i \ con_{i_k r}$)
for all selector elements $s_i \in S_i$ corresponding to sel_i and all its context
pairs $(c_{i_k l}, c_{i_k r}) \in U_i$ (represented by $con_{i_k l}$ and $con_{i_k r}$). This means the

---

[1]Here, NP_a stands for an accusative noun phrase and NP_d stands for a dative one.

identification of a derivation step of a CG.

So that context–selector pairs can be identified after the elimination of contexts in the input string, the two phases run intertwined. The input string in which $con_{i_k l}$ and $con_{i_k r}$ are eliminated not only continues the iteration in phase two for each successfully applied rule $(sel_i \longrightarrow con_{i_k l}\ sel_i\ con_{i_k r})$, but also reruns the two phases. It is important to note that the numbering in the newly built input strings remains the same so that both phases can reuse all previously computed results. The strings to be erased are only marked to be empty by the new terminal symbol ϵ. The rules $(q \longrightarrow \epsilon q)$ and $(q \longrightarrow q\epsilon)^2$ eliminate all occurences of ϵ and represent that q covers the eliminated context as well.

We show that in the worst case the space complexity of an acceptor is $O(n^4)$ and the run time is $O(n^6)$ for the ordinary recursion definition *in*. The computation according to Mg and Ml and parsing according to *in*, Mg and Ml costs at most $O(n^6)$ space and $O(n^9)$ time units. The parser is available online at the following address:

www.uni–koblenz.de/~harbusch/CG-PARSER/welcome-cg.html.

As the definitions of CGs can be skipped in this volume[3], the next section addresses the linguistic relevance of Contextual Grammar with context–free selectors. In section 3, a polynomial parser for CGs is presented. Finally, some open questions and possibilities for future work are discussed.

2   The Linguistic Relevance of CGs

[13] and [8] have explained that CGs are appropriate for describing natural languages largely because the grammar writer can straightforwardly describe all the usual restrictions that appear in natural languages and which lead to their non-context-freeness: *reduplication, crossed dependencies*, and *multiple agreement*. For instance, the non–context–free language L_1 (*crossed dependencies*) mentioned in the introduction circumscribes phenomena in a dialect spoken around Zurich (Switzerland) [14]. L_1 can be specified by $G_1 = (\{a, b, c\}, \{acbcacb\}, \{(\{acb^+ca\}, \{(a,a)\}), (\{bca^+cb\}, \{(b,b)\})\})$.

However, the introduction also mentioned a surprising limitation of Contextual Grammars with maximal global use of selectors. The language $L_2=\{a^n cb^m cb^m ca^n | n, m \geq 1\}$ (*center-embedded structures*) cannot be generated. Note that this is a linear language in Chomsky's sense and that it

---

[2]Notice that both rules are required to detect empty selector elements (i.e. $= \lambda$) at the beginning and at the end of the input string.

[3]In the following, we adopt the terminology of [8]. For an overview of the recent literature, see e.g. [7], [13], [2], [11] or [9]. For our purposes, the following definition is added. For all selectors $s_i \in S_i$, all contextual pairs $(c_{i_k l}, c_{i_k r}) \in C_i$ $(1 \leq k \leq$ cardinality of $C_i)$ and (S_i, C_i) a production *(context–selector pair)* of the CG G, $(s_i, (c_{i_k l}, c_{i_k r}))$ is called a *context–selector pair–element*; $c_{i_k l}$ is called a *left context*; and $c_{i_k r}$ a *right context* of the *selector element* s_i.

belongs to the families $CL_{in}(FIN)$ and $CL_{Ml}(FIN)$. However, L_2 is not in the family $CL_{Mg}(REG)$. As we use a polynomial parser for CGs with context–free selectors, here the class of languages with context–free selectors is investigated with respect to the question whether it generates L_2.

The Contextual Grammar with context–free selectors G_2, the alphabet $\{a, b, c\}$, the axiom bcb, and the context–selector pairs:

$$(\{a^n cb^* cb^* ca^n | n \geq 0\}, \{(a, a)\}), (\{b^m cb^m | n \geq 0\}, \{(b, b), (ac, ca)\})$$

generates L_2 under the assumption of Mg. For instance, all accepted input strings up to length 11 are produced in the following manner. The axiom bcb derives $bbcbb$ or $acbcbca$ according to in, Mg, Ml because $b^1 cb^1$ is the only applicable selector. The string $bbcbb$ derives $bbbcbb$ and $acbbcbbca$ according to Ml, Mg (here $x_1 = \lambda$, $x_2 = b^1 cb^1$, $x_3 = \lambda$; $x_1' = b$, $x_2' = b^0 cb^0$, $x_3' = b$, which would produce the same string, is suppressed; according to in, $bbacccabb, bacbcbccab$ (i.e. $x_1' = bb$, $x_2' = b^0 cb^0$, $x_3' = bb$, $x_1' = b$, $x_2' = b^1 cb^1$, $x_3' = b$) would also be produced; for reasons of space, we give them no further consideration here). The string $acbcbca$ derives $aacbcbcaa$ according to Mg (here $x_1 = \lambda, x_2 = a^1 cb^1 cb^1 ca^1, x_3 = \lambda$; $x_1' = a, x_2' = a^0 cbcbca^0, x_3' = a$, which would produce the same string, is suppressed; according to in, $acbbcbbca, acbacccabca, acacbcbcaca$ would be also produced (i.e. $x_1' = ac, x_2' = b^1 cb^1, x_3' = ca, x_1' = acb, x_2' = b^0 cb^0, x_3' = bca$); again, we give them no further consideration). Consequently, for the three strings $bbbcbbb, acbbcbbca$ and $aacbcbcaa$ the following holds:

- The string $bbbcbbb$ derives $bbbbcbbbb$ and $acbbbcbbbca$ according to Mg with the argumentation as above.

- The string $acbbcbbca$ derives $aacbbcbbcaa$ according to Mg (here $x_1 = \lambda, x_2 = a^1 cb^2 cb^2 ca^1, x_3 = \lambda$; e.g. $x_1' = a, x_2' = a^0 cb^2 cb^2 ca^0, x_3' = a$ or $x_1' = aacb, x_2' = b^2 cb^2, x_3' = bcaa$ are suppressed as are the strings $acbbacccabbca, acbacbcbcabca, acacbbcbbcaca, acbbbcbbbca$ produced according to in).

- The string $aacbcbcaa$ derives $aaacbcbcaaa$ according to Mg ($x_1 = \lambda, x_2 = a^2 cbcbca^2, x_3 = \lambda$; e.g. $x_1' = a, x_2' = a^1 cbcbca^1, x_3' = a, x_1' = aa, x_2' = a^0 cbcbca^0, x_3' = aa$ or $x_1' = aac, x_2' = bcb, x_3' = caa$).

Again, the string $bbbbcbbbb$ derives $bbbbbcbbbbb$ and $acbbbbcbbbbca$ under the assumption of Mg according to the argumentation as above.

This example shows that the class Contextual Grammars with context-free selectors according to the definition Mg satisfies all the requirements of natural language grammars. It should be noticed here that the parser allows the grammar writer to specify suitably complex selectors, e.g. finite selectors according to L_1 and context–free selectors according to L_2. This suitability constraint imposed on the process of grammar writing reduces the average runtime of the parser. In our current work, we are concentrating on how to obtain natural–language Contextual Grammars with context-free selectors by automatically processing corpora where this constraint is automatically satisfied (cf. section 4).

3 CGP — A Polynomial Parser for CGs

Below, we first present an acceptor for $CL_{in}(F)$, where $F \in \{FIN, REG, CF\}$ and then we extend this module according to Mg and Ml and the explicit representation of derivations. Finally, we address the correctness and the time complexity of all variants.

Now, we present the two Earley–based[4] passes $FRAG$ (*compute FRAG-ments*) and $PROCO$ (*Production Combination*) of CGP, the *Contextual-Grammar Parser*. Loosely speaking, $FRAG$ identifies all the individual contexts and selectors in the input string basically by the context–free rules $(sel_i \longrightarrow s_i)$, $(con_{i_k l} \longrightarrow c_{i_k l})$, $(con_{i_k r} \longrightarrow c_{i_k r})$ for all $(s_i, (c_{i_k l}, c_{i_k r}))$ context–selector pair–elements. In $PROCO$, the rules $(sel_p \longrightarrow con_{p_k l}\ sel_p\ con_{p_k r})$ identify applicable productions of the CG. The rules in $FRAG$ $(a \longrightarrow \epsilon a)$ and $(a \longrightarrow a\epsilon)$ are required to rerun this phase with eliminated contexts according to $PROCO$ where these contexts are set ϵ.[5] CGP finally accepts iff $j = (sel_p, con_{p_k l}\ sel_p\ con_{p_k r}, \lambda, 0) \in I_n$ exists where $s_p \in A_{cg}$ or $\lambda \in A_{cg}$ for $w = \lambda$.

3.1 The Two Phases $FRAG$ and $PROCO$ of $CGP_{in-accept}$

For all computations in the first phase $FRAG$, for a given CG $G_{cg} = (V_{cg}, A_{cg}, \{(S_1, C_1), ..., (S_m, C_m)\})$, $m \geq 1$, the context–free grammar $G_{cfI} = (N_{cfI}, T_{cfI}, P_{cfI}, S_{cfI} := N_{cfI})$ is constructed:

- $N_{cfI} := V_{cg} \cup \{sel_i | \forall s_i \in S_i \text{ where } |s_i| \leq n = |input\ sting|\}^6 \cup \{con_{i_k l}| \text{ for each } c_{i_k l}\} \cup \{con_{i_k r}| \text{ for each } c_{i_k r}\}$,

- $T_{cfI} := \{a'^7 | \forall a \in V_{cg}\} \cup \{\epsilon | \{\epsilon\} \cap V_{cg} = \emptyset\}$ and

---

[4] In order to avoid normal–form transformations, we adopt the Earley algorithm [1] for the context–free analyses in the following. Furthermore, the representation of infinite derivations according to rules of the form $X \longrightarrow \lambda$ is adequate for the frequently specified empty selectors and contexts.

[5] The following context-selector pairs provide an example of this necessity: $(\{bd\}, \{(a, \lambda)\})$, $(\{d\}, \{(c, e)\})$. A possible derivation is: $bd \Longrightarrow abd \Longrightarrow abcde$. Here, the selector element bd is not a substring of the input string $abcde$. So the proper items which are constructed in the first phase are missing, i.e. a rerun is required.

[6] Here, a restriction on the length of the selector elements reduces the possibly infinite language to the finite subset of applicable selector elements. The online parser defines a rewriting process of the infinite selector languages (cf. [4]), which reduces the number of essential grammar rules. For our purposes here, the two methods do not need to be differentiated. Only the internal processing in $FRAG$ differs according to the two individual grammars.

[7] "a'" means a concatenated with " $\prime$ ". The rules $(a \longrightarrow a')$, which uniquely map terminals in the input string to nonterminals ($\in V_{cg}$), are necessary for technical

- $P_{cfI} := \{(a \longrightarrow a')\} \cup \{(a \longrightarrow \epsilon a)\} \cup \{(a \longrightarrow a\epsilon)\} \cup$
  $\{(sel_i \longrightarrow s_i)\} \cup \{(con_{i_k l} \longrightarrow c_{i_k l})\} \cup \{(con_{i_k r} \longrightarrow c_{i_k r})\}$

$\forall a \in V_{cg}$, $\forall s_i \in S_i$ where $|s_i| \leq |n|$, $\forall (c_{i_k l}, c_{i_k r}) \in C_i$ ($1 \leq k \leq$ cardinality of C_i), $\forall (S_i, C_i)$ ($1 \leq i \leq$ number of context–selector pairs).

This grammar and the input string $w = w_1...w_n \in V_{cg}$ ($n \geq 0$) where all terminals are concatenated with a prime in order to become terminals of G_{cfI} (cf. footnote 7) are handed to the first pass $FRAG$. An ordinary Earley acceptor runs the modified input. Finally, the acceptance or rejection of the parser is ignored. Here, only the following items representing individual contexts and selectors are of interest:

$$j = (sel_i, s_i, \lambda, k), \ j' = (con_{i_k l}, c_{i_k l}, \lambda, k') \text{ and } j'' = (con_{i_k r}, c_{i_k r}, \lambda, k'')$$

in any item list I_m ($0 \leq m \leq n$), i.e. only completely analysed fragments are handed to phase two. All other items of $FRAG$ can be ignored.

In $PROCO$ all fragments computed by $FRAG$ are combined according to the context–free rule–type ($sel_i \longrightarrow con_{i_k l} \ sel_i \ con_{i_k r}$) which represents a derivation step in the CG. In terms of the Earley algorithm, the component $PROCO$ performs all the possible completion steps throughout the item lists according to the context–free rules specified above.

The context–free grammar for the Earley processing in $PROCO$ is called $G_{cfII} = (N_{cfII}, T_{cfII}, P_{cfII}, S_{cfII})$, where:

- $N_{cfII} := N_{cfI}$,
- $P_{cfII} := \{(sel_i \longrightarrow con_{i_k l} \ sel_i \ con_{i_k r})\}$,
- $T_{cfII} := \emptyset$,
- $S_{cfII} := N_{cfII}$.

Here, $con_{i_k l}$ corresponds to $c_{i_k l}$, $con_{i_k r}$ to $c_{i_k r}$ and sel_i to s_i (as defined in $FRAG$) and $(c_{i_k l}, c_{i_k r}) \in C_i$ ($1 \leq k \leq$ cardinality of C_i) and (S_i, C_i) is a context–selector pair ($1 \leq i \leq$ number of context–selector pairs).

In phase two, the grammar G_{cfII} and the preset item lists initialise the Earley acceptor $PROCO$. Since $T_{cfII} := \emptyset$, any input is ignored. Thus, the Earley parser can be slightly modified so that no scanning and no prediction steps are performed at all. Accordingly, the component only performs completion for each item list until no changes take place in item list I_i.

It is important to notice here that $PROCO$ initialises the rerun of CGP by performing the assignment (*) for successfully completed productions, i.e. items of the form $j = (sel_i, con_{i_k l} sel_i con_{i_k r}, \lambda, t) \in I_m$, where sel_i corresponds to the selector s_i, and $con_{i_k l}$ to the left context $c_{i_k l}$ and $con_{i_k r}$ to the right context $c_{i_k r}$:

(*) if $\neg((c_{i_k l} = \lambda) \wedge (c_{i_k l} = \lambda))$, i.e. no new neighbors are found.

---

reasons. The elements of V_{cg} should occur on the left–hand side of a context–free rule, e.g. $(a \longrightarrow \epsilon a)$, and thus must be in N_{cfI}. Here, ϵ — a terminal according to G_{cfI} — denotes elements of the input that belong to eliminated contexts. These signs are removed by scanning them in the ordinary manner and marking the input as successfully consumed (if $j = (a, \lambda, \epsilon a, p) \in I_p$ then $j' = (a, \epsilon, a, p) \in I_{p+1}$). So, the corresponding selector element becomes adjacent to new input signs (cf. footnote 5).

Then run CGP with the original input string $w := w_1 \ldots w_n$, where:

$$w_p := \epsilon, p = t, \ldots, m - |s_i| + 1 \text{ and } w_{m-|s_i|+2}\ldots w_{m+1} := s_i.{}^8$$

By the assignment (*) in $PROCO$, the overall system runs a new input string where the currently identified contexts are marked to be eliminated. This enables all the intermediate results already provided by the parser to be reused. As described for $FRAG$, the rules $(a \longrightarrow \epsilon a)$ and $(a \longrightarrow a\epsilon)$ eliminate an arbitrary number of ϵs and represent that a finally a covers the corresponding input string. Consequently, new adjacent terminals (represented by their corresponding nonterminals in N_{cfI}) are inspected and new selectors and contexts can be produced. As the rerun is initiated by CGP, $PROCO$ checks for context–selector pairs and further reruns can be initialised recursively.

Now the pure acceptor is extended towards a parser. Furthermore the recursion definitions Ml and Mg are imposed on the algorithm. For reasons of space, the following extensions are presented on an informal level. For more details see [3].

3.2   Adapting CPG for Mg and Ml and Parsing

In order to extend $CPG_{in-accept}$ to test for Mg or Ml, only the component $PROCO$ needs to be extended to compare complete context–selector pair-elements. Such elements are represented by items of the form $j = (sel_i, con_{i_k l} \, sel_i \, con_{i_k r}, \lambda, k)$ in an item list I_p $(0 \leq p \leq n)$. According to the definition of the two maximal modes, all context–selector pair–elements must be compared. If we presuppose that the two additional positions of each item sb (selector–begin) and se (selector–end) are computed in the following manner:[9]

1. $j = (sel_i, con_{i_k l}, sel_i con_{i_k r}, k, sb := p, se \text{ not yet defined}) \in I_p$ and

2. $j' = (sel_i, con_{i_k l} sel_i, con_{i_k r}, k', sb = p, se := q) \in I_q$.

All items of the form $j = (sel_p, con_{p_k l} \, sel_p \, con_{p_k r}, \lambda, k, sb_j, se_j) \in I_m$ can be straightforwardly compared to all existing items of the same form. However, the corresponding eliminations work too locally. An item to be erased can have caused completions. Thus, we finally test whether a complete derivation still exists. Obviously, this test is identical to the computation of all parses.

---

[8]Note that it is not necessary to identify the actual boundaries between contexts and selector. According to the rules $(a \longrightarrow \epsilon a)$ and $(a \longrightarrow a\epsilon)$, it suffices to initialise the string $w_t\ldots w_{m+1} = \epsilon^p s_i$ where $p = m - |s_i| - t + 1$ in order to represent that the string from t to $m+1$ is derivable by s_i. Obviously, an explicit representation (cf. sb, se in section 3.2) would also work. Actually, this behaviour is realized in the online parser for reasons of readability by the user.

[9]This processing increases the time and space complexity only constantly.

This check at the end of CGP runs through the item lists in an inverse manner (cf. the computation of the derivations in the basic Earley algorithm [5]), i.e. for all accepting items $j = (lhs, \alpha, \lambda, k)$ in I_n, the initally empty variable der is set to der "+" name of the applied context–selector pair–element. Recursively all imposed completions and scannings are collected and realized by a product. All alternatives on the same level of recursion are summed up. The result is a regular expression in which the individual derivations are the result of computing all pure products. But this could cost exponential time and space whereas the computation and storing of der costs polynomial time, because it is – in a sense — another representation of the item lists.

Now the correctness and the time complexity of CGP is informally addressed. For more details, see [3].

3.3   Correctness and Time Complexity

The correctness of the individual passes directly follows from the correctness of the Earley parsing. In $FRAG$, $(con_{i_kl}, c_{i_kl}, \lambda, s)$, $(sel_i, sel_i, \lambda, s')$ and $(con_{i_kr}, c_{i_kr}, \lambda, s'')$ represent individual contexts and selectors according to a context–selector pair–element $(s_i, (c_{i_kl}, c_{i_kr})) \in (S_i, C_i)$. Since these items are used in $PROCO$ and the context–free rule type $(sel_i \longrightarrow con_{i_kl}, sel_i con_{i_kr})$ directly corresponds to the derivation definition of a CG, all context–selector pairs in the given input are found. Using the same argument, both phases can be rerun if (con_{i_kl}, con_{i_kr}) are eliminated (that is to say, if they are scanned in $FRAG$). Since contexts and selectors can only be directly identified in the input or constructed by eliminating contexts, all the correct derivations of a given CG will have been checked after at most n repetitions of context eliminations. The correctness of the computation of all derivations follows from the inverse traversal of all items. The correctness of the derivations according to Mg and Ml is directly obvious.

The time complexity during the first run of $FRAG$ is $O(n^2)$ and the space complexity is $O(n)$ because only a finite set of finite strings (contexts) is identified in a given input string of length n. During the reruns, the same holds for a new input but it requires up to n repetitions of the rules $(a \longrightarrow \epsilon a)$, $(a \longrightarrow a\epsilon)$. Accordingly, $FRAG$'s time complexity is $O(n^3)$. Since there exist at most $O(n^3)$ — the number of new items in phase two — new input strings which are handed to phase one, the overall space complexity is $O(n^4)$ and its overall time complexity is (number of inputs $(O(n^3))$ × execution time for each input $(O(n^3))$).

The time and space complexity of the algorithm $PROCO$ is $O(n^3)$ as it is a context–free acceptor. Since it is activated for all new items here the overall time complexity is (number of items $(O(n^4))$ × execution time for each item $O(n^2))$. Accordingly, the overall processing can take at most

$O(n^6)$ steps for the recursion definition *in*.

The computation of the variable *der* — and accordingly the computation of Mg and Ml — causes some extra costs. The length of *der* can become $O(n^6)$ because it revisits all items and in the worst case stores a pointer to each item in a sum or product ($O(n^2)$ for the number of candidates in $I_n \times O(n^4)$ for the number of items). This processing can be interpreted as an inverse Earley parse, i.e. the completions are inspected from the last completion (i.e. the item in the item list I_n) back to the first completion, which must be related to an item in I_0. Accordingly, the processing takes $O(n^3)$ steps and the overall time complexity is $O(n^9)$.

4 Final Discussion

In this paper, we have described a fully implemented polynomial parser for a CG with $CL_\alpha(F)$ ($F \in \{FIN, REG, CF\}$, $\alpha \in \{in, Ml, Mg\}$) (cf. www.uni–koblenz.de/~harbusch/CG-PARSER/welcome-cg.html).

In the future, we shall focus on the following two questions. Since we are especially interested in natural language parsing with CGs, we are going to build a Contextual Grammar with context–free selectors for English and German. At present we are investigating how to extract context–selector pairs from corpora. The heads are specified as features for selectors. The patterns, i.e. the contexts, are extracted according to the significant examples in the corpus. This work has just begun and as yet we have no results. On the theoretical side, the formal properties of CGs with context–free selectors will be further investigated.

References

[1] J. Earley, An efficient context–free parsing algorithm, *Communications of the ACM*, 13.2 (1970), 94–102.

[2] A. Ehrenfeucht, Gh. Păun and G. Rozenberg, Contextual grammars and formal languages. In G. Rozenberg and A. Salomaa (eds.), *Handbook of Formal Languages*. Springer, Berlin, 1997, vol. 2, 237–293.

[3] K. Harbusch, A polynomial parser for contextual grammars with linear, regular and context–free selectors. In *Proceedings of the Sixth Meeting on Mathematics of Language*. University of Central Florida, Orlando, Fl., 1999, 323–335.

[4] K. Harbusch, An online parser for contextual grammars with context–free selectors. Submitted; see www.uni–koblenz.de/~harbusch/online-parser.ps

[5] J.E. Hopcroft and J.D. Ullman, *Introduction to Automata Theory, Languages and Computation*. Addison-Wesley, Reading, Mass., 1979.

[6] S. Marcus, Contextual grammars, *Revue Roumaine des Mathématiques Pures et Appliquées*, 14 (1969), 1525–1534.

[7] S. Marcus, Contextual grammars and natural language. In G. Rozenberg and A. Salomaa (eds.), *Handbook of Formal Languages*. Springer, Berlin, 1997, vol. 2, 215–235.

[8] S. Marcus, C. Martín-Vide and Gh. Păun, Contextual grammars as generative models of natural languages, *Computational Linguistics*, 24.2 (1998), 245–274.

[9] C. Martín–Vide, Contextual automata. In *Proceedings of the Sixth Meeting on Mathematics of Language*. University of Central Florida, Orlando, Fl., 1999, 315–321.

[10] C. Martín–Vide, A. Mateescu, J. Miquel–Vergés and Gh. Păun, Internal contextual grammmars: minimal, maximal and scattered use of selectors. In M. Koppel and E. Shamir (eds.), *Proceedings of The Fourth Bar–Ilan Symposium on Foundations of Artificial Intelligence*. AAAI Press, Menlo Park, Ca., 1995, 159–168.

[11] C. Martín–Vide and Gh. Păun, Structured contextual grammars, *Grammars*, 1.1 (1998), 91–101.

[12] Gh. Păun, Marcus contextual grammmars: after 25 years, *Bulletin of the European Association for Theoretical Computer Science*, 52 (1994), 263–273.

[13] Gh. Păun, *Marcus Contextual Grammmars*. Kluwer, Dordrecht, 1997.

[14] S.M. Shieber, Evidence against the context–freeness of natural language, *Linguistics and Philosophy*, 8 (1985), 333–343.

Linguistic Grammar Systems: A Grammar Systems Approach for Natural Language[1]

M. Dolores Jiménez-López

Research Group on Mathematical Linguistics
Rovira i Virgili University
Tarragona, Spain
mdjl@astor.urv.es

Abstract. The aim of this paper is to point out the possible suitability of Grammar Systems Theory in Linguistics. After reviewing some of the features that make Grammar Systems potentially suitable for accounting for linguistic issues, we introduce a new variant: *Linguistic Grammar Systems*.

1 Introduction

Grammar Systems is a new branch of the field of Formal Languages, which aims to describe multi-agent complex systems. While in classical formal language and automata theory, each grammar (or automata) works individually to generate (or recognize) a language, in this new field a *set* of grammars following a specified protocol work together to generate a language. There are two main types of Grammar Systems: namely *Cooperating Distributed Grammar Systems* (CDGS), which work sequentially, and *Parallel Communicating Grammar Systems* (PCGS), which work in parallel.

Since 1988 –the year in which CDGS were introduced to model syntactic aspects of the blackboard model of problem solving– the theory of Grammar

---

[1]Research supported by an FI Fellowship from the Direcció General de Recerca/CIRIT, Generalitat de Catalunya.

Systems has developed considerably and new variants have been added to the two basic types (for an exhaustive bibliography, see [7]). Research in this area has also been carried out into finding possible applications of the theory ([1], [3], [4], [9], [10], [11], [12], [13]).

In this paper we shall point out some of the features of Grammar Systems that make them potentially suitable for accounting for natural language issues. In order to do so, we have divided the paper into two main sections: the first one outlines some of the features of Grammar Systems that fit with features of what may be a model for natural language; the second one introduces a new variant of Grammar Systems –*Linguistic Grammar Systems*– to show the possible applicability of Grammar Systems Theory to Linguistics.

2   Grammar Systems Suitability in Linguistics

One of the most persistent issues in Linguistics is the question of determining where natural languages are located in the 'Chomsky hierarchy.' The debate started several years ago, but the question still open in Linguistics: can natural languages be characterized using context-free grammars or can't they? Several attempts have been made to prove the non-context-freeness of natural languages. Many authors have attempted to demonstrate the non-context-freeness of natural languages by providing examples of structures that are present in natural languages but which cannot be described using a context-free grammar.

The difficulty of working with context-sensitive grammars has forced researchers to look for ways of generating the non-context-free structures present in natural languages while using context-free rules. This idea has led to *Regulated Rewriting* and to the so-called *Mildly Context-Sensitive Devices* in Formal Languages and Linguistics, respectively.

We would like to propose Grammar Systems as a new tool that may allow us to produce the non-context-free structures present in natural languages using context-free grammars.

2.1   Modularity

The idea of modularity has proved to be very important in a wide range of fields. Cognitive Science, Natural-Language Processing, Computer Science and, of course, Linguistics are, among others, examples of fields in which modular models have been proposed.

If we concentrate on the field of Linguistics, we can see that the modular approach of grammar has been shown to have important consequences on the study of language. This has led many grammatical theories to use modular models. The idea of having a system made up of several independent components (syntax, semantics, phonology, morphology, etc.) –each one governed by its own rules and with its own primitives, all of which interact with one another– seems to be a good way of accounting for linguistic issues. We could cite several modular approaches in grammar theory, from Chomsky's Generative Grammar to Autolexical Syntax ([18]) or Jackendoff's view of Architecture of the Language Faculty ([8]).

It seems, then, that the idea of modularity may be considered as an important feature of language. The same idea of modularity can be found in Grammar Systems Theory. Therefore, 'modularity' is an important feature in the study of language that is also present in Grammar Systems Theory and that may somehow support our initial thesis about the possible adequacy of Grammar Systems Theory in the field of Linguistics.

2.2   Parallelism & Interaction

Traditionally, Linguistics has portrayed a hierarchical view of grammar, where the different organizational dimensions of language are seen as 'levels' obtainable from one another in a certain fixed order. A similar situation can be found in Natural Language Processing where serial models of human sentence processing were used. These models, however, have certain problems that suggest that it would be better to postulate systems with parallel, autonomous components working independently to account for the several organizational dimensions that expressions of natural language manifest. In fact, many recent theories approach language in a parallel way. Sadock's Autolexical Syntax ([18]), Jackendoff's Architecture of Language Faculty ([8]) or parallel interactive models of syntactic processing are just some examples.

If we approach Grammar Systems Theory with the ideas of *parallelism* and *interaction* in mind, we will realize that this new theory can provide a good framework to account for *interaction* between modules and their *parallel* functioning in the language system.

The easy generation of *non-context-free structures* using context-free rules, *modularity, parallelism* and *interaction* are just some of the reasons that, we think, could support the suitability of Grammar Systems in Linguistics. But we could adduce other defining features of such systems that could enable them to be applied to a large range of topics in Linguistics, since Pragmatics to Language Change. *Cooperation, communication, distribution, interdependence among parts, emergent behaviour,* etc. are –in addition to the above-mentioned– features that make Grammar Systems a potentially good candidate to account for linguistic matters.

The section below introduces a new variant of Grammar Systems that may account for the way in which different modules in natural language grammar work and interact with one another.

3   Linguistic Grammar Systems. A Preview

In order to show the applicability of Grammar Systems Theory in Linguistics, we would like to extend the notion of Grammar System to the notion of *Linguistic Grammar System*. A Linguistic Grammar System should be understood as a grammar system whose components are not grammars, but grammar systems[2]. So, we are speaking about a macro-grammar system (i.e. Linguistic Grammar System) composed of several micro-grammar systems (components of Linguistic Grammar System) that, in turn, are composed of several grammars.

We would like to present Linguistic Grammar Systems as a generic tool that includes the main features of what we consider to be a good model for natural language, namely:

- *Modularity.* We need several modules (syntax, semantics, pragmatics, phonology, etc.) working independently, with different alphabets (N, V, Adj... in syntax; Argument, Predicate... in semantics; stems, derivational morphemes... in morphology, etc.), different rules and generating different types of structures (syntactic structure, semantic structure...).

- *Parallelism.* Modules should work simultaneously, in a parallel way.

- *Interaction* among modules. Although modules are independent and work in parallel, they need to interact in order to share information and reach their common goal: generation of language.

- *Coordinator element.* If we stop here, and simply have several modules working independently, and generating different structures, we would get a sequence of structures of different types. But since we want the output of our system to be a *single* language, we need an element that can coordinate the work of all the modules. This special component could be the 'master'.

Taking into account all the above, a Linguistic Grammar System is a *macro-Parallel Communicating Grammar System with separate alphabets* ([14]) where components work independently and in parallel, interacting with one another from time to time.

The modules that build up a Linguistic Grammar System could be Cooperating Distributed Grammar Systems, in which different components work sequentially, cooperating to generate the language of the system (i.e. the corresponding syntactic, semantic, phonological or whatever structure).

Taking into account that a Linguistic Grammar System is a PCGS with separate alphabets, and given the need for modules to communicate with

---

[2]In [15] we find a similar idea in the so-called *hierarchical grammar systems*.

one another, we need a way of *translating* the information from one module into information understandable for another. This problem can be solved by using *PCGS with Renaming* introduced in [16].[3]

Defining *Linguistic Grammar Systems* as PCGS poses another problem: what kind of communication should we use? There are different types of communication within PCGS, in particular we can refer to *communication by request* (defined in the initial model) and to *communication by 'command'* (introduced in [5]). In a very informal way, we can say that while the former type of communication allows components to use 'query symbols' anytime they feel they need the help of another component, in the latter, components can send information when their strings fit with requirements of other modules. So, while in the first type, modules 'ask' for information, in the second type, modules 'give' information (without having been asked for it). In the case of Linguistic Grammar Systems we think it would be best to have a combination of both modes of communication (request and command).

According to what has been said up to now, we have a set of components that generate independently (with or without communication, with communication by command or by request) and in parallel a set of structures (syntactic, semantic, phonological, etc.). The result, then, is an n-tuple of structures $< Syntax, Semantics, Phonology... >$. But somehow we have to unify the different structures in such a way that we have only one language: the language of the system. This is where the 'master' comes in. The 'master' is a special component without *axiom* that can start its work just when it receives strings from every component that make up the system. It has to unify all that information and generate the language of the system. In this way, the language generated by a Linguistic Grammar System will be the result of putting in correspondence (via master) all the structures generated by components of the system. Summing up, a Linguistic Grammar System is a PCGS with renaming, with separate alphabets, composed of CDGS (with output and input filters), non-returning, non-centralized and having two types of communication (request and command.) Before embarking on the formal definition of a Linguistic Grammar System, let us see how it looks in a picture:

---

[3]In PCGS with Renaming we add 'weak codes' to the basic model that allow us to translate the strings generated by a component before communicating them to another module.

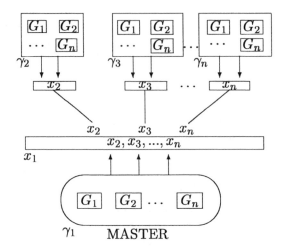

3.1   An Attempt to Formally Define Linguistic Grammar Systems

We assume that the reader is familiar with the basics of Formal Language Theory. For more information we refer to [19], [17].

Definition 1 *A Linguistic Grammar System of degree $n+m$, with $n, m \geq 1$, is an $(n+m+1)-tuple$:*

$$\Gamma = (K, (\gamma_1, I_1,), (\gamma_2, I_2, O_2), ..., (\gamma_n, I_n, O_n), h_1, ..., h_m)$$

where:

- $K = \{Q_1, ..., Q_n, q_1, ..., q_n\}$ *are query symbols, their indices $1, .., n$ pointing to $\gamma_1, ..., \gamma_n$ components, respectively. $Q_1, ..., Q_n$ refers to the whole string of the i-th component, while $q_1, ..., q_n$ refer to a substring of the i-th component.*

- $(\gamma_1, I_1,), (\gamma_2, I_2, O_2), ..., (\gamma_n, I_n, O_n)$ *are the components of the system:*

  - $\gamma_1 = (N_1, T_1, G_1, ..., G_k, f_1)$, *is the 'master' of the system, where:*
    * N_1 *is the non-terminal alphabet.*
    * T_1 *is the terminal alphabet.*
    * *Without axiom.*
    * $G_r = (N_1, T_1, P_r)$, *for $1 \leq r < k$, is a usual Chomsky grammar, where:*
      · N_1 *is the non-terminal alphabet.*
      · T_1 *is the terminal alphabet.*

- P_r are finite sets of rewriting rules over $N_1 \cup T_1 \cup K \cup K'$, where $K' = \{[h_j, Q_i] \mid 1 \le i \le n, 1 \le j \le m\}$, and every $[h_j, Q_i]$ is a symbol.
    * $f_1 \in \{*, t, \le k, \ge k, = k\}$ is the derivation mode of γ_1.
- $\gamma_i = (N_i, T_i, S_i, G_1, ..., G_k, f_i)$, for $2 \le i \le n$, is a CD Grammar System where:
    * N_i is the non-terminal alphabet.
    * T_i is the terminal alphabet.
    * S_i is the axiom.
    * $G_r = (N_i, T_i, P_r)$, for $1 \le r \le k$, is a usual Chomsky grammar, where:
        - N_i is the non-terminal alphabet.
        - T_i is the terminal alphabet.
        - P_r are finite set of rewriting rules over $N_i \cup T_i \cup K \cup K'$, where $K' = \{[h_j, Q_i] \mid 1 \le i \le n, 1 \le j \le m\}$, and every $[h_j, Q_i]$ is a symbol.
    * $f_i \in \{*, t, \le k, \ge k, = k\}$ is the derivation mode of γ_i.
- $I_i \subseteq \bigcup_{j=2}^{n} T_j^*$, $i = 1$ is the input filter of the master.
- $I_i \subseteq \bigcup_{j=1}^{n} (N_j \cup T_j)^*$, $2 \le i \le n$ is the input filter of the i-th component.
- $O_i \subseteq \bigcup_{j=1}^{n} (N_j \cup T_j)^*$ $2 \le i \le n$ is the output filter of the i-th component.

- $h_j : \bigcup_{i=1}^{n} (N_i \cup T_i)^* \to \bigcup_{i=1}^{n} (N_i \cup T_i)^*$, $1 \le j \le m$ are weak codes such that:

    - $h_j(A) = A$, for $A \in N$.
    - $h_j(a) \in T \cup \{\lambda\}$, for $a \in T$.

We write $V_i = N_i \cup T_i \cup K \cup K'$ and $V_\Gamma = \bigcup_{i=1}^{n} (N_i \cup T_i) \cup K \cup K'$. Sets N_i, T_i, K, K' are mutually disjoint for any i, $1 \le i \le n$. We do not require $N_i \cap T_j = \emptyset$ for $1 \le i, j \le n$, $i \ne j$.

Definition 2 *Given a Linguistic Grammar System* $\Gamma = (K, (\gamma_1, I_1,), (\gamma_2, I_2, O_2), \ldots, (\gamma_n, I_n, O_n), h_1, \ldots, h_m)$, *its state is described at any moment by an* n-*tuple* $(x_1, \ldots, x_n)$, *where each* $x_i \subseteq V_i^*$, $1 \le i \le n$, *represents the string that is available at node* i *at that moment.*

Definition 3 *Given a Linguistic Grammar System* $\Gamma = (K, (\gamma_1, I_1), (\gamma_2, I_2, O_2), \ldots, (\gamma_n, I_n, O_n), h_1, \ldots, h_m)$, *for two* n-*tuples* $(x_1, x_2, \ldots, x_n)$, $(y_1, y_2, \ldots, y_n)$, $x_i, y_i \in V_i^*$, $1 \le i \le n$, *we write* $(x_1, \ldots, x_n) \Rightarrow (y_1, \ldots, y_n)$ *if one of the following cases holds:*

1. $|x_i|_K = 0$, $|x_i|_{K'} = 0$, $x_i \notin O_i$, and for each i, $1 \le i \le n$, we have $x_i \Rightarrow y_i$ in the CDGS γ_i, or $x_i \in T_i^*$ and $x_i = y_i$.

   For each γ_i, with $x_i, y_i \in V_i^*$ we write $x_i \Rightarrow_{G_r}^k y_i$, iff $\exists x_1, x_2, \ldots, x_{k+1}$ such that:

   - $x_i = x_1$, $y_i = x_{k+1}$.
   - $x_j \Rightarrow_{G_r} x_{j+1}$, i.e. $x_j = x_j' A_j x_j''$, $x_{j+1} = x_j' w_j x_j''$, $A_j \to w_j \in P_r$, $1 \le j \le k$.

2. There is an i, $1 \le i \le n$, such that $|x_i|_K > 0$, then for each such i we write $x_i = z_1 Q_{i_1} z_2 Q_{i_2} \ldots z_t Q_{i_t} z_{t+1}$, $t \ge 1$, for $z_j \in V_\Gamma^*$, $|z_j|_K = 0$, $1 \le j \le t+1$;if $|x_{i_j}|_K = 0$, $1 \le j \le t$, then $y_i = z_1 x_{i_1} z_2 x_{i_2} \ldots z_t x_{i_t} z_{t+1}$ providing that $y_i \in V_i^*$; when for some j, $1 \le j \le t$, $|x_{i_j}|_K \ne 0$, then $y_i = x_i$; for all i, $1 \le i \le n$, for which y_i is not specified above, we have $y_i = x_i$.

3. There is an i, $1 \le i \le n$, such that $|x_i|_{K'} > 0$, then for each such i we write $x_i = z_1 [h_j, Q_{i_1}] z_2 [h_j, Q_{i_2}] \ldots z_t [h_j, Q_{i_t}] z_{t+1}$, $t \ge 1$, for $z_j \in V_\Gamma^*$, $|z_j|_{K'} = 0$, $1 \le j \le t+1$; if $|x_{i_j}|_{K'} = 0$, $1 \le j \le t$, then $y_i = z_1 h_j(x_{i_1}) z_2 h_j(x_{i_2}) \ldots z_t h_j(x_{i_t}) z_{t+1}$ providing that $y_i \in V_i^*$; when for some j, $1 \le j \le t$, $|x_{i_j}|_{K'} \ne 0$, then $y_i = x_i$; for all i, $1 \le i \le n$, for which y_i is not specified above, we have $y_i = x_i$.

4. $(x_1, \ldots, x_n) \vdash (y_1, \ldots, y_n)$ iff $y_i = x_i \bigcup (I_i \cap (\bigcup_{j=1, i \ne j}^n O_j \cap x_j))$, for $i = 1, \ldots, n$.

Point 1 defines a *rewriting step*, whereas points 2, 3 and 4 define *communication steps*. In 1 no query symbol Q_i, q_i, or $h_j(Q_i)$ is present in the current string and the string doesn't match the output filter of the CDGS, so no communication (by request or command) can be done. In this case we perfom a *rewriting step*. In 2 we define a *communication step by request without renaming*. Some query symbols, say $Q_{i_1} \ldots Q_{i_l}$ (or $q_{i_1} \ldots q_{i_l}$) appear in a string x_i. In this case rewriting stops and some communication steps are performed. Every symbol Q_{i_j} (or q_{i_j}), $1 \le j \le l$, must be replaced with the current string (or substring) of the component γ_{i_j}, say x_{i_j}, assuming that no x_{i_j}, $1 \le j \le l$, contains a query symbol. If one of the strings x_{i_j}, $1 \le j \le l$, also contains query symbols these symbols must be replaced with the requested strings before communicating that string. In 3, we define a *communication step by request with renaming*. In this case some query symbols $[h_j, Q_{i_1}] \ldots [h_j, Q_{i_l}]$ appear in a string x_i. Everything works as in the case of the communication by request without renaming with the only difference that here $[h_j, Q_{i_1}] \ldots [h_j, Q_{i_l}]$ must be replaced not by the string x_{i_j}, but by $h_j(x_{i_j})$. And finally, in 4, we define a *communication step by command*. In this case, copies of those strings which are able to pass the output filter of some γ_j and the input filter of some γ_i $(i \neq j)$ join (concatenated in the order of system components) the string present at γ_i.

Definition 4 *The language generated by a Linguistic Grammar System as above is:*

$$L(\Gamma) = \{x \in T_1^* \mid (\emptyset, S_2, \ldots, S_3) \Rightarrow (\emptyset, \alpha_2^{(1)}, \ldots, \alpha_n^{(1)}) \vdash (\emptyset, y_2^{(1)}, \ldots, y_n^{(1)})$$
$$\Rightarrow (\emptyset, \alpha_2^{(2)}, \ldots, \alpha_n^{(2)}) \vdash (\emptyset, y_2^{(2)}, \ldots, y_n^{(2)}) \Rightarrow \ldots \Rightarrow (\emptyset, \alpha_2^{(s)}, \ldots, \alpha_n^{(s)}) \vdash$$
$$(\alpha_2^{(s)}, \ldots \alpha_n^{(s)}, \alpha_2^{(s)}, \ldots, \alpha_n^{(s)}) \Rightarrow \ldots \Rightarrow (x, \alpha_2^{(s)}, \ldots, \alpha_n^{(s)}), s \geq 1, \alpha_i^{(s)} \in T_i^*, 2 \leq i \leq n\}.$$

Notice that we start with the set of axioms of the components $\gamma_2, \ldots, \gamma_n$ of a Linguistic Grammar System and with an empty set in the master module (γ_1). We perform derivation and communication steps until the master (which has not axiom) produces a terminal string, x.

4   Final Remarks

Our aim in this paper was to show the possible adequacy of Grammar Systems Theory in the field of Linguistics. We have presented some traits that may justify applying this theory to the study of language. *Modularity, easy generation of non-context-free structures, parallelism, interaction, co-operation, distribution* and other features have been adduced as important notions for language that can be captured by using Grammar Systems to study linguistic matters. By no means was our purpose to make an exhaustive revision of all the aspects that prove that this theory is suitable in Linguistics. Nevertheless, we have attempted to show its applicability by means of a new variant of grammar systems –the so-called *Linguistic Grammar Systems*– which we have introduced to show how the different modules that make up a grammar work and interact with one another to generate an acceptable language structure. We have informally presented the model and attempted to give a formal definition.

We know how difficult introducing a new theory in the study of natural language can be but, taking into account the properties of Grammar Systems, we think it would be worthwhile trying to apply this theory in Linguistics. This paper is just one small example of how valuable the application of Grammar Systems Theory in Linguistics could be.

References

[1] E. Csuhaj-Varjú, Grammar systems: a multi-agent framework for natural language generation. In Gh. Păun (ed.), *Mathematical Aspects of*

Natural and Formal Languages. World Scientific, Singapore, 1994, 63–78.

[2] E. Csuhaj-Varjú, J. Dassow, J. Kelemen and Gh. Păun, *Grammar Systems: A Grammatical Approach to Distribution and Cooperation*. Gordon and Breach, London, 1994.

[3] E. Csuhaj-Varjú and M.D. Jiménez-López, Cultural eco-grammar systems: a multi-agent system for cultural change. In A. Kelemenová (ed.), *Proceedings of the MFCS'98 Satellite Workshop on Grammar Systems*, Silesian University, Opava, Czech Republic, 1998, 165–182.

[4] E. Csuhaj-Varjú, M.D. Jiménez-López and C. Martín-Vide, Pragmatic eco-rewriting systems': pragmatics and eco-rewriting systems. In Gh. Păun and A. Salomaa (eds.), *Grammatical Models of Multi-Agent Systems*. Gordon and Breach, London, 1999, 262–283.

[5] E. Csuhaj-Varjú, J. Kelemen and Gh. Păun, Grammar systems with WAVE-like communication. *Computers and AI*, 15/5 (1996), 419–436.

[6] J. Dassow, Gh. Păun, *Regulated Rewriting in Formal Language Theory*. Springer, Berlin, 1989.

[7] J. Dassow, Gh. Păun and G. Rozenberg, Grammar systems. In G. Rozenberg, A. Salomaa (eds.), *Handbook of Formal Languages*. Springer, Berlin, 1997, vol. 2, 155–213.

[8] R. Jackendoff, *The Architecture of Language Faculty*. MIT Press, Cambridge, 1997.

[9] M.D. Jiménez-López, Sistemas de gram'aticas y lenguajes naturales: ideas intuitivas al respecto. In C. Martín-Vide (ed.), *Lenguajes Naturales y Lenguajes Formales XII*. PPU, Barcelona, 1996, 223–236.

[10] M.D. Jiménez-López, Cultural eco-grammar systems: agents between choice and imposition. A preview. In G. Tatai & L. Gulyás (eds.), *Agents Everywhere*. Springer, Budapest, 1999, 181–187.

[11] M.D. Jiménez-López, *Grammar Systems: A Formal-Language-Theoretic Framework for Linguistics and Cultural Evolution*, PhD Dissertation, Universitat Rovira i Virgili, Tarragona, 2000.

[12] M.D. Jiménez-López and C. Martín-Vide, Grammar systems for the description of certain natural language facts. In Gh. Păun, A. Salomaa (eds.), *New Trends in Formal Languages*. Springer, Berlin, 1997, 288–298.

[13] M.D. Jiménez-López and C. Martín-Vide, Grammar Systems and Autolexical Syntax: Two Theories, One Single Idea. In R. Freund & A. Kelemenová (eds.), *Grammar Systems 2000*. Silesian University, Opava, 2000, 283–296

[14] V. Mihalache, PC grammar systems with separated alphabets. *Acta Cybernetica*, 12/4 (1996), 397–409.

[15] V. Mitrana, Gh. Păun and G. Rozenberg, Structuring grammar systems by priorities and hierarchies. *Acta Cybernetica*, 11/3 (1994), 189–204.

[16] Gh. Păun, PC grammar systems and natural languages, private communication.

[17] G. Rozenberg and A. Salomaa (eds.), *Handbook of Formal Languages*. Springer, Berlin, 1997.

[18] J.M. Sadock, *Autolexical Syntax. A Theory of Parallel Grammatical Representations*. University of Chicago Press, Chicago, 1991.

[19] A. Salomaa, *Formal Languages*. Academic Press, New York, 1973.

Multi-Bracketed Contextual Rewriting Grammars with Obligatory Rewriting

Martin Kappes

Fachbereich Informatik
Johann Wolfgang Goethe University
Frankfurt am Main, Germany
kappes@psc.informatik.uni-frankfurt.de

Abstract. We study the generative capacity and closure properties of multi-bracketed contextual rewriting grammars with obligatory rewriting. This model is a generalization of multi-bracketed contextual rewriting grammars. They possess an induced Dyck-structure to control the derivation process and to provide derivation trees. It will be shown that this class of grammars is closed under intersection with regular sets.

1 Motivation and Introduction

Contextual grammars were introduced by Marcus in [8]. They are a formalization of the linguistic idea that more complex well formed strings are obtained by inserting contexts into already well formed strings. Therefore, these grammars are based on the principle of adjoining.

Multi-bracketed contextual grammars were introduced in [5]. Generally speaking, this is a class of contextual grammars working on an induced Dyck-structure which controls the derivation process and also provides context-free-like derivation trees. They were generalized to multi-bracketed contextual rewriting grammars in [7].

Here we will study a further generalization of this model called multi-bracketed contextual rewriting grammars with obligatory adjoining. In the models studied in [5] and [7], each string derived by a grammar G in a finite number of derivation steps belongs to the language generated by G. We will now "filter" these strings by imposing the restriction that only certain generated strings are in the language of G as only strings consisting exclusively of terminal symbols are in the language generated by a Chomsky grammar. We will briefly study the generative capacity of these grammars. Our main result is that, in contrast to the classes investigated in [5] and [7], the classes of languages generated by those grammars are closed under intersection with regular sets.

All the models investigated in this paper are based on so-called internal contextual grammars which were introduced by Păun and Nguyen in [11]. Detailed information on contextual grammars can be found in the monograph [10]; a more compressed source of information is [1]. The first approach to induce a bracket structure in contextual grammars was so called bracketed contextual grammars, introduced by Martín-Vide and Păun in [9]; the generative capacity of this class was studied in [6].

2 Definitions

Let Σ^* denote the free monoid generated by the finite alphabet Σ and $\Sigma^+ = \Sigma^* - \{\lambda\}$, where λ denotes the empty word. FIN, REG, CF and CS denote the families of finite, regular, context-free and context-sensitive languages. We assume the reader is familiar with the common notions of formal language theory as presented in [3].

For an alphabet Γ we define the projection to Γ via $pr_\Gamma(\sigma) = \sigma$, if $\sigma \in \Gamma$, and λ otherwise. Let Δ denote a finite set of indices. By $B_\Delta = \{[_A,]_A \mid A \in \Delta\}$ we define the bracket alphabet induced by Δ. By D_Δ we denote the Dyck-Language over B_Δ (see [2]).

Let Σ and Δ denote two alphabets. The set of all Dyck-covered words over Σ with respect to the index alphabet Δ is given by:

$$\mathrm{DC}_\Delta(\Sigma) \;=\; \{\alpha \in (\Sigma \cup B_\Delta)^+ \mid \alpha = [_A\alpha']_A \text{ for an } A \in \Delta,\ pr_{B_\Delta}(\alpha') \in D_\Delta$$
$$\text{and for all decompositions } \alpha = \alpha_1[_B\alpha_2]_B\alpha_3 \text{ with } \alpha_2 \in \Sigma^*$$
$$\text{we have } \alpha_2 \neq \lambda\} \cup \{\lambda\}.$$

Furthermore, for each $A \in \Delta$ we define $\mathrm{DC}_\Delta^A(\Sigma) = \{\alpha \in \mathrm{DC}_\Delta(\Sigma) \mid \alpha = [_A\alpha']_A\}$. Throughout the paper we always assume $\Sigma \cap B_\Delta = \emptyset$. Notice that the first and the last symbol of each non-empty Dyck-covered word is a pair of brackets belonging together.

It is easy to see that each $\alpha \in \mathrm{DC}_\Delta(\Sigma)$ can be interpreted as unique encoding for a tree, where Δ is the label alphabet for the internal nodes and Σ is the label alphabet for the leaf nodes in the following way: a string

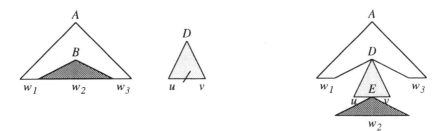

Figure 1: The derivation process in an MBICR-grammar: a context (μ, ν) may be adjoined to an $\alpha = \alpha_1[_B\alpha_2]_B\alpha_3$ yielding a tree $\alpha_1\mu[_E\alpha_2]_E\nu\alpha_3$ if and only if there is an $(S, C, K, H) \in P$ such that $pr_\Sigma(\alpha_2) \in S$, $(\mu, \nu) \in C$, $[_B\alpha_2]_B \in DC_\Delta^B(\Sigma)$, $B \in K$ and $E \in H$. In the above figure, we have $\alpha \in DC_\Delta^A(\Sigma)$, $\mu\nu \in DC_\Delta^D(\Sigma)$ $pr_\Sigma(\alpha_1) = w_1$, $pr_\Sigma(\alpha_2) = w_2$, $pr_\Sigma(\alpha_3) = w_3$, $pr_\Sigma(\mu) = u$ and $pr_\Sigma(\nu) = v$.

$[_A\alpha]_A \in DC_\Delta(\Sigma)$ is identified with a tree where the root is labelled by A, and the subtrees of the root are determined by the unique decomposition of $\alpha = \alpha_1\alpha_2 \ldots \alpha_n$ such that $\alpha_i \in (DC_\Delta(\Sigma) - \{\lambda\}) \cup \Sigma$, $1 \leq i \leq n$.

A multi-bracketed contextual rewriting grammar (MBICR) is a tuple $G = (\Sigma, \Delta, \Omega, P)$, where Σ is a finite set of terminals, Δ is a finite set of indices, $\Omega \subseteq DC_\Delta(\Sigma)$ is a finite set of axioms and P is a finite set of tuples (S, C, K, H), where $S \subseteq \Sigma^+$, $K, H \subseteq \Delta$ and C is a finite subset of $(\Sigma \cup B_\Delta)^+ \times (\Sigma \cup B_\Delta)^+$ such that, for all $(\mu, \nu) \in C$, we have $\mu\nu \in DC_\Delta(\Sigma)$. The derivation relation "$\Longrightarrow_G$" on $(\Sigma \cup B_\Delta)^*$ is defined by $\alpha \Longrightarrow_G \beta$ if and only if $\alpha = \alpha_1[_A\alpha_2]_A\alpha_3$, $\beta = \alpha_1\mu[_D\alpha_2]_B\nu\alpha_3$, $[_A\alpha_2]_A \in DC_\Delta(\Sigma)$ and there is an $(S, C, K, H) \in P$, such that $pr_\Sigma(\alpha_2) \in S$, $(\mu, \nu) \in C$, $A \in K$ and $B \in H$. The structure language generated by G (or strong generative capacity of G) is $T(G) = \{\beta \in (\Sigma \cup B_\Delta)^* \mid$ there is an $\alpha \in \Omega$ such that $\alpha \overset{*}{\Longrightarrow}_G \beta\}$, where "$\overset{*}{\Longrightarrow}_G$" denotes the reflexive transitive closure of the derivation relation. The language generated by G (or weak generative capacity of G) is $L(G) = \{pr_\Sigma(\beta) \mid \beta \in T(G)\}$. An MBICR-grammar $G = (\Sigma, \Delta, \Omega, P)$ is with F-choice for a family of languages F, if $S \in F$ for all $(S, C, K, H) \in P$. By MBICR(F) we denote the set of all languages which can be generated by an MBICR(F)-grammar. The derivation process in an MBICR-grammar is illustrated in Figure 1.

A special case occurs in MBICR(F)-grammars, where for all $(S, C, K, H) \in P$ there is an $A \in \Delta$ such that $K = H = \{A\}$ and $\mu\nu \in DC_\Delta^A(\Sigma)$ for all $(\mu, \nu) \in C$. These grammars are called MBIC-grammars and were studied in [5].

In MBICR-grammars, we have $\beta \in T(G)$ for each β such that there is an $\alpha \in \Omega$ with $\alpha \overset{*}{\Longrightarrow}_G \beta$ and $pr_\Sigma(\beta) \in L(G)$ for each $\beta \in T(G)$. In this sense, each derivation step immediately yields a string in $T(G)$ and $L(G)$.

Analogously to the concept of terminal symbols in Chomsky grammars, we will now impose the restriction that only some indices $\Upsilon \subseteq \Delta$ are considered "valid" and that a string β derived from an axiom in a finite number of derivation steps is only in $T(G)$ if $\beta \in (B_\Upsilon \cup \Sigma)^*$, i.e. it only contains brackets with "valid" indices.

A multi-bracketed contextual rewriting grammar with obligatory adjoining (MBICRO) is a tuple $G = (\Sigma, \Delta, \Upsilon, \Omega, P)$, where Σ, Δ, Ω and P are defined as in an MBICR-grammar and $\Upsilon \subseteq \Delta$ is a set of permitted indices. The derivation process is defined as in an MBICR-grammar but the strong and weak generative capacities of G are given by $T(G) = \{\beta \in DC_\Upsilon(\Sigma) \mid$ there is an $\alpha \in \Omega$ such that $\alpha \stackrel{*}{\Longrightarrow}_G \beta\}$ and $L(G) = \{pr_\Sigma(\beta) \mid \beta \in T(G)\}$. Thus, all brackets indexed by symbols from $\Delta - \Upsilon$ have to be replaced during the derivation process in order to obtain a string in $T(G)$.

Let us consider the following example:

$$G = (\{a, b, c, d, e\}, \{A, B\}, \{A\}, \{[_A a[_B bc]_B d]_A\}, \{\pi_1, \pi_2\}),$$

where:

$$\begin{aligned} \pi_1 &= (\Sigma^+, \{([_A a[_B b, c]_B d]_A)\}, \{B\}, \{A\}), \\ \pi_2 &= (\Sigma^+, \{([_A e, e]_A)\}, \{B\}, \{A\}). \end{aligned}$$

It is not difficult to see that using π_1 i times yields a derivation:

$$[_A a[_B bc]_B d]_A \stackrel{i}{\Longrightarrow}_G ([_A a)^{i+1}[_B b([_A b)^i(c]_A)^i c]_B(d]_A)^{i+1}.$$

In order to obtain a string in $T(G)$ we have to use production π_2 exactly once to remove the pair of brackets indexed by B from the sentential form. After applying π_2 once, no further derivation steps are possible. Hence $L(G) = \{a^n e b^n c^n e d^n \mid n \geq 1\}$.

3   Generative Capacity

We will now investigate the generative capacity of MBICRO-grammars.

Theorem 1 *The diagram in Figure 2 represents the relation between the various language classes, where a (dashed) arrow indicates a (not necessarily) strict inclusion and families not linked by a path in the diagram are not necessarily incomparable.*

Proof.   For the results about MBIC- and MBICR-Grammars we refer the reader to [5] and [7]. Since each language generated by an MBICR-grammar possesses the so-called internal bounded step-property (cf. [10]) and since the language generated in the above example does not possess this property, we have MBICR(F) $\subset$ MBICRO(F) for all families of languages F with $\Sigma^+ \in F$.

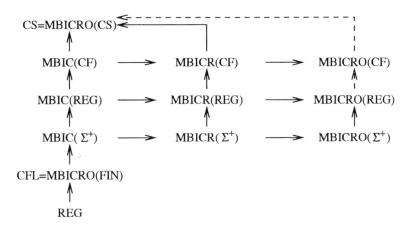

Figure 2: Generative capacity of MBICRO-Grammars.

Lemma 6 in [5] proves that CF $\subseteq$ MBIC(FIN). Lemma 4.5 in [7] proves that for each MBICR(FIN) grammar G there is a context free grammar G' with $L(G') = T(G)$. Clearly, for each MBICRO(F) grammar $G = (\Sigma, \Delta, \Upsilon, \Omega, P)$ we have $T(G) = T(G'') \cap (B_\Upsilon \cup \Sigma)^*$ for the MBICR(F)-grammar $G'' = (\Sigma, \Delta, \Omega, P)$. As context-free languages are closed under intersection with regular languages and morphisms, we also have MBICRO(FIN) $\subseteq$ CF. Hence MBICRO(FIN) = CF.

In Lemma 4.6 of [7] it was shown that there is a language which can be generated by an MBIC(REG)-grammar but there is no so-called tree adjoining grammar (c.f. [4] for details) generating this language. Furthermore, Lemma 3.2 in [7] presents a construction to convert a given MBICR(Σ^+)-grammar into a tree adjoining grammar generating the same language. It is straightforward to modify this construction so that it yields the same result for MBICRO(Σ^+)-grammars. Therefore, MBIC(REG) $-$ MBICRO(Σ^+) $\neq \emptyset$.

Let $L \subseteq \Sigma^*$ be a context-sensitive language. We construct the MBICRO-grammar $G = (\Sigma, \{A, B\}, \{A\}, \Omega, \{\pi\} \cup \{\pi_\sigma \,|\, \sigma \in \Sigma\})$, where $\Omega = \{\lambda \,|\, \lambda \in L\} \cup \{[_Ax]_A \,|\, x \in L, |x| = 1\} \cup \{[_B\sigma]_B \,|\, \sigma \in \Sigma\}$, $\pi = (\Sigma^+, \{([_B\sigma, \lambda]_B) \,|\, \sigma \in \Sigma\}, \{B\}, \{A\})$ and $\pi_\sigma = (\{x \in \Sigma^+ \,|\, \sigma x \in L\}, \{([_A\sigma, \lambda]_A)\}, \{B\}, \{A\})$. Since the family of context-sensitive languages is closed under quotient with singleton sets, all selector languages are context-sensitive, and it is not difficult to prove $L(G) = L$. Hence MBICRO(CS) = CS.   □

4   Closure Properties

It is easy to prove that MBICRO(F) is closed under union, concatenation and Kleene-star for all families of languages F with FIN $\subseteq F$ by using the same constructions as for MBIC and MBICR-grammars (cf. [5] for details).

The classes MBIC(F) and MBICR(F) are not closed under intersection with regular languages if $\Sigma^+ \in F$. For MBICR-grammars this result follows from the introductory example given in this paper: the language $L = \{a^n \sigma b^n c^n \sigma c^n \mid n \geq 1, \sigma \in \{\lambda, e\}\}$ can be generated by an MBICR(Σ^+)-grammar. However, $L \cap a^* e b^* c^* e d^* = \{a^n e b^n c^n e d^n \mid n \geq 1\}$ cannot be generated by any MBICR-grammar.

In what follows, we will give a construction to prove that MBICRO(F) is closed under intersection with regular languages for arbitrary families of languages F.

Theorem 2 *For all families of languages F, MBICRO(F) is closed under intersection with regular languages.*

Proof. Let $G'' = (\Sigma, \Delta, \Upsilon, \Omega'', P'')$ be an arbitrary MBICRO(F)-grammar and R a regular language. Without loss of generality, we assume $\Sigma \cap \Delta = \emptyset$. In order to simplify the construction we will first transform G'' into a normal form. Consider the mapping $t : \Sigma \cup B_\Delta \to (\Sigma \cup B_{\Delta \cup \Sigma})^*$ defined by $t(\sigma) = \sigma$ if $\sigma \in B_\Delta$ and $[_\sigma \sigma]_\sigma$ if $\sigma \in \Sigma$. It is easy to see that applying the homomorphic extension of t to an $\alpha \in DC_\Delta(\Sigma)$ yields a string $t(\alpha) \in DC_{\Delta \cup \Sigma}(\Sigma)$, where each symbol $\sigma \in \Sigma$ is replaced by $[_\sigma \sigma]_\sigma$. Furthermore, for the MBICRO(F)-grammar $G = (\Sigma, \Delta \cup \Sigma, \Upsilon \cup \Sigma, \Omega, P)$, where $\Omega = \{t(\alpha) \mid \alpha \in \Omega''\}$ and:

$$P = \{(S, \{(t(\mu), t(\nu)) \mid (\mu, \nu) \in C\}, K, H) \mid \text{ there is an } (S, C, K, H) \in P''\},$$

we have $\alpha \in T(G)$ if and only if there is an $\alpha'' \in T(G'')$ such that $t(\alpha'') = \alpha$. Hence $L(G) = L(G'')$. Figure 3 shows an example for this transformation.

Since R is regular, there exists a deterministic finite automaton $M = (Q, \Sigma, \delta, q_0, F)$ with $L(M) = R$ (cf. [3] for notational details). We construct the index-set $\Phi = \{(A, [p, q], [r, s]) \mid A \in \Delta \cup \Sigma, p, q, r, s \in Q\}$. For an intuitive explanation, let us take a look at the tree interpretation: If a node is labelled by $(A, [p, q], [r, s])$, then $[p, q]$ is a value propagated from the immediate predecessor of the node stating that this node is supposed to generate a yield w such that $\delta(p, w) = q$. The tuple $[r, s]$ denotes that the immediate successors of the node are supposed to generate a yield w such that $\delta(r, w) = s$. Now we construct a mapping *convert* which relabels the indices of the brackets in all possible ways such that for the resulting strings α the following properties hold:

(1) For each partition $\alpha = \alpha_1 \alpha_2 \alpha_3$ such that $\alpha_2 \in t(DC_\Delta(\Sigma)) - \{\lambda\}$ we have $\alpha_2 = [_X \gamma_1 \ldots \gamma_n]_X$, where $X = (A, [p, q], [p_0, p_n])$, $\gamma_i \in t(DC_\Delta(\Sigma) \cup \Sigma) - \{\lambda\}$ and $\gamma_i = [_{Y_i} \gamma_i']_{Y_i}$, $Y_i = (B_i, [p_{i-1}, p_i], [r_i, s_i])$, $1 \leq i \leq n$. See Figure 4 for an illustration.

(2) For each partition $\alpha = \alpha_1 \alpha_2 \alpha_3$ such that $\alpha_2 \in t(\Sigma)$ we have $\alpha_2 = [_X \sigma]_X$, where $X = (\sigma, [p, q], [r, s])$ and $\delta(r, \sigma) = s$.

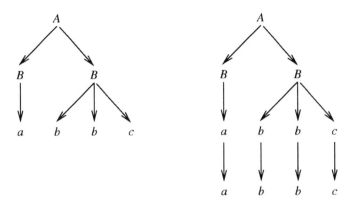

Figure 3: Example for the normal form used to prove the closure under intersection with regular sets: the tree $\alpha = [_A[_Ba]_B[_Bbbc]_B]_A$ is converted into the normal form tree $t(\alpha) = [_A[_B[_aa]_a]_B[_B[_bb]_b[_bb]_b[_cc]_c]_B]_A$. Notice that $pr_\Sigma(\alpha) = pr_\Sigma(t(\alpha))$.

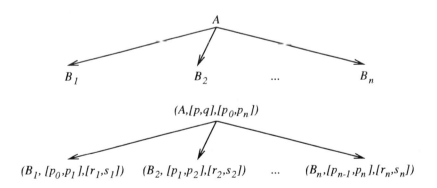

Figure 4: Example for the mapping *convert*. Here, a node is shown with its immediate nonterminal successors $B_1, \ldots, B_n$. Applying the mapping *convert* with $[p, q]$ as set of states to the above tree yields exactly all trees of the above form for arbitary $p_i, r_i, s_i \in Q$, $0 \leq i \leq n$.

Formally we define the mapping $convert : t(DC_\Delta(\Sigma) \cup \Sigma) \times (Q \times Q) \to 2^{DC_\Phi(\Sigma)}$ as follows: if $\alpha = [_A\alpha_1 \dots \alpha_n]_A$ with $\alpha_i \in t(DC_\Delta(\Sigma) \cup \Sigma) - \{\lambda\}, 1 \leq i \leq n$, then:

$$convert(\alpha, [p,q]) = \{[_X\beta_1 \dots \beta_n]_X \mid \quad X = (A, [p,q], [p_0, p_n]),$$
$$\beta_i \in convert(\alpha_i, [p_{i-1}, p_i]), 1 \leq i \leq n,$$
$$p_i \in Q, 0 \leq i \leq n\}.$$

If $\alpha = [_\sigma\sigma]_\sigma$ for a $\sigma \in \Sigma$, then:

$$convert(\alpha, [p,q]) = \{[_X\sigma]_X \mid X = (\sigma, [p,q], [r,s]) \text{ and } \delta(r, \sigma) = s \text{ for } r, s \in Q\}.$$

Furthermore, $convert(\lambda, [p,q]) = \lambda$ if $\lambda \in R$ and $\emptyset$ else.

Notice that there is exactly one decomposition of the above kind for each $\alpha \in t(DC_\Delta(\Sigma) \cup \Sigma)$. Analogously, we also have to define a mapping $convert2$ which relabels the contexts of the grammar G: for an arbitrary (μ, ν) such that $\mu\nu \in t(DC_\Delta(\Sigma))$ and arbitrary $[p,q], [p',q'] \in Q \times Q$, we define the mapping $convert2(\mu, \nu, [p,q], [p',q']) = \{([_X\beta_1 \dots \beta_{k-1}\xi, \rho\beta_{k+1} \dots \beta_n]_X) \mid X = (A, [p,q], [p_0, p_n]), \beta_i \in convert(\alpha_i, [p_{i-1}, p_i]), 1 \leq i \neq k \leq n :$

$$(\xi, \rho) \in \begin{cases} convert2(\alpha'_k, \alpha''_k, [p_{k-1}, p_k], [p', q']) & \text{if } \alpha'_k\alpha''_k \neq \lambda \\ \{(\lambda, \lambda)\} & \text{if } \alpha'_k\alpha''_k = \lambda \text{ and } [p_{k-1}, p_k] = [p', q'] \\ \emptyset & \text{if } \alpha'_k\alpha''_k = \lambda \text{ and } [p_{k-1}, p_k] \neq [p', q'], \end{cases}$$

where $\mu = [_A\alpha_1 \dots \alpha_{k-1}\alpha'_k, \nu = \alpha''_k\alpha_{k+1} \dots \alpha_n]_A$ with $\alpha_i \in t(DC_\Delta(\Sigma) \cup \Sigma) - \{\lambda\}, 1 \leq i \neq k \leq n, \alpha'_k\alpha''_k \in t(DC_\Delta(\Sigma) \cup \Sigma), \alpha'_k, \alpha''_k \notin t(DC_\Delta(\Sigma) \cup \Sigma) - \{\lambda\}$ and $p_i \in Q, 0 \leq i \leq n$. Notice that there is exactly one such decomposition for each (μ, ν) such that $\mu\nu \in t(DC_\Delta(\Sigma)) - \{\lambda\}$.

We define the grammar $G' = (\Sigma, \Phi, \Phi', \Omega', P')$, where:

$$\begin{aligned} \Omega' &= \{convert(\alpha, [q_0, f]) \mid \alpha \in \Omega, f \in F\}, \\ \Phi' &= \{(A, [p,q], [p,q]) \in \Phi \mid A \in \Upsilon \cup \Sigma, p, q \in Q\}, \end{aligned}$$

and:

$$\begin{aligned} P' = \{(S, C', K', H') \mid \quad &\text{there is an } (S, C, K, H) \in P, \\ &[p,q], [p',q'], [r,s] \in Q \times Q \text{ such that} \\ &C' = \{convert2(\mu, \nu, [p,q], [p',q']) \mid (\mu, \nu) \in C\}, \\ &K' = \{(A, [p,q], [r,s]) \mid A \in K\}, \\ &H' = \{(B, [p',q'], [r,s]) \mid B \in H\}\}. \end{aligned}$$

It is not difficult to prove that, if $\alpha \in \Omega$ and $\alpha \overset{*}{\Rightarrow}_G \beta$, then for all $\beta' \in convert(\beta, [q_0, f])$ for an $f \in F$ there is an $\alpha' \in convert(\alpha, [q_0, f])$ such that $\alpha' \in \Omega'$ and $\alpha' \overset{*}{\Rightarrow}_{G'} \beta'$. On the other hand, if $\alpha' \in \Omega'$ and $\alpha' \Rightarrow_{G'} \beta'$, then there is an $f \in F$, an $\alpha \in \Omega$ and a β such that $\alpha' \in convert(\alpha, [q_0, f])$, $\beta' \in convert(\beta, [q_0, f])$ and $\alpha \overset{*}{\Rightarrow}_G \beta$. Furthermore, since properties (1) and

(2) hold for all strings derived in G', it is not difficult to see that for each $\beta' \in T(G')$ we have $pr_\Sigma(\beta') \in R$. Also, for each $\beta \in T(G)$ such that $pr_\Sigma(\beta) \in R$ there is a $\beta' \in convert(\beta, [q_0, \delta(q_0, pr_\Sigma(\beta))])$ such that $\beta' \in T(G')$.

Using these facts we can show that $L(G') = L(G) \cap R = L(G'') \cap R$. Since the selector languages of G'' and G' are identical, the claim follows.

□

5 Conclusion

We have studied the generative capacity and the closure properties of multi-bracketed contextual rewriting grammars with obligatory rewriting. The family of languages generated by an MBICRO(F)-grammar is closed under intersection with regular languages for arbitrary families of languages F. The questions whether MBICRO(REG) $\subset$ MBICRO(CF) and whether MBICRO(CF) $\subset$ CS or not remain open.

References

[1] A. Ehrenfeucht, Gh. Păun and G. Rozenberg, Contextual grammars and formal languages. In G. Rozenberg and A. Salomaa (eds.), *Handbook of Formal Languages*. Springer, Berlin, 1997, vol. 2, 237–294.

[2] M.A. Harrison, *Introduction to Formal Language Theory*. Addison-Wesley, Reading, Mass., 1978.

[3] J.E. Hopcroft and J.D. Ullman, *Introduction to Automata Theory, Languages and Computation*. Addison-Wesley, Reading, Mass., 1979.

[4] A.K. Joshi and Y. Schabes, Tree-adjoining grammars. In G. Rozenberg and A. Salomaa (eds.), *Handbook of Formal Languages*. Springer, Berlin, 1997, vol. 3, 69–123.

[5] M. Kappes, Multi-bracketed contextual grammars, *Journal of Automata, Languages and Combinatorics*, 3.2 (1998), 85–103.

[6] M. Kappes, On the generative capacity of bracketed contextual grammars, *Grammars*, 1.2 (1998), 91–101.

[7] M. Kappes, Multi-bracketed contextual rewriting grammars, *Fundamenta Informaticae*, 38.8 (1999), 257–280.

[8] S. Marcus, Contextual grammars, *Revue Roumaine des Mathématiques Pures et Appliquées*, 14.10 (1969), 1525–1534.

[9] C. Martín-Vide and Gh. Păun, Structured contextual grammars, *Grammars*, 1.1 (1998), 33–55.

[10] Gh. Păun, *Marcus Contextual Grammars*. Kluwer, Dordrecht, 1997.

[11] Gh. Păun and X.M. Nguyen, On the inner contextual grammars, *Revue Roumaine des Mathématiques Pures et Appliquées*, 25.4 (1980), 641–651.

Semi-Top-Down Syntax Analysis

Jaroslav Král

Michal Žemlička

Faculty of Mathematics and Physics
Charles University
Prague, Czech Republic
kral@ksi.mff.cuni.cz
zemlicka@ksi.ms.mff.cuni.cz

Abstract. We show that the parsers developed for LR grammars can be modified so that they can produce outputs intuitively near to top-down parses for LR grammars. The parsers can be implemeted by recursive procedures, i.e. they can behave like recursive descent parsers. We present a class of grammars with left recursive symbols that are well suited to this technique.

1 Introduction

Many modern compilers use parsers that are implemented as a collection of recursive procedures (procedure driven parsers, PDP) because they have many software engineering advantages. Such parsers can be well understood by human beings, and they can be easily documented, and manually modified. The manual modification is often necessary to support effectiveness, error recovery, pecularities of semantics, etc. PDP can be very effective.

In the theory of parsing, procedure driven parsing is an implementation of recursive descent parsing. Recursive descent parsing works, however, for LL grammars only. The grammars of programming languages are not LL.

PDP in compilers must therefore be adapted by ad hoc modifications so that
they can work for non LL grammars and languages.

On the other hand, the output of LR parsers in compilers is something
more than bottom-up parses in the classical sense, i.e. more than merely
the reversed right-most derivations. We show that the output of LR parsers
can have a form that contains the maximal possible amount of on-line information about the syntactical structure of the input string and that LR
parsers can be implemented as PDP. In order to simplify the explanation,
we shall assume a lookahead of length one. We use, unless stated otherwise,
the notation from [7] and [1].

All the grammars discussed below are LR(1). The generalization to general LR(k) case is straightforward. $G, G0, G1, G2, \ldots$ will denote grammars.
N, T, P are nonterminal and terminal alphabets and the set of productions,
respectively. $p_1, p_2, \ldots$ are productions. $V = N \cup T$. We write P_G, T_G, N_G, V_G
if G cannot be understood implicitly.

The procedures in procedure driven parsers (PDP) usually contain calls
of semantic routines. The actions of any PDP are read and lexical operations
of tokens from input and the calls of semantic routines are activated from
the PDP procedures. It leads us to the following convention.

The output of the parser $\mathcal{P}$ for the input $x = x_1 x_2 \ldots x_n$ is the string
$out_{\mathcal{P}}(x) = \sigma_0 x_1 \sigma_1 x_2 \sigma_2 \ldots$, where x_i are input tokens and σ_i are sequences
of *semantic symbols* for $i = 0, 1, 2, \ldots, n$. The semantic symbols are new
symbols that are different from symbols in $N \cup T$.

We shall further assume without any substantial loss of generality and/or
applicability of the results given below that the grammars contain no rules
of the form $A \to \lambda$, where λ is an empty string. We further assume, that the
initial symbol S' is the left hand side of just one production $S' \to \vdash S \dashv$ and
that the grammars are reduced, i.e. for each $A \in V^*$ there are strings x_1, x_2
from V^* and z from T^* such that $S' \Rightarrow^* x_1 A x_2$ and $A \Rightarrow^* z$. Under these
conditions any grammar is completely defined by the set of its productions.

We assume that the productions are unambiguously numbered. The productions will often be given in a modified Backus-Naur normal form. In order
to present all the relevant information, the production numbers inclusive, we
use the notation clear from the following example of the grammar $G0$[1]:

$$G0 :: \quad 1 : S' \to \vdash S \dashv, \ 2.3 : S \to Ac \mid Bb, \ 4.5 : B \to aB \mid a,$$
$$6.7 : A \to aA \mid a$$

The full syntax grammar $Syn(G)$ of the grammar G is defined by the
productions $\{i : A \to A_{i.0} x_1 A_{i,1} x_2 A_{i,2} \ldots \ x_n A_{i,n} \mid i : A \to x_1 x_2 \ldots x_n \in
P_G\}$. $A_{i,k}$ are new terminal symbols called *syntactical symbols*. In order to
enhance legibility we often write $<_{A.i}$ instead of $A_{i,0}$ and $>_{A.i}$ instead of
$A_{i,n}$. The symbols $<_{A.i}$ and $>_{A.i}$ are *left and right s-brackets respectively.*

---

[1]$G0$ is somewhat ugly because it has to show all the cases and peculiarities of the
algorithms given below in very little space.

LB_G is the set of left s-brackets of G and RB_G the set of right s-brackets of G.

A homomorphism h on a language K is the *semantic homomorphism* for G if $h(a) = a$ for all a in T_G and $h(K) = L(G)$. Let M be a set of symbols. Then h_M is the homomorphism for which $h(a) = a$ for $a \in M$, $h(a) = \lambda$ otherwise.

For every $x \in L(G)$ the *complete parse* $Syn_G(x)$ of x is the string $y \in L(Syn_G)$ such that $h_T(y) = x$.

Obviously for every $x \in L(G)$, $h_{LB_G}(Syn(x))$ corresponds to (codes) the left-most derivation of x, i.e. the top-down parse of x. Similarly, h_{RB_G} produces a bottom-up parse, i.e. reversed right-most derivations, in an appropriate coding.

2   Syntax Directed Semantics

Semantics Sem on $L(G)$ is a function from $L(G)$ into a language (set of strings) $Sem(G)$ such that $h_T(Sem(x)) = x$ for every $x \in L(G)$. Syn_G is semantics by definition. A semantics $sem1$ on $L(G)$ covers a semantics $sem2$ on $L(G)$ if there is a semantic homomorphism h such that for every $x \in L(G)$, $sem2(x) = h(sem1(x))$. A semantics sem is *syntax directed by a grammar G* if it is covered by $Syn(G)$. We say that $sem1$ covers $sem2$ via h. A grammar G *strongly covers* a grammar $G1$ if $L(G) = L(G1)$ and Syn_G covers Syn_{G1}.

The relation *strongly covers* is incomparable with the relations *left covers* and *right covers* [6].

A semantics sem on $L(G)$ is *on line computable* if $x - x_1 x_2 \ldots x_s u \in L(G)$ and $y = x_1 x_2 \ldots x_s w \in L(G)$, $s > 0$, $x_i \in T$ for $i = 1, 2, \ldots, s$, implies $sem(x) = \sigma_0 x_1 \sigma_1 x_2 \ldots \sigma_s x_s u'$ and $sem(y) = \sigma_0 x_1 \sigma_1 x_2 \ldots \sigma_s x_s w'$; i.e. σ_i depends only on the left context (read off prefix) and the coming symbol. On line computable semantics can be generated on line with input. A production $i{:}A \to w$ is left recursive if $w \Rightarrow^* Ay$ for some $y \in V^*$. The s-bracket $<_{A.i}$ is then left recursive.

Theorem 1 [Král [6]] Syn_G *is on-line computable iff G is LL. If a semantics sem is directed by a grammar G via a semantic homomorphism h then $h(b) = \lambda$ for any left recursive s-bracket b and the semantics can be directed by a non left recursive grammar G'. G' can be constructed by a slight modification of the left factorization algorithm.*

The proofs can be found in [6]. We shall study the case of on line computable syntax. Due to the previous theorem we can assume without any

Table 1: The standard LR parser $\mathcal{P}_{G0}$ with added output.

S		Set	Action	Go	Output
1	1:	$S' \to . \vdash S \dashv, \lambda$	read	2	$<_{S',1}\vdash$
2	1:	$S' \to \vdash .S \dashv, \lambda$		3	
	2:	$S \to .Ac, \dashv$		4	
	3:	$S \to .Bb, \dashv$		6	
	4.5.6.7:	$B \to .aB, b$	read	8	$S'_{1,1}\{<_{S.2}<_{A.6}, <_{S.2}<_{A.7},$ $<_{S.3}<_{B.4}, <_{S.3}<_{B.5}\}a$
3	1:	$S' \to \vdash S. \dashv, \lambda$	read	11	$S'_{1,2}\dashv$
4	2:	$S \to A.c, \dashv$	read	5	$S_{2,1}c$
5	2:	$S \to Ac., \dashv$	RD2		$>_{S.2}$
6	3:	$S \to B.b, \dashv$	read	7	$S_{3,1}b$
7	3:	$S \to Bb., \dashv$	RD3		$>_{S.3}$
8	4:	$B \to a.B, b$		9	
	5:	$B \to a., b$	RD5		$>_{B.5}$
	6:	$A \to a.A, c$		10	
	7:	$A \to a., c$	RD7		$>_{A.7}$
	4.5.6.7:	$B \to .aB, b$	read	8	$\{B_{4,1}<_{B.4}, B_{4,1}<_{B.5},$ $A_{6,1}<_{A.6} A_{6,1}<_{A.7}\}a$
9	4:	$B \to .aB, b$	RD4		$>_{B.4}$
10	6:	$A \to .aA, c$	RD6		$>_{A.6}$
11	1:	$S' \to \vdash S \dashv ., \lambda$	accept		$>_{S',1}$

loss of generality that all the grammars discussed are left recursion free.[2]

3 Almost Top-Down Parsing of LR Grammars

Syn_G is not on-line computable for general LR grammars. It cannot be produced on line by standard LR parsers. We show, however, that the standard LR parser can produce substantially more information than mere bottom-up parses (the reversed right-most derivations).

A semi-top-down parse of $std_G(\vdash x_1 x_2 \ldots x_n \dashv)$, $x_i \in T$ for $i = 1, 2, \ldots x_n$ is the string $<_{S',1}\vdash \tilde{D}_0 x_1 \tilde{D}_1 x_2 \tilde{D}_2 \ldots x_n \tilde{D}_n \dashv >_{S',1}$, where $\tilde{D}_i$ is for every i a uniquely formed expression (string) of the language of the theory of sets defining the set $\{d_{i,0}, d_{i,1}, \ldots d_{i,n}\}$[3]. The set specified by $\tilde{D}_i$ is the set $D_i = \{d \mid \text{there are } y, z \text{ such that } h(y) = x, y = d_0 \vdash d_1 x_1 \ldots x_i d x_{i+1} z \in L(Syn(G))\}$. We treat all possible expressions for a singleton $\{x\}$ as equal to x, i.e. $x \approx \{x\} \approx \{x, x\}$ etc.

$std_{G0}(\vdash aab\dashv) = <_{S',1}\vdash S'_{1,1}\{<_{S.2}<_{A.6}, <_{S.2}<_{A.7}, <_{S.3}<_{B.4}, <_{S.3}<_{B.5}\}a\{$ $B_{4,1} <_{B.4}, B_{4,1}<_{B.5}, A_{6,1}<_{A.6}, A_{6,1}<_{A.7}\}a >_{B.5}>_{B.4} S_{3,1}b > S'_{1,2}\dashv, >_{S',1}.$

---

[2]The construction of G' is internal matter of the construction of the parser and need not be visible for parser users. Another solution is in [3] or in [8]. The second case is discussed below.

[3]It suffices to assume that $d_{i,j}$ are lexicographically ordered.

Table 2: Run of the parser $\mathcal{P}_{G0}$.

Input	Stack	Action	Output generated by the Action
$\vdash aab\dashv$	1	read $\vdash$	$<_{S'.1}\vdash$
$aab\dashv$	$1 \vdash 2$	read a	$S'_{1,1}\{<_{S.2}<_{A.6}, <_{S.2}$
			$<_{A.7}, <_{S.3}<_{B.4}, <_{S.3}<_{B.5}\}a$
$ab\dashv$	$1 \vdash 2$ a8	read a	$\{B_{4,1}<_{B.4}, B_{4,1}<_{B.5}, A_{6,1}<_{A.6}$
			$A_{6,1}<_{A.7}\}a$
$b\dashv$	$1 \vdash 2$ a8 **a8**	RD5	$>_{B.5}$
$b\dashv$	$1 \vdash 2$ **a8 B9**	RD4	$>_{B.4}$
$b\dashv$	$1 \vdash 2$ B6	read b	$S_{3,1}b$
$\dashv$	$1 \vdash 2$ **B6 b7**	RD3	$>_{S.3}$
$\dashv$	$1 \vdash 2$ S3	read$\dashv$	$S'_{1,2}\dashv$
	$1 \vdash 2$ **S3** $\dashv$11	RD1	$>_{S'.1}$
	1 S'	accept	

In order to simplify the explanation we assume that any expression of the language of the theory of sets is a syntactic sugar only for the basic form equivalent to it, i.e. $<_{S.3}\{<_{A.6}, <_{A.7}\}$ is an abbreviation of $\{<_{S.3}<_{A.6}, <_{S.3}<_{A.7}\}$.

The semantic *sem* on $L(G)$ is *semidirected* by G if there is a semantic homomorphism h such that, for every $x \in L(G)$, $sem(x) = h(std_G(x))$. Note that all $h(\tilde{D}_i)$ must be singletons.

Theorem 2 [Král [6]] *Let G be an LR(1) grammar and $\mathcal{P}$ the standard LR parser for it. It is possible to add an output function to $\mathcal{P}$ such that $\mathcal{P}$ generates $std_G(x)$ for every $x \in L(G)$.*

The formal proof is rather cumbersome [6], but the principle, demonstrated in tables 1 and 2, is quite simple. Knowledge of the construction of LR parsers is, however, necessary (see e.g. [1]). An item is the 3-tuple *production number, production* and *context*. We use the following convention. $(m : A \to w_1.w_2, a/b/c)$ denotes the items of the parser with the lookaheads a, b, c, and the production $m : A \to w_1.w_2$. The expression $m_1.m_2.\ldots.m_s : A_1 \to .w_1, a/b/c$ is used as an abbreviation for the s lines with the items $m_i : A_i \to .w_i, a/b/c, i = 1, 2, \ldots, s$, having the same Action, Go, and Outputs.

4   Recursive Descent Parsing by LR-Like Parsers

Let $out_\mathcal{P}(x)$ denote the output of the parser $\mathcal{P}$ for the input x. The parser $\mathcal{P}1$ covers the parser $\mathcal{P}$ if there is a semantic homomorphism h such that for every string $x \in L(G)$ $Out_\mathcal{P}(x) = h(Out_{\mathcal{P}1}(x))$.

Theorem 3 [Král [6]] *For every LR(1) parser $\mathcal{P}$ there is a parser $\mathcal{PP}$ covering $\mathcal{P}$ such that $\mathcal{PP}$ can be implemented as a procedure driven parser.*

The construction of $\mathcal{PP}$ goes via a *normalized parser $\mathcal{PN}$*. Let S be a state of $\mathcal{P}$. Slices $_1S,\ _2S,\ \ldots,\ _kS$ of S are defined as follows:

1. The slice $_1S$ is the kernel of S, i.e. $_1S$ contains all the items in S of the form $A \to x_1.Bx_2, M$ where x_1 is not an empty string.

2. For $i = 2, 3, \ldots, k,\ _iS = \{(t : B \to .y,\ M) \mid (t : B \to .y,\ M) \in S,$ there is a $(r : A \to x_1.Bx_2,\ M') \in\ _{i-1}S\}$.

As we assume that all the grammars discussed are left recursion free, every state has only a limited number of slices.

A state S is *normalized* if all its slices are disjoint. A parser is normalized if all its states are normalized. Note that $\mathcal{P}_{G0}$ is normalized:

We show how to construct a normalized parser $\mathcal{PN}$ for a standard LR parser $\mathcal{P}$ such that $\mathcal{PN}$ covers $\mathcal{P}$.

A state S is normalized by the following algorithm:

Input: S
Output: Normalized state $S\mathcal{N}$

```
FOR i:=1 TO number of slices in the state S DO
  WHILE there is  p = (k:B -> x.y, u) in Si and p is in Sj for a
        j>i DO
  BEGIN
    replace p in  Si by (r:B -> x.Q,u) and add (Q ->y,u) to S,
    Q is a new nonterminal and r a new production number
  END
```

We shall illustrate the operation of normalization on the example of the grammar. The construction of $\mathcal{PN}$ proceeds in the same way as the construction of a standard LR parser with the difference that any state is normalized immediately after it is created. The process is shown on the example of the following grammar:

$$G1 ::\ 1 : S' \to\ \vdash S \dashv,\ 2.3.4.5 : S \to Aa \mid Bc \mid aBa \mid aAc,$$
$$6 : B \to Aq,\ 7.8 : A \to bA \mid b$$

The construction of $\mathcal{PN}$ for the grammar $G1$ is in table 3, and an example of how it works in table 4. Note, however, that normalization is an internal operation on the data structures to be used by the constructed parser. It is *not* any operation that changes the underlying grammar:

$$out_{\mathcal{PN}}(\vdash bba \dashv) =<_{S'.1}\dashv S'_{1,2}\{<_{S.20}<_{Q.30},\ <_{S.3}<_{B.6}\}\{<_{A.7},\ <_{A.8}\}bA_{7,1}$$
$$\{<_{A.7},\ <_{A.0}\}b >_{A.8} >_{A.1} Q_{30.1}u >_{Q.30} >_{S.20} S''_{1,2} \dashv >_{S'.1}.$$

$\mathcal{PN}_G$ covers $\mathcal{P}_G$ if G is LR(1). In our example it suffices to perform the following changes: $<_{S.20} \to <_{S.2},\ >_{S.20} \to >_{S.2},\ <_{S.50} \to <_{S.5},\ >_{S.50} \to >_{S.5}$

Table 3: The normalized parser $\mathcal{PN}_{G1}$. Added items have numbers greater than 22. S denotes state, s slice.

S	s	Set normalized		Action	Go	Output
1	1	1:	$S' \to .\vdash S \dashv, \lambda$	read	2	$<_{S'.1}\vdash$
2	1	1:	$S' \to \vdash .S \dashv, \lambda$		19	
	2	20:	$S \to .Q, \dashv$		21	
		3:	$S \to .Bc, \dashv$		14	
		5.4:	$S \to .aAc, \dashv$	read	3	$S_{1,1}\{<_{S.4}, <_{S.5}\}a$
	3	30.6:	$Q \to .Aa, \dashv$		16	
	4	7.8:	$A \to .bA, a/q$	read	4	$S'_{1,1}\{<_{S.20}<_{Q.30}, <_{S.3}<_{B.6}\}$
						$\{<_{A.7}, <_{A.8}\}b$
						$\{<_{A.7}, <_{A.8}\}b$
3	1	4:	$S \to a.Ba, \dashv$		10	
		50:	$S \to a.R, \dashv$		21	
	2	60.6:	$R \to .Ac, \dashv$		12	
	3	7.8:	$A \to .bA, c/q$	read	4	$S_{4,1}\{<_{B.6}, S_{5,1}<_{R.60}\}$
						$\{<_{A.7}, <_{A.8}\}b$
						$\{<_{A.7}, <_{A.8}\}b$
4	1	8:	$A \to b., c/q/a$	RD8		$>_{A.8}$
		7:	$A \to b.A, c/q/a$		5	
	2	7.8:	$A \to .bA, c/q/a$	read	4	$A_{7,1}\{<_{A.7}, <_{A.8}\}b$
5	1	7:	$A \to bA., c/q/a$	RD7		$>_{A.7}$
8	1	6:	$B \to Aq., \dashv$	RD6		$>_{B.6}$
10	1	4:	$S \to aB.a, \dashv$	read	11	$S_{4,2}a$
11	1	4:	$S \to aBa., \dashv$	RD4		$>_{S.4}$
12	1	6:	$B \to A.q, c/a$	read	8	$B_{6,1}q$
		60:	$R \to A.c, \dashv$	read	13	$R_{60,1}c$
13	1	60:	$R \to Ac., \dashv$	RD60		$>_{R.60}$
14	1	3:	$S \to B.c, \dashv$	read	15	$S_{3,1}c$
15	1	3:	$S \to Bc., \dashv$	RD3		$>_{S.3}$
16	1	30:	$Q \to A.a, \dashv$	read	17	$Q_{30,1}a$
		6:	$B \to A.q, c/a$	read	8	$B_{6,1}q$
17	1	30:	$Q \to Aa., \dashv$	RD30		$>_{Q.30}$
19	1	1:	$S' \to \vdash S. \dashv, \lambda$	read	22	$S'_{1,2}\dashv$
20	1	20:	$S \to Q., \dashv$	RD20		$>_{Q.20}$
21	1	50:	$S \to aR., \dashv$		RD50	$>_{S.50}$
22	1	1:	$S' \to \vdash S \dashv ., \lambda$	RD1	accept	$>_{S'.1}$

Table 4: Run of the parser $\mathcal{PN}_{G1}$

Input	Stack	Action	Output during Action
$\vdash bba \dashv$	1	read $\vdash$	$<_{S'.1}\vdash$
$bba\dashv$	$1 \vdash 2$	read b	$S'_{1,1}\{<_{S.20}<_{Q.30}, <_{S.3}<_{B.6}\}\{<_{A.6}, <_{A7}\}b$
$ba\dashv$	$1 \vdash 2$ b4	read b	$A_{7,1}\{<_{A.6}, <_{A.7}\}b$
$a\dashv$	$1 \vdash 2$ b4 **b4**	RD8	$>_{A.8}$
$a\dashv$	$1 \vdash 2$ **b4 A5**	RD7	$>_{A.7}$
$a\dashv$	$1 \vdash 2$ A16	read a	$Q_{30,1}a$
$\dashv$	$1 \vdash 2$ **A16 a17**	RD30	$>_{Q.30}$
$\dashv$	$1 \vdash 2$ **Q21**	RD20	$>_{S.20}$
$\dashv$	$1 \vdash 2$ S19	read $\dashv$	$S'_{1,2}\dashv$
	$1 \vdash 2$ S3 $\dashv$11	RD1	$>_{S'.1}$

, $R_{60.1} \to S_{5.2}$. The symbols $<_{Q.30}, >_{Q.30}, <_{R.60}, >_{R.60}$ are replaced with empty strings.

The parser $\mathcal{PN}$ can be mechanically transformed into a parser $\mathcal{PP}$ with the same functionality, implementable as a PDP. The stack symbols of $\mathcal{PP}$ are the symbols of slices. $n.m$ is the symbol of the m-th slice of the state n.

If $\mathcal{PN}$ performs a read action according to the item $i : A \to x_1.ax_2$, u belonging to the slice $(n.m)$ and $\mathcal{PN}$ moves to the state t, then $\mathcal{PP}$ pushes the symbols $n.1$, $n.2$, ..., $n.m$ on the stack replaces the top symbol $m.n$ by the symbol $t.1$ and gives an output that is identical to the output of $\mathcal{PN}$. The actions can be viewed as m call operations of procedures slicen_1, slicen_2, ..., slicen_m and the first action of the procedure slicen_m.

If $\mathcal{PN}$ performs the reduce move RDi for the item that has a production with the left hand side A, then $\mathcal{PP}$ pops the top stack symbol. Now let the new top stack symbol of $\mathcal{PP}$ be the symbol $k.s$ and let the Go move under A in $\mathcal{PN}$ lead from state k to state t. Then the top symbol $k.s$ in $\mathcal{PP}$ is replaced by $t.1$ and the symbol $>_{A.i}$ is output. This can be interpreted as the operation RTi, which is equivalent to the return from a procedure and the first operation after return. At the start of $\mathcal{PP}$, the stack of $\mathcal{PP}$ contains only the symbol 1.1, the input string is on the input tape.

In order to make it easier to implement $\mathcal{PP}$ in the form of a PDP we modify the way the output of $\mathcal{PP}$ is computed. We assume that $calPP$ has a memory $\mathcal{M}$. Let the content of $\mathcal{M}$ be $\mathcal{S}(\mathcal{M})$. With each call and read action A of $\mathcal{PP}$ in the state S the set $\mathcal{S}_A$ is associated. Before the action A the operation $\mathcal{S}(\mathcal{M}) \circ \mathcal{S}_A$ is performed. The content of $\mathcal{M}$ is output during the read operation and $\mathcal{M}$ is emptied. The closing action of the read operation is the output of the just read up symbol. The read operation is implemented as the function **read** returning the read symbol. Let a be the symbol on input. The $\circ$ operation on the expressions of the sets theory language defining the sets M and K produces the expression defining the set $M \circ K = \{xz \mid x \in M, x = x' <_{U.n}, z = <_{B.m}\in K$, there is $n : U \to .Bz \in K, a \in FIRST(w)$, there is item $m : B \to w, u\}$. The principle is clear from tables 5 and 6.

Table 5: A modification $\mathcal{PP}$ of the normalized parser for $G1$ implementable as PDP. S is the state number, s is the slice number.

State	Slice		Set normalized	Action	Go	Memory modifier
1	1	1:	$S' \rightarrow . \vdash S \dashv, \lambda$	read	2.1	$<_{S'.1}\dashv$
2	1	1:	$S' \rightarrow \vdash .S \dashv, \lambda$	call 2.2	19.1	$S'_{1,1}$
	2	20:	$S \rightarrow .Q, \dashv$	call 2.3	21.1	$\{<_{S.20}, <_{S.3}\}$
		3:	$S \rightarrow .Bc, \dashv$	call 2.3	14.1	$\{<_{S.20}, <_{S.3}\}$
		5.4:	$S \rightarrow .aAc, \dashv$	read	3.1	$\{<_{S.4}, <_{S.5}\}a$
	3	30.6:	$Q \rightarrow .Aa, \dashv$	call 2.4	16.1	$\{<_{Q.30}, <_{B.6}\}$
	4	7.8:	$A \rightarrow .bA, a/q$	read	4.1	$\{<_{A.7}, <_{A.8}\}b$
3	1	4:	$S \rightarrow a.Ba, \dashv$	call 3.2	10.1	$\{S_{4,1}, S_{5,1}\}$
		50:	$S \rightarrow a.R, \dashv$	call 3.2	21.1	$\{S_{4,1}, S_{5,1}\}$
	2	60.6:	$R \rightarrow .Ac, \dashv$	call 3.3	12.1	$\{<_{B.6}, <_{R.60}\}$
	3	7.8:	$A \rightarrow .bA, c/q$	read	4.1	$\{<_{A.7}, <_{A.8}\}b$
4	1	8:	$A \rightarrow b., c/q/a$	RT8		$>_{A.8}$
		7:	$A \rightarrow b.A, c/q/a$	call 4.2	5.1	$A_{7,1}$
	2	7.8:	$A \rightarrow .bA, c/q/a$	read	4.1	$\{<_{A.7}, <_{A.8}\}b$
5	1	7:	$A \rightarrow bA., c/q/a$	RT7		$>_{A.7}$
8	1	6:	$B \rightarrow Aq., \dashv$	RT6		$>_{B.6}$
10	1	4:	$S \rightarrow aB.a, \dashv$	read	11.1	$S_{4,2}a$
11	1	4:	$S \rightarrow aBa., \dashv$	RT4		$>_{S.4}$
12	1	6:	$B \rightarrow A.q, c/a$	read	8.1	$B_{6,1}q$
		60:	$R \rightarrow A.c, \dashv$	read	13.1	$R_{60,1}c$
13	1	60:	$R \rightarrow Ac., \dashv$	RT60		$>_{R.60}$
14	1	3:	$S \rightarrow B.c, \dashv$	read	15.1	$S_{3,1}$
15	1	3:	$S \rightarrow Bc., \dashv$	RT3		$>_{S.3}$
16	1	30:	$Q \rightarrow A.a, \dashv$	read	17.1	$Q_{30,1}a$
		6:	$B \rightarrow A.q, c/a$	read	8.1	$B_{6,1}q$
17	1	30:	$Q \rightarrow Aa., \dashv$	RT30		$>_{Q.30}$
19	1	1:	$S' \rightarrow \vdash S. \dashv, \lambda$	read	22.1	$S'_{1,2}\dashv$
20	1	20:	$S \rightarrow Q., \dashv$	RT20		$>_{S.20}$
21	1	50:	$S \rightarrow aR., \dashv$	RT50		$>_{S.50}$
22	1	1:	$S' \rightarrow \vdash S \dashv ., \lambda$	accept		$>_{S'.1}$

The parser $\mathcal{PP}$ can be easily, even mechanically, rewritten in the procedure driven form. An example of a procedure for **slice 2.2** in the Pascal-like code follows:

```
FUNCTION slice2_2:char;
  BEGIN
    IF on_input('a') THEN
      BEGIN (* 3.1 *) (* R or B *)
        remember('{<_{S.4}, <_{S.5}}'); read; remember('{S_{4,1}, S_{5,1}}');
        CASE slice3_2 OF
          'R': (* 21.1 *)
            IF on_input('⊣') THEN BEGIN out('>_{S.50}'); return('S') END
                         ELSE error;
          'B': BEGIN (* 10.1 *)
            remember('{S_{4,1}}'); read; (* 11.1 *)
            IF on_input('⊣') THEN BEGIN out('>_{S.4}'); return('S') END
                         ELSE error
        END
```

Table 6: Run of the parser $\mathcal{PP}_{G1}$.

Input	Stack	Action	Memory content	Output
⊢bba⊣	1.1	read ⊢	$<_{S'.1}$	$<_{S'.1}⊢$
bba⊣	2.1	call 2.2	$S'_{1,1}$	
bba⊣	2.1 2.2	call 2.3	$S'_{1,1}\{<_{S.20},$ $<_{S.3}\}$	
bba⊣	2.1 2.2 2.3	call 2.4	$S'_{1,1}\{<_{S.20}<_{Q.30},$ $<_{S.3}<_{B.6}\}$	
bba⊣	2.1 2.2 2.3 2.4	read b	$\emptyset$	$S'_{1,2}\{<_{S.20}<_{Q.30},$ $<_{S.3}<_{B.6}\}$ $\{<_{A.7},<_{A.8}\}b$
ba⊣	2.1 2.2 2.3 4.1	call 4.2	$A_{7.1}$	
ba⊣	2.1 2.2 2.3 4.1 4.2	read b	$\emptyset$	$A_{7.1}\{<_{A.7},<_{A.8}\}b$
a⊣	2.1 2.2 2.3 4.1 4.1	RT8	$\emptyset$	$>_{A.8}$
a⊣	2.1 2.2 2.3 5.1	RT7	$\emptyset$	$>_{A.7}$
a⊣	2.1 2.2 16.1	read a	$\emptyset$	$Q_{30.1}a$
⊣	2.1 2.2 17.1	RT30	$\emptyset$	$>_{Q.30}$
⊣	2.1 20.1	RT20	$\emptyset$	$>_{S.20}$
⊣	19.1	read ⊣	$\emptyset$	$S'_{1,2}⊣$
	22.1	accept	$\emptyset$	$>_{S'.1}$

```
            END
          END
        ELSE
          IF on_input('b') THEN
            BEGIN
...
```

on_input('a') returns **true** if the character a is equal to the just read symbol. remember(S) performs the ∘ operation of S with memory contents. The description of **read** is given above.**out** outputs the string given by the argument.

The lookahead in $\mathcal{PP}$ is given statically in the procedure bodies. It is also possible to pass it into the procedure via a parameter with the form of a list of pairs: nonterminal, its possible right contexts. The procedure code is needed only once for a given LR(0) kernel (see parsers for LALR grammars [1]). The possible right context is computed dynamically at the place of the call.

5 Kind Grammars

The PDP for general LR grammars are quite complex, not easy to modify manually and not easy to be extensible in the sense proposed in [8]. The grammars of programming languages are "almost" LL. The main problem is that the grammars have left recursive symbols that are usually used to

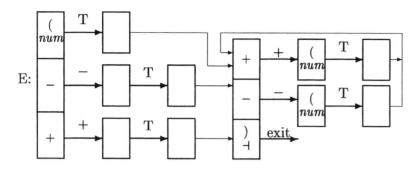

Figure 1: Parsing structure for E-productions.

define the syntax of arithmetic expressions and lists. A closer look shows that a limited form of left recursiveness is used (see the grammar G3):

G3:: S'→⊢E⊣, 2.3.4.5.6: E→E+T | E−T | T | +T| −T,
    7.8.9: T→T∗F | T/F | F, 13.11.12: F→(E) | id | num

This observation led to the definition of *kind grammars*.

$$DLRP_G = \{A \to A\alpha | A \to A\alpha \in P_G\}$$
$$NLRF_G(k, A) = \{a_1 \ldots a_k \in Pref(k, \beta Follow(B)) \mid$$
$$B \to \alpha A\beta \in P_G \wedge ((\alpha \neq \varepsilon) \vee (B \neq A))\}$$

Definition 1 (k-kind grammar) *A context-free grammar G that only has productions without left recursion and productions with direct left recursion is called* kind *if:*

1. *for every two productions $A \to \alpha X\beta$, $A \to \alpha Y\gamma \in P_G$ and $X \neq Y$ then $Pref(k,X\beta) \cap Pref(k,Y\gamma) = \emptyset$, and*

2. *for every nonterminal A $DLRF_G^k(A) \cap NLRF_G^k(A) = \emptyset$.*

Productions are grouped by the nonterminal on the left hand side of each production. Kind grammars produce LL languages such as those shown in [11], which also shows how to mechanically generate graphs that are easily transformable into programs. An example for E-group is in Figure 1.

For this structure it is easy to generate a parsing procedure like in the following listing:

```
PROCEDURE Parse_E;
  BEGIN
    CASE LookAhead OF
      term_num, term_left: Parse_T;
      term_minus: BEGIN ReadTerm(term_minus); Parse_T END;
    END;
    WHILE LookAhead IN [term_plus,term_minus] DO
      CASE LookAhead OF
```

```
        term_minus: BEGIN ReadTerm(term_minus); Parse_T END;
        term_plus: BEGIN ReadTerm(term_plus); Parse_T END;
        END;
    END;
```

Note that the program contains loops. This is a substantial difference from the previous case.

Kind grammars form a proper superclass of LL grammars. It is possible to design an extensible parser for kind grammars that may be extended during parsing time — even by a parallel process.

6   Conclusions

LR parsing is implementable in the procedure driven form and can produce information close to top-down parses. So top-down parsing is sometimes possible using tools derived for bottom-up parsing. We prepare an implementation of a dynamically modifiable parser based on the concept of kind grammars.

References

[1] A.V. Aho, R. Sethi and J.D. Ullman, *Compilers: Principles, Techniques, and Tools.* Addison-Wesley, Reading, Mass., 1986.

[2] J. Drózd, Semi-Top-Down Parsing, Master thesis, Faculty of Mathematics and Physics, Charles University, Prague, 1985 (in Czech).

[3] J. Drózd, Recursive Descent Parsing for LR(k) Grammars, PhD dissertation, Faculty of Mathematics and Physics, Charles University, Prague, 1990 (in Czech).

[4] J. Král, Almost top-down analysis for generalized LR (k) grammars. In *Trudy Vsesojuznogo sympoziuma po metodam realizacii algoritmičeskich jazykov I*, SO AN SSSR Novosibirsk, 1976, 230–247.

[5] J. Král, A top-down no backtrack parsing of general context-free languages. In *Proceedings of Mathematical Foundations of Computer Science, MFCS'77*, Springer, Berlin, 1977, 333–341.

[6] J. Král, Parsing and Syntax Directed Compiling, Technical Report, Institute of Computing Techniques, Prague, 1982 (in Czech).

[7] Gh. Păun, *Marcus Contextual Grammars.* Kluwer, Dordrecht, 1998.

[8] M. Žemlička, Extensible Language Compiler, Master thesis, Faculty of Mathematics and Physics, Charles University, Prague, 1994 (in Czech).

[9] M. Žemlička, Extensible LL(1) parsing. In *Proceedings of SOFSEM'95*, Milovy, 1995.

[10] M. Žemlička, Parsing of Extensible Languages, Technical Report, Faculty of Mathematics and Physics, Charles University, Prague, 1996 (in Czech).

[11] M. Žemlička and J. Král, Run-time extensible deterministic top-down parsing, *Grammars*, 2.3 (1999), 283–293.

Descriptional Complexity of Multi-Parallel Grammars with Respect to the Number of Nonterminals

Alexander Meduna

Dušan Kolář

Department of Computer Science and Engineering
Technical University of Brno
Czech Republic
{meduna,kolar}@dcse.fee.vutbr.cz

Abstract. The present paper discusses multi-parallel grammars and their descriptional complexity with respect to the number of nonterminals. It proves that eight-nonterminal multi-parallel grammars characterize the family of recursively enumerable languages. It also summarizes all the important results on the descriptional complexity of multi-grammars with respect to the number of nonterminals.

1   Introduction

The theory of selective substitution grammars classified multi grammars into three basic types –multi-sequential grammars, multi-continuous grammars, and multi-parallel grammars (see [2]). The descriptional complexity of the first two types with respect to the number of nonterminals was investigated in [4] and [5]. The present paper completes this investigation by discussing this complexity regarding multi-parallel grammars.

More specifically, the present paper proves that eight-nonterminal multi-parallel grammars characterize the family of recursively enumerable languages. It also concludes by summarizing all the important results on the descriptional complexity of multi-grammars with respect to the number of nonterminals.

2   Definitions

This paper assumes that the reader is familiar with formal language theory, including selective substitution grammars (see [6] and Chapter 10 in [1]).

Let Σ be an alphabet. The cardinality of Σ is denoted by $card(\Sigma)$. Σ^* represents the free monoid generated by Σ under the operation of concatenation. The unit of Σ^* is denoted by ε. Let $\Sigma^+ = \Sigma^* - \{\varepsilon\}$; algebraically, Σ^+ is the free semigroup generated by Σ under the operation of concatenation. For $w \in \Sigma^*$, $|w|$ denotes the length of w.

Next, we give the definition of a multi-parallel grammar. Compared to the definition of a multi-parallel grammar given in [2], the following definition is simpler; however, it is easy to see that both definitions are equivalent.

Let m be a positive integer. An *m-parallel grammar*, G, is a quintuple:

$$G = (\Sigma, P, S, T, K),$$

where Σ is an alphabet, $T \subseteq \Sigma$, $S \in \Sigma - T$, P is a finite substitution on Σ^*, and:

$$K = \{\pi_1, \ldots, \pi_n\},$$

where for $i = 1, \ldots, n$, π_i is a language of the form:

$$\pi_i = F_1 F_2 \ldots F_m,$$

with:

$$F_h \in \{W^+ : W \subseteq \Sigma, W \neq \emptyset\},$$

for all $h = 1, \ldots, m$. Let $u, v \in \Sigma^*$. G *directly derives* v *from* u, symbolically denoted as:

$$u \Rightarrow v,$$

if either $u = S$ and $v \in P(S)$ or there exists a natural number, n, so:

1. $u = a_1 \ldots a_n$ with $a_i \in V$ for all $i = 1, \ldots, n$;

2. $u \in \pi_j$ for some $\pi_j \in K$, where $j \in \{1, \ldots, n\}$;

3. $v = x_1 \ldots x_n$ with $x_i \in P(a_i)$.

Instead of $x \in P(a)$, this paper writes $a \to x$ hereafter. In the standard manner, extend $\Rightarrow$ to $\Rightarrow^n$ (where $n \geq 0$), $\Rightarrow^+$, and $\Rightarrow^*$. The *language* of G, $L(G)$, is defined as:

$$L(G) = \{w \in T^* : S \Rightarrow^* w\}.$$

G is a *multi-parallel grammar* if G represents an m-parallel grammar for some $m \geq 1$.

A *queue grammar* (see [3]) is a sixtuple, $Q = (V, T, W, F, R, g)$, where V and W are alphabets satisfying $V \cap W = \emptyset$, $T \subseteq V$, $F \subseteq W$, $R \in (V - T)(W - F)$, and $g \subseteq (V \times (W - F)) \times (V^* \times W)$ is a finite relation such that for any $a \in V$, there exists an element $(a, b, x, c) \in g$. If there exist $u, v \in V^*W$, $a \in V$, $r, z \in V^*$, and $b, c \in W$ such that $(a, b, z, c) \in g$, $u = arb$, and $v = rzc$, then Q directly derives v from u, denoted by $u \Rightarrow v$. In the standard manner, define $\Rightarrow^n$, $\Rightarrow^+$, and $\Rightarrow^*$. A derivation of the form $R \Rightarrow^* wf$ with $w \in T^*$ and $f \in F$ is a successful derivation. The language of Q, $L(Q)$, is defined as $L(Q) = \{w \in T^* : R \Rightarrow^* wf, \text{ where } f \in F\}$.

3 Results

The present section demonstrates that the family of recursively enumerable languages is equal to the family of languages generated by eight-nonterminal multi-parallel grammars.

Lemma 1 *Let:*

$$Q = (V, T, W, F, R, g)$$

be a queue grammar. Then, there exists an eight-nonterminal multi-parallel grammar, G, satisfying:

$$L(G) = L(Q)$$

Proof. Let:

$$Q = (V, T, W, F, R, g)$$

be a queue grammar. Without any loss of generality, assume that:

$$(V \cup W) \cap \{0, 1, 2, 3, 4, 5, 6, X\} = \emptyset$$

.

Construction:
For a natural number, $n \geq 2^{card(V \cup W)}$, introduce the following four mappings — χ, τ, β, and δ:

1. Define an injection, χ, from $(V \cup W)$ to $(\{4, 5\}\{3\})^n$. In the standard manner, extend χ so it is defined from $(V \cup W)^*$ to $((\{4, 5\}\{3\})^n)^*$.

2. Define the bijection, τ, from $\{4, 5, 3\}$ to $\{0, 1, 3\}$ as $\tau(4) = 0$, $\tau(5) = 1$, and $\tau(3) = 3$. In the standard manner, extend τ so it is defined from $\{4, 5, 3\}^*$ to $\{0, 1, 3\}^*$.

3. Define the injection, β, from $(V \cup W)$ to $(\{0, 1\}\{3\})^n$ so that for every $a \in (V \cup W)$, $\beta(a) = \tau(\chi(a))$. In the standard manner, extend χ so it is defined from $(V \cup W)^*$ to $((\{0, 1\}\{3\})^n)^*$.

4. Define the relation, δ, from $(V \cup W \cup \{X\})$ to $(T\{X\} \cup (\{4,5\}\{3\})^n)$ so that:

for every $a \in ((V \setminus T) \cup W), \delta(a) = \{\chi(a)\}$;

for every $a \in T, \delta(a) = \{\chi(a), aX\}$;

$\delta(X) = \{X\}$.

In the standard manner, extend δ so it is defined from $(V \cup W \cup \{X\})^*$ to $((T\{X\} \cup (\{4,5\}\{3\})^n))^*$.

Let m be any natural number satisfying:

$$m \geq \max\{|\chi(abxc)| : (a,b,x,c) \in g\} + 3.$$

Construct the following m-parallel grammar:

$$G = (T \cup \{0,1,2,3,4,5,6,X\}, P, 2, T, K),$$

with:

$P = \{2 \to \chi(b)2\chi(a)X^m 2 : a \in V - T, b \in W - F, ab = R\}$
$\cup \{a \to a : a \in T \cup \{4,5\}\}$
$\cup \{3 \to \chi(c)2 : c \in W\}$
$\cup \{i \to \tau(i) : i = 3,4,5\}$
$\cup \{a \to \varepsilon : a \in \{X,0,1,2,3\}\}$
$\cup \{2 \to X^m \delta(x)6 : x \in V^*, (a,b,x,c) \in g \text{ for some } a,c \in W, b \in V\}$
$\cup \{6 \to X^m 2\}$.

Initially, set $K = \emptyset$; then, extend K in the following three-step way:

A. For every $(a,b,x,c) \in g$, where $b,c \in W, a \in V$, and $x \in V^*$, add:

$\{ \{b_1\}^+\{3\}^+ \dots \{b_n\}^+\{3\}^+\{2\}^+\{a_1\}^+\{3\}^+ \dots \{a_n\}^+\{3\}^+$
$(\{4,5,3,X\} \cup T)^+ H_1 \dots H_j\{2\}^+ \}$

to K, where:

$a_i, b_i \in \{3,4\}$ for $i = 1, \dots, n$

$a_1 3 \dots a_n 3 = \chi(a)$

$b_1 3 \dots b_n 3 = \chi(b)$

$j = m - 4n - 3$

$H_h = \{X\}^+$ for all $h = 1, \dots, j$

B.   For every $(a, b, x, c) \in g$, where $b, c \in W$, $a \in V$, and $x \in V^*$, and every $y \in \delta(x)$, add:

$$\{ \{b_1\}^+\{3\}^+ \ldots \{b_n\}^+\{3\}^+\{a_1\}^+\{3\}^+ \ldots \{a_n\}^+\{3\}^+$$
$$\{c_1\}^+\{3\}^+ \ldots \{c_n\}^+\{3\}^+\{2\}^+$$
$$(\{4,5,3,X\} \cup T)^+ H_1 \ldots H_j\{d_1\}^+\{f_1\}^+ \ldots \{d_r\}^+\{f_r\}^+\{6\}^+ \}$$

to K, where:

$a_i, b_i \in \{0, 1\}$ for $i = 1, \ldots, n$

$c_q \in \{4, 5\}$ for $q = 1, \ldots, n$

$a_1 3 \ldots a_n 3 = \beta(a)$

$b_1 3 \ldots b_n 3 = \beta(b)$

$c_1 3 \ldots c_n 3 = \chi(b)$

$r = \frac{|y|}{2}$

$d_k \in \{4, 5\} \cup T$ for $j = 1, \ldots, r$

$f_l \in \{X, 3\}$ for $l = 1, \ldots, r$

$d_1 f_1 \ldots d_r f_r = y$

$j = m - 6n - |y| - 3$

$H_h = \{X\}^+$ for all $h = 1, \ldots, j$

C.   For every $b \in F'$, add:

$$\{ \{b_1\}^+\{3\}^+ \ldots \{b_n\}^+\{3\}^+\{2\}^+(\{X\} \cup T)^+ H_1 \ldots H_j\{2\}^+ \}$$

to K, where:

$b_1 3 \ldots b_n 3 = \beta(b)$ with $b \in F$

$j = m - 2n - 3$

$H_h = \{X\}^+$ for all $h = 1, \ldots, j$

The construction of G is completed.

Next, we outline a proof demonstrating $L(G) \subseteq L(Q)$, but leave it up to the reader to provide a rigorous version of this proof. Examine the construction of G to make the following three observations:

I. G derives no sentential form containing two consecutive identical nonterminals from $\{0, 1, 2, 3, 4, 5, 6\}$.

II. If some terminals precede some occurrences of symbols 4, 5, or 3 in a sentential form, f, derived by G, then these occurrences can never be removed, so G cannot derive a member of $L(G)$ from f at this point.

III. G uses a selector introduced in C only during the last derivation step of any successful derivation.

Based on properties A through C, observe that every successful derivation actually simulates a successful derivation in Q. To give an insight into this simulation in greater detail, consider:

$$avb \Rightarrow vxc$$

according to $(a, b, x, c) \in g$ in Q. By using selectors constructed in A and B, G simulates $avb \Rightarrow vxc$ by making these two steps:

$$\chi(b)2\chi(a)\delta(vX^m)2 \Rightarrow \beta(ba)\chi(c)2\delta(vX^m x)6 \Rightarrow \chi(c)2\delta(vxX^m)2$$

(X acts as a "filling" symbol). More precisely, every successful derivation, $2 \Rightarrow^* v$ with $v \in T^*$, has this form:

$$
\begin{array}{lllll}
2 & \Rightarrow & x_1 & \Rightarrow & y_1 \\
  & \Rightarrow & x_2 & \Rightarrow & y_2 \\
  & \vdots & & & \\
  & \Rightarrow & x_h & \Rightarrow & y_h \\
  & \Rightarrow & x_{h+1} & \Rightarrow & v
\end{array}
$$

where:

$h \geq 1$, and:

for $j = 1, \ldots, h$:

- $x_j, y_j \in \{0, 1, 2, 3, 4, 5, 6, X\}^* (\{X\} \cup T)^* \{2, 6\}^*$,
- $x_j = \chi(b_j)2\delta(a_j z_j X^m)2$ and
  $y_j = \beta(b_j a_j)\chi(c_j)2\delta(w_j X^m q_j)6$
  for some $(a_j, b_j, q_j, c_j) \in R$ so that
  $\quad w_j, z_j \in V^*, b_{j+1} = c_j,$
  $\quad w_j$ begins with a_{j+1}, and $a_{j+1} z_{j+1} = w_j q_j,$
- $x_1 = \chi(b_1)2\chi(a_1)X^m2$ with $a_1 b_1 = R,$
- $x_{h+1} = \beta(b_{h+1})2z_{h+1}X^m2$ with $z_{h+1} = v$ and $b_{h+1} \in F.$

Then, in Q:

$$a_1 b_1 \quad \Rightarrow \quad a_2 z_2 b_2$$

$$\vdots$$

$$\Rightarrow \quad a_h z_h b_h$$

$$\Rightarrow \quad v b_{h+1}$$

Therefore, $L(G) \subseteq L(Q)$. A proof demonstrating $L(Q) \subseteq L(G)$ is left to the reader.

As $L(G) \subseteq L(Q)$ and $L(Q) \subseteq L(G)$, $L(Q) = L(G)$. Because G has only eight nonterminals $-0, 1, 2, 3, 4, 5, 6$, and $X-$, Lemma 1 holds.   □

Theorem 1 *The family of languages generated by eight-nonterminal multi-parallel grammars coincides with the family of recursively enumerable languages.*

Proof. Obviously, every language generated by an eight-nonterminal multi-parallel grammar represents a recursively enumerable language.

By Lemma 1, for every queue grammar, Q, there exists an eight-nonterminal multi-parallel grammar, G, satisfying $L(G) = L(Q)$. Recall that the family of languages generated by queue grammars coincides with the family of recursively enumerable languages (see [3]). Consequently, every recursively enumerable language is generated by an eight-nonterminal multi-parallel grammar.

Therefore, Theorem 1 holds.   □

The following theorem summarizes all the fundamental results on the descriptional complexity of multi-grammars with respect to the number of nonterminals.

Theorem 2 *The following four families are identical:*

A. *the family of languages generated by eight-nonterminal multi-parallel grammars;*

B. *the family of languages generated by six-nonterminal multi-continuous grammars;*

C. *the family of languages generated by eight-nonterminal multi-sequential grammars;*

D. *the family of recursively enumerable languages.*

Proof. This theorem follows from Theorem 1 above, Theorem 1 in [4], and Theorem 1 in [5].   □

References

[1] J. Dassow and Gh. Păun, *Regulated Rewriting in Formal Language Theory.* Springer, New York, 1989.

[2] H.C.M. Kleijn and G. Rozenberg, Multi grammars, *International Journal of Computer Mathematics*, 12 (1983), 177–201.

[3] H.C.M. Kleijn and G. Rozenberg, On the generative power of regular pattern grammars, *Acta Informatica*, 20 (1983), 391–411.

[4] A. Meduna, Eight-nonterminal multi-sequential grammars characterize the family of recursively enumerable languages, *International Journal of Computer Mathematics*, 65 (1997), 179–189.

[5] A. Meduna, Descriptional complexity of multi-continues grammars, *Acta Cybernetica*, 1998, 375–384.

[6] A. Meduna, *Automata and Languages: Theory and Applications.* Springer, London, 2000.

On the Generative Capacity of Parallel Communicating Extended Lindenmayer Systems[1]

György Vaszil

Computer and Automation Research Institute
Hungarian Academy of Sciences
Budapest, Hungary
`vaszil@luna.aszi.sztaki.hu`

Abstract. We investigate the generative power of parallel communicating systems with extended Lindenmayer systems as components. First we prove that, like the context-free case, non-returning systems can be simulated by returning systems. Then we demonstrate the power of non-returning systems by showing that they are able to generate the twin-shuffle language over arbitrary alphabets.

1   Introduction

Parallel communicating Lindenmayer systems were introduced in [3], the paper which initiated the study of their generative power (see also [1]). It was shown that three, or even two, components are enough for a system to be strictly more powerful then the components it contains.

The study of parallel communicating extended Lindenmayer systems continued in [9], where it was shown that systems that have components with or without tables generate the same class of languages. A normal form for the production rules was also given. Different modes of derivation were investigated in [7] and in [10], and the equivalence of several different ways of communication was shown in [8].

---

[1] Research supported by the Hungarian Scientific Research Fund "OTKA" Grant no. T 029615.

In this paper we continue to investigate the generative power of parallel communicating extended Lindenmayer systems. We show that non-returning systems, with or without tables, can be simulated by returning systems, and we demonstrate the power of non-returning systems by showing that they are able to generate the twin-shuffle language over arbitrary alphabets.

2   Preliminaries

The reader is assumed to be familiar with the basics of formal language theory. Further details can be found in [6].

An alphabet V is a finite set of symbols and a string over V is a finite sequence of elements of V. The set of all nonempty strings over V is denoted by V^+. The empty string is denoted by ε, V^* stands for the union of V^+ and $\{\varepsilon\}$. A language L over V is a subset of V^*. $|w|$ and $|w|_X$ denotes the length of a word w and the number of occurences of symbols from set X in w. For details about Lindenmayer systems, consult [5]. Bellow only the basic definitions about extended Lindenmayer systems are presented.

An *E0L system* is a quadruple $G = (N, T, P, \omega)$, where N and T are disjoint sets; N is the nonterminal alphabet and T is the terminal alphabet. P is a finite set of rewriting rules (productions) over $(N \cup T)$ with rules of the form $a \to v$, where $a \in (N \cup T)$ and $v \in (N \cup T)^*$, and $\omega \in (N \cup T)^*$ is the axiom of the system. Furthermore, P is complete, that is, all letters of $(N \cup T)$ can be rewritten by at least one rule of P.

A string x *directly derives* a string y in G, denoted by $x \Rightarrow_G y$, if $x = a_1 a_2 \ldots a_n$, $y = \alpha_1 \alpha_2 \ldots \alpha_n$, and for every i, $a_i \in (N \cup T)$, $\alpha_i \in (N \cup T)^*$ and $a_i \to \alpha_i \in P$, $1 \le i \le n$.

The *language* generated by an *E0L* system G is $L(G) = \{x \in T^* \mid \omega \Rightarrow_G^* x\}$, where $\Rightarrow_G^*$ denotes the reflexive and transitive closure of $\Rightarrow_G$.

A *tabled E0L system* or an *ET0L system* is a quadruple $G = (N, T, P, \omega)$, where N and T are the disjoint alphabets of nonterminals and terminals, $\omega \in (N \cup T)^*$ is the axiom and P is a finite set of *E0L* production sets over $(N \cup T)$, $P = \{P^{(1)}, \ldots, P^{(t)}\}$. The elements of P are called *tables*. In a direct derivation step only productions from one of the tables in P can be chosen, but in different steps of the same derivation, different production sets can be used.

PC grammar systems with Chomsky type components were introduced in [4], with Lindenmayer type components in [3].

Definition 1 *A* parallel communicating extended Lindenmayer system *with* n *components (a PC E0L or PC ET0L system in short)*, $n \ge 1$, *is an* $(n+3)$-*tuple* $\Gamma = (N, K, T, G_1, \ldots, G_n)$, *where* N *is a* nonterminal alphabet, T *is a* terminal alphabet, *and* $K = \{Q_1, Q_2, \ldots, Q_n\}$ *is an alphabet of query symbols*. N, T, *and* K *are pairwise disjoint sets*. $G_i = (N \cup K, T, P_i, \omega_i)$, $1 \le$

$i \leq n$, *called a* component *of* Γ, *is an extended Lindenmayer system as above,* G_1 *is said to be the* master *of* Γ.

An *n-tuple* $(x_1, \ldots, x_n)$, *where* $x_i \in (N \cup T \cup K)^*$, $1 \leq i \leq n$, *is called a* configuration *of* Γ, $(\omega_1, \ldots, \omega_n)$ *is said to be the* initial configuration.

PC Lindenmayer systems change their configurations by performing direct derivation steps.

Definition 2 *Let* $\Gamma = (N, K, T, G_1, \ldots, G_n)$, $n \geq 1$, *be a parallel communicating Lindenmayer system. We say that a configuration* $(x_1, \ldots, x_n)$ *directly derives* $(y_1, \ldots, y_n)$, *denoted by* $(x_1, \ldots, x_n) \Rightarrow (y_1, \ldots, y_n)$, *if one of the following two cases holds:*

1. There is no x_i *which contains any query symbol, that is,* $x_i \in (N \cup T)^*$ *for* $1 \leq i \leq n$. *Then for each* i, $1 \leq i \leq n$, $x_i \Rightarrow_{G_i} y_i$ *(y_i is obtained from x_i by a direct derivation step in G_i).*

2. There is some x_i, $1 \leq i \leq n$, *which contains at least one occurrence of a query symbol. In this case* $(y_1, \ldots, y_n)$ *is obtained from* $(x_1, \ldots, x_n)$ *as follows:*

For each x_i *with* $|x_i|_K \neq 0$ *we write* $x_i = z_1 Q_{i_1} z_2 Q_{i_2} \ldots z_t Q_{i_t} z_{t+1}$, *where* $z_j \in (N \cup T)^*$, $1 \leq j \leq t+1$, *and* $Q_{i_l} \in K$, $1 \leq l \leq t$. *If* $|x_{i_l}|_K = 0$ *for each* i_l, $1 \leq l \leq t$, *then* $y_i = z_1 x_{i_1} z_2 x_{i_2} \ldots z_t x_{i_t} z_{t+1}$ *and in returning systems* $y_{i_l} = \omega_{i_l}$, *in non-returning systems* $y_{i_l} = x_{i_l}$, $1 \leq l \leq t$. *If* $|x_{i_l}|_K \neq 0$ *for some* i_l, $1 \leq l \leq t$, *then* $y_i = x_i$. *For all* j, $1 \leq j \leq n$, *for which* y_j *is not specified above,* $y_j = x_j$.

Let $\Rightarrow^*$ *denote the reflexive and transitive closure of* $\Rightarrow$.

The first case is the description of a rewriting step. If no query symbol is present in any of the sentential forms, then each component uses its rewriting rules. The second case describes a communication. If a query symbol appears in a sentential form, the rewriting process is interrupted and one or more communication steps must be performed. Each query symbol must be replaced by the sentential form of the component with the same index, provided that the replacing strings do not contain further query symbols. If this condition cannot be fulfilled, a circular query has appeared and the derivation is blocked. In *returning* systems, after they have communicated their sentential form to another one, components must return to their axioms and begin to generate a new string. In *non-returning* systems they continue to rewrite the current string. Let $\Rightarrow_{rew}$ and $\Rightarrow_{com}$ denote a rewriting and a communication step, respectively.

Definition 3 *The* language *generated by a parallel communicating system of extended Lindenmayer systems* $\Gamma = (N, K, T, G_1, \ldots, G_n)$, $G_i = (N \cup K, T, P_i, \omega_i)$, $1 \leq i \leq n$, *is:*

$$L(\Gamma) = \{\alpha_1 \in T^* \mid (\omega_1, \ldots, \omega_n) \Rightarrow^* (\alpha_1, \ldots, \alpha_n)\},$$

where G_1 is the master of Γ.

Thus, the generated language consists of terminal strings appearing as sentential forms of the master.

Let us denote the classes of languages generated by returning and non-returning PC Lindenmayer systems with at most n components of type $X \in \{E0L, ET0L\}$ by $\mathcal{L}(PC_n X)$ and $\mathcal{L}(NPC_n X)$, respectively. When an arbitrary number of components is considered, we use $*$ in the subscript instead of n. Let also $\mathcal{L}(X)$, X as above, denote the class of languages generated by extended Lindenmayer systems, and $\mathcal{L}(RE)$ the class of recursively enumerable languages.

3 About Generative Capacity

First we recall a theorem from [9], which shows that returning or non-returning PC $E0L$ systems and PC $ET0L$ systems generate the same class of languages.

Theorem 1 [9] $\mathcal{L}(X_* E0L) = \mathcal{L}(X_* ET0L)$, $X \in \{PC, NPC\}$.

Now we continue by showing that non-returning PC $E0L$ or PC $ET0L$ systems can be simulated by returning systems.

Theorem 2 $\mathcal{L}(NPC_* X) \subseteq \mathcal{L}(PC_* Y)$, $X, Y \in \{E0L, ET0L\}$.

Proof. By Theorem 1 above, it is sufficient to show that non-returning PC $E0L$ systems can be simulated by returning PC $E0L$ systems. Let $\Gamma = (N, K, T, G_1, \ldots, G_n)$ be a non-returning PC $E0L$ system with $G_i = (N \cup K, T, P_i, \omega_i)$, $1 \leq i \leq n$. Let us also assume that $\omega_i \notin T^*$, $1 \leq i \leq n$. This does not involve any loss in generality, since a nonterminal S can be added to N and new rules $S \to \omega_i$ can be added to P_i, $1 \leq i \leq n$, without changing the generated language. Now we construct a returning PC $E0L$ system Γ' which generates the same language as Γ. Let:

$$\Gamma' = (N', K', T, G'_1, \ldots, G'_{2n}, G_{11}, \ldots, G_{1n}, \ldots, G_{n1}, \ldots, G_{nn}, G_a),$$

where G'_1 is the master grammar, $G'_i = (N' \cup K', T, P'_i, S_i)$, $1 \leq i \leq 2n$, $G_{ij} = (N' \cup K', T, P_{ij}, S_{ij})$, $1 \leq i, j \leq n$, $G_a = (N' \cup K', T, P_a, S)$, and:

$$\begin{aligned} N' = & \{x, [x], [y] \mid x \in N, y \in T\} \cup \\ & \{S, S', S'', S_i, S_{jk}, S'_{jk} \mid 1 \leq i \leq 2n, \ 1 \leq j, k \leq n\}. \end{aligned}$$

The set of productions are as follows. If rules for a symbol $x \in N' \cup T$ are not explicitly given, we assume the presence of "chain" rules, $x \to x$.

$$\begin{aligned} P'_i = & \{S_i \to Q'_{n+i}, S_i \to [\omega_i], S'_{jk} \to Q'_i \mid 1 \leq j, k \leq n\} \cup \\ & \{x \to [x], [x] \to Q'_i \mid x \in N \cup T\}, \end{aligned}$$

for $1 \leq i \leq n$,

$$
\begin{aligned}
P_i' \;=\; & \{S_i \to Q_{i-n}'\} \cup \{[x] \to \alpha \mid x \to \alpha \in P_{i-n}, \alpha \in (N \cup T)^*\} \cup \\
& \{[x] \to \alpha_0 Q_{ij_1}\alpha_1 \ldots Q_{ij_t}\alpha_t \mid x \to \alpha_0 Q_{j_1}\alpha_1 \ldots Q_{j_t}\alpha_t \in P_{i-n}, \\
& \alpha_l \in (N \cup T)^*, \; 0 \leq l \leq t\},
\end{aligned}
$$

for $n + 1 \leq i \leq 2n$,

$$
\begin{aligned}
P_{ij} \;=\; & \{S_{ij} \to S_{ij}', S_{ij}' \to Q_{j+n}', S_{ij} \to Q_{j+n}'\} \cup \\
& \{[x] \to Q_{ij} \mid x \in N \cup T\},
\end{aligned}
$$

for $1 \leq i, j \leq n$, and:

$$
\begin{aligned}
P_a \;=\; & \{S \to S', S' \to S'', S'' \to Q_{11}Q_{12}\ldots Q_{nn}S'\} \cup \\
& \{[x] \to Q_a \mid x \in N \cup T\}.
\end{aligned}
$$

Let us now follow the derivations in Γ' to see that it generates the same language as Γ. In the following, for any string $\alpha = x_1 x_2 \ldots x_t \in (N \cup T)^*$, $x_i \in N \cup T$, $1 \leq i \leq t$, $[\alpha]$ denotes the string $[x_1][x_2]\ldots[x_n]$. If $(\omega_1, \ldots, \omega_n) \Rightarrow_{rew} (\alpha_1, \ldots, \alpha_n)$ is the initial step of Γ, then in Γ' after the first rewriting step we get:

$$
\begin{aligned}
& ([\omega_1], \ldots, [\omega_n], Q_1', \ldots, Q_n', S_{11}', \ldots, S_{nn}', S') \Rightarrow_{com} \\
& (S_1, \ldots, S_n, [\omega_1], \ldots, [\omega_n], S_{11}', \ldots, S_{nn}', S') \Rightarrow_{rew} \\
& (Q_{n+1}', \ldots, Q_{2n}', \alpha_1', \ldots, \alpha_n', Q_{n+1}', \ldots, Q_{2n}',, \ldots, Q_{n+1}', \ldots, Q_{2n}', S''),
\end{aligned}
$$

where α_i' and α_i, $1 \leq i \leq n$ differ only in the indices of the query symbols they contain; if α_i contains Q_j then α_i' contains Q_{ij}. Otherwise, the system is blocked after the next rewriting step.

If no communication follows the first rewriting step in Γ, then the α_i' sentential forms are sent to $G_1', \ldots, G_n'$ and the the simulation of the next rewriting step starts.

If $(\alpha_1, \ldots, \alpha_n) \Rightarrow_{com}^* (\beta_1, \ldots, \beta_n)$ holds in Γ, then this communication is simulated with the aid of the components $G_{11}, \ldots, G_{nn}$ in the following way. If for some j, $1 \leq j \leq n$, a sentential form α_j' does not contain query symbols, it is transmitted to G_j' and also to the components G_{ij}, $1 \leq i \leq n$, where n copies are saved. The k-th of these saved copies can then be transmitted to a sentential form α_k', if it contains Q_{kj} (Q_{kj} is contained by α_k', if α_k in Γ contains Q_j). This way none of the query symbols in the α_i', $1 \leq i \leq n$, sentential forms are replaced by start symbols, but they all receive a different copy of the string they have requested. This way we get:

$$
(\beta_1, \ldots, \beta_n, S_{n+1}, \ldots, S_{2n}, \delta_{11}, \ldots, \delta_{1n}, \ldots, \delta_{n1}, \ldots, \delta_{nn}, S'')
$$

through a series of communication steps. The δ_{jk}, $1 \leq j, k \leq n$ "garbage" will be removed after the next rewriting step by G_a.

The further rewriting steps and communications of Γ are simulated in a similar way. Let us assume that $(\alpha_1, \ldots, \alpha_n) \Rightarrow_{rew} (\beta_1, \ldots, \beta_n)$ holds in Γ. Now Γ' starts from a configuration:

$$(\alpha_1, \ldots, \alpha_n, S_{n+1}, \ldots, S_{2n}, \delta_{11}, \ldots, \delta_{1n}, \ldots, \delta_{n1}, \ldots, \delta_{nn}, \delta S''),$$

where $\delta_{jk} \in (N' \cup T)^+$, $1 \leq j, k \leq n$. After a rewriting and a communication step we get:

$$(S_1, \ldots, S_n, [\alpha_1], \ldots, [\alpha_n], S_{11}, \ldots, S_{1n}, \ldots, S_{n1}, \ldots, S_{nn}, \delta' S'),$$

and then:

$$(Q'_{n+1}, \ldots, Q'_{2n}, \beta'_1, \ldots, \beta'_n, Q'_{n+1}, \ldots, Q'_{2n}, \ldots, Q'_{n+1}, \ldots, Q'_{2n}, \delta' S''),$$

where β'_i and β_i, $1 \leq i \leq n$, differ only in the indices of the query symbols they contain. Otherwise, the system is blocked after the next rewriting step.

If $(\beta_1, \ldots, \beta_n) \Rightarrow^*_{com} (\gamma_1, \ldots, \gamma_n)$ holds in Γ then it is simulated in the same way as explained above, and we get:

$$(\gamma_1, \ldots, \gamma_n, S_{n+1}, \ldots, S_{2n}, \delta_{11}, \ldots, \delta_{1n}, \ldots, \delta_{n1}, \ldots, \delta_{nn}, \delta' S'').$$

Now the simulation of the next rewriting step can start.   □

Next we focus on the power of non-returning systems. We recall a lemma from [2], which states that every recursively enumerable language is the morphic image of the intersection of the so called twin-shuffle language over some alphabet and a regular language.

First we recall the shuffle operation and the notion of the twin-shuffle language. Let V be a finite alphabet and let $\alpha, \beta \in V^*$. The *shuffle* of strings α and β, denoted by $\alpha \amalg \beta$, is the set $\alpha \amalg \beta = \{\alpha_1 \beta_1 \ldots \alpha_n \beta_n \mid \alpha = \alpha_1 \ldots \alpha_n,\ \beta = \beta_1 \ldots \beta_n,\ \alpha_i, \beta_i \in V^*,\ 1 \leq i \leq n\}$. Consider now an alphabet V, denote by $\bar{V}$ the set $\bar{V} = \{\bar{x} \mid x \in V\}$, and define the coding $h : V^* \longrightarrow \bar{V}^*$ by $h(x) = \bar{x}$, $x \in V$. Let $h(w)$, $w \in V^*$ be denoted by $\bar{w}$. The *twin-shuffle* language over alphabet V, denoted by $TS(V)$, is $TS(V) = \bigcup_{w \in V^*} (w \amalg \bar{w})$.

In [2] the following characterisation of recursively enumerable languages is given.

Lemma 1 [2] *For every recursively enumerable language L, there is a twin-shuffle language $TS(V)$, a regular language R, and a weak coding h, such that $L = h(TS(V) \cap R)$.*

Based on this lemma, we obtain the following theorem.

Theorem 3 *For every recursively enumerable language L, there is a non-returning PC E0L system Γ, a regular language R, and a weak coding h, such that $L = h(L(\Gamma) \cap R)$.*

Proof.   By Lemma 1, it is sufficient to construct a non-returning PC E0L system Γ which generates $TS(V)$, the twin-shuffle language over an arbitray alphabet V. Let $V = \{a^{(1)}, a^{(2)}, \ldots, a^{(m)}\}$ be an alphabet with m elements and let $\bar{V} = \{\bar{a}^{(1)}, \bar{a}^{(2)}, \ldots, \bar{a}^{(m)}\}$. Let:

$$\Gamma = (N, K, T, G, G_0, G_1, \ldots, G_m, G_{m+1}, G_{m+2}),$$

where $G = (N \cup K, T, P, S)$ is the master grammar, the other components are $G_0 = (N \cup K, T, P_0, A\bar{A}[A\bar{A}])$, $G_i = (N \cup K, T, P_i, S)$, $1 \leq i \leq m + 2$, with:

$$N = \{A, \bar{A}, [A\bar{A}], [\bar{A}A], S, A_j, \bar{A}_j, [A\bar{A}]_j, [\bar{A}A]_j, S_j,$$
$$x_j \mid x \in V \cup \bar{V}, \ 1 \leq j \leq 2\},$$
$$T = V \cup \bar{V}.$$

The production sets are as follows. If for a symbol $x \in N \cup T$ no rule is explicitly given, we assume the presence of "chain" rules, $x \to x$:

$$P = \{S \to S, S \to Q_0\} \cup$$
$$\{x_j \to \varepsilon \mid x \in \{A, \bar{A}, [A\bar{A}], [\bar{A}A]\}, \ 1 \leq j \leq 2\} \cup$$
$$\{x_j \to x \mid x \in V \cup \bar{V}, \ 1 \leq j \leq 2\},$$

$$P_0 = \{x \to x_1 \mid x \in \{A, \bar{A}, [A\bar{A}]\}\} \cup \{S_2 \to Q_0\} \cup$$
$$\{[A\bar{A}]_1 \to Q_i, [\bar{A}A]_1 \to Q_i \mid 1 \leq i \leq m + 2\} \cup$$
$$\{x_1 \to \varepsilon \mid x \in \{A, \bar{A}\} \cup V \cup \bar{V}\} \cup$$
$$\{x_2 \to x_1 \mid x \in \{A, \bar{A}, [A\bar{A}], [\bar{A}A]\} \cup V \cup \bar{V}\}.$$

Let also:

$$P_i = \{x_1 \to x_2 \mid x \in \{[A\bar{A}], [\bar{A}A]\} \cup V \cup \bar{V}\} \cup$$
$$\{A_1 \to a_2^{(i)} A_2, \bar{A}_1 \to \bar{a}_2^{(i)} \bar{A}_2\} \cup$$
$$\{x_2 \to \varepsilon \mid x \in \{A, \bar{A}\} \cup V \cup \bar{V}\} \cup$$
$$\{x \to Q_0 \mid x \in \{S, [A\bar{A}]_2, [\bar{A}A]_2\}\},$$

for $1 \leq i \leq m$, and:

$$P_{m+1} = \{S \to S_1, S \to Q_0, S_1 \to S_2, S_2 \to S_1, S_2 \to Q_0\} \cup$$
$$\{A_1 \to \varepsilon, \bar{A}_1 \to \bar{A}_2 A_2, [A\bar{A}]_1 \to [\bar{A}A]_2, [\bar{A}A]_1 \to Q_{m+1}\} \cup$$
$$\{x_1 \to x_2 \mid x \in V \cup \bar{V}\} \cup$$
$$\{x_2 \to \varepsilon \mid x \in \{A, \bar{A}\} \cup V \cup \bar{V}\} \cup$$
$$\{[\bar{A}A]_2 \to S_1, [\bar{A}A]_2 \to Q_0\},$$

$$P_{m+2} = \{S \to S_1, S \to Q_0, S_1 \to S_2, S_2 \to S_1, S_2 \to Q_0\} \cup$$
$$\{\bar{A}_1 \to \varepsilon, A_1 \to A_2 \bar{A}_2, [\bar{A}A]_1 \to [A\bar{A}]_2, [A\bar{A}]_1 \to Q_{m+2}\} \cup$$
$$\{x_1 \to x_2 \mid x \in V \cup \bar{V}\} \cup$$
$$\{x_2 \to \varepsilon \mid x \in \{A, \bar{A}\} \cup V \cup \bar{V}\} \cup$$
$$\{[A\bar{A}]_2 \to S_1, [A\bar{A}]_2 \to Q_0\}.$$

Let us now follow the derivations in this system to see how it generates the twin-shuffle language over V. It starts with:

$$(S, A\bar{A}[A\bar{A}], S, \ldots, S, S, S),$$

and after a rewriting step we get:

$$(\delta, A_1\bar{A}_1[A\bar{A}]_1, Q_0, \ldots, Q_0, \delta_{m+1}, \delta_{m+2}),$$

where δ is either S or Q_0, δ_j, $m + 1 \leq j \leq m + 2$ is either S_1 or Q_0. If $\delta = Q_0$ the system generates ε after the next rewriting step, so let us assume that $\delta = S$ in order to continue. The string $A_1\bar{A}_1[A\bar{A}]_1$ is transferred to components $G_1, \ldots, G_m$, and possibly to components G_{m+1}, or G_{m+2}. We get:

$$(S, A_1\bar{A}_1[A\bar{A}]_1, A_1\bar{A}_1[A\bar{A}]_1, \ldots, A_1\bar{A}_1[A\bar{A}]_1, \delta'_{m+1}, \delta'_{m+2}),$$

where δ'_j, $m + 1 \leq j \leq m + 2$ is either S_1 or $A_1\bar{A}_1[A\bar{A}]_1$. Now each G_i starts deriving a different word of the twin shuffle language by rewriting A_1 and $\bar{A}_1$ to $a_2^{(i)} A_2$ and $\bar{a}_2^{(i)} \bar{A}_2$, $1 \leq i \leq m$. If G_{m+1} has received $A_1\bar{A}_1[A\bar{A}]_1$, it changes the order of the two "key" As and the nonterminal marking this order by producing $\bar{A}_2 A_2[\bar{A}A]_2$ (if G_{m+2} has received $A_1\bar{A}_1[A\bar{A}]_1$ the system is going to be blocked; this component is designed to change the order of the As in the opposite way). Meanwhile, G_0 erases its string and introduces a query symbol Q_i, for some i, $1 \leq i \leq m + 2$. This process produces:

$$(S, Q_i, a_2^{(1)} A_2\bar{a}_2^{(1)} \bar{A}_2[A\bar{A}]_2, \ldots, a_2^{(m)} A_2\bar{a}_2^{(m)} \bar{A}_2[A\bar{A}]_2, \delta''_{m+1}, S_2),$$

where $1 \leq i \leq m + 2$ and δ''_{m+1} is either $\bar{A}_2 A_2[\bar{A}A]_2$ or S_2. Now one of the $m + 2$ strings is received by G_0, then $G_1, \ldots, G_{m+2}$ erase their sentential forms and introduce Q_0 (or possibly S_1 in the case of G_{m+1}, G_{m+2}). Thus, after the next rewriting step we get:

$$(S, \delta_0, Q_0, \ldots, Q_0, \delta_{m+1}, \delta_{m+2}),$$

where δ_0 is $a_1^{(i)} A_1\bar{a}_1^{(i)} \bar{A}_1[A\bar{A}]_1$ for some i, $1 \leq i \leq m$, or $\bar{A}_1 A_1[\bar{A}A]_1$, and δ_j, $m + 1 \leq j \leq m + 2$, is S_1 or Q_0. Now the process can continue in the same way by adding more symbols to the subwords of the twin-shuffle string in $G_1, \ldots, G_m$, or by changing the order of A and $\bar{A}$ in G_{m+1}, G_{m+2}. If G, the master, introduces Q_0, it receives the string generated so far, erases the nonterminals $A_j, \bar{A}_j$, $[A\bar{A}]_j$, or $[\bar{A}A]_j$, $1 \leq j \leq 2$, and produces a word of the twin-shuffle language over V.

Any other way of functioning leads to a blocking configuration, so our proof is complete. $\square$

Corollary 1 *For each family of languages $\mathcal{L}$, with $\mathcal{L} \subset \mathcal{L}(RE)$ and $\mathcal{L}$ being closed under arbitrary morphisms and intersection with regular languages $\mathcal{L}(NPC_*X) - \mathcal{L} \neq \emptyset$, $X \in \{E0L, ET0L\}$ holds.*

Proof. By Theorem 3 and the properties of $\mathcal{L}$ the inclusion $\mathcal{L}(NPC_*X)$ $\subseteq \mathcal{L}$, $X \in \{E0L, ET0L\}$ would imply $\mathcal{L}(RE) \subseteq \mathcal{L}$, a contradiction.   □

Since $\mathcal{L}(ET0L)$ is closed under intersection with regular languages and arbitrary morphisms (see [5]), this corollary combined with Theorem 1 again implies the inclusion $\mathcal{L}(ET0L) \subset \mathcal{L}(NPC_*X)$, $X \in \{E0L, ET0L\}$, which is also the consequence of earlier results from [3] (see also [1]) combined with Theorem 1.

References

[1] E. Csuhaj-Varjú, J. Dassow, J. Kelemen and Gh. Păun, *Grammar Systems: A Grammatical Approach to Distribution and Cooperation.* Gordon and Breach, London, 1994.

[2] J. Engelfriet and G. Rozenberg, Fixed point languages and representations of recursively enumerable languages. *Journal of the ACM*, 27/3 (1980), 499–518.

[3] Gh. Păun, Parallel communicating systems of L systems. In G. Rozenberg and A. Salomaa (eds.), *Lindenmayer Systems: Impacts on Theoretical Computer Science, Computer Graphics, and Developmental Biology.* Springer, Berlin, 1992, 405–418.

[4] Gh. Păun and L. Sântean, Parallel communicating grammar systems: The regular case. *Annals of the University of Bucharest, Mathematics-Informatics Series*, 38/2 (1989), 55–63.

[5] G. Rozenberg and A. Salomaa, *The Mathematical Theory of L Systems.* Academic Press, New York, 1980.

[6] A. Salomaa, *Formal Languages.* Academic Press, New York, 1973.

[7] Gy. Vaszil, Parallel communicating grammar systems without a master. *Computers and Artificial Intelligence*, 15/2-3 (1996), 185–198.

[8] Gy. Vaszil, Communication in parallel communicating Lindenmayer systems. *Grammars*, 1/3 (1999), 255–270.

[9] Gy. Vaszil, On parallel communicating Lindenmayer systems. In Gh. Păun and A. Salomaa (eds.), *Grammatical Models of Multi-Agent Systems.* Gordon and Breach, London, 1999, 99–112.

[10] Gy. Vaszil, Further remarks on parallel communicating grammar systems without a master. *Journal of Automata, Languages and Combinatorics*, accepted.

II

AUTOMATA

Cellular Automata and Probabilistic L Systems: An Example in Ecology

Manuel Alfonseca

Alfonso Ortega

Alberto Suárez

Department of Computer Science Engineering
Autonomous University of Madrid
Spain
{manuel.alfonseca, alfonso.ortega}@ii.uam.es

Abstract. This paper revisits the formal definition of deterministic and probabilistic cellular automata, with special attention to the problem of updating the probabilistic information of each automaton in the grid. An example is given. On the other hand, we introduce a formal notation for probabilistic L systems and the language generated by them. Several examples are given. We propose a new equivalence between both fields: the step-equivalence between a probabilistic L system and a probabilistic cellular automaton. The paper includes a constructive proof of this result and its application to a bi dimensional probabilistic cellular automaton that models an ecosystem.

1 Cellular Automata

A cellular automaton is defined as six-fold (G, G_0, N, Q, f, T), where G is a matrix of automata. G_0 is the initial state of the grid and is a mapping $G_0 : G \to Q$, an injective function that assigns an initial state to each automaton in the grid. N (neighborhood) is a function that assigns to each automaton in the grid the set of its neighbors. Q is the set of possible states of every automaton in the grid. f is the transition mapping $f : Q \times Q^n \rightarrow Q$, where $f(q_0, (q_1, ..., q_n)), n = \#Q$ is the next state of any automaton in the grid if its current state is q_0 and whose neighborhood's states are $(q_1, ..., q_n)$. $T \subseteq Q$ is the set of final or target states. Every automaton in the grid has the same number of neighbors, transition mappings and set of possible and final states.

It is obvious that each finite automaton in the grid is defined by $a = (Q^n, Q, f, G_0(a), (T))$.

2   Probabilistic Cellular Automata

Cellular automata are probabilistic if each automaton in the grid is a probabilistic finite automaton.

In probabilistic cellular automata, the automata on the grid choose their next state from a set of options by assigning probabilities to each transition while the pure non-deterministic approach only establishes the set of options.

A probabilistic cellular automaton is the six-fold (G, G_0, N, Q, M, T), where G, N, Q, T are defined as in a cellular automaton, and G_0 is both the initial state of the grid and the mapping:

$$G_0 : G \rightarrow \left([0,1] \bigcap \Re\right) | \forall x \in G \Rightarrow \sum_{i=1}^{\#Q} \prod_i (G_0(x)) = 1$$

This mapping is an injective function that assigns an initial state vector to each automaton in the grid. The state vector of an automaton shows the probability of the automaton being in each state. The following notations will be used indistinctly in the following pages:

$\prod_i (G_0(x)) = \prod_{q_i} (G_0(x))$ =probability that the automaton x is in state q_i at the initial moment.

M is the transition matrix, a matrix of probabilities of transition between states, with dimension $\#Q^n \times \#Q \times \#Q$. In order to simplify the notation, that M is considered a family of $\#Q^n$ square matrices $\#Q \times \#Q$ (there is a matrix for each particular neighborhood configuration). Each finite probabilistic automaton a in the grid is defined as $a = (Q^n, Q, M, G_0(a), T)$.

Example 1 *Assume an infinite square grid. The concatenation of the row and column indices identifies the automaton at this position in the grid. The Von Neumann neighborhood will be used. Automata are binary, that is $Q = \{0, 1\}$ The cellular automaton is $pca_1 - (G_1, G_0, N_{v_N}, Q, M, T)$, where $G_1 \in M_{Z \times Z}$ is an infinite square matrix of automata around position (0,0). $G_0(x) = 0.5 \forall x \in Q$, i.e. each state is initially equiprobable. $N_{v_N} : F \rightarrow$*

$F^4 | \forall (i,j) \in Z \times Z, N_{v_N} (G[i,j]) = (G[i-1,j], G[i,j+1], G[i+1,j],$
$G[i,j-1]),$

$$M = \left\{ \begin{array}{cccc} \begin{bmatrix} 0.1 & 0.9 \\ 0.8 & 0.2 \end{bmatrix} & \begin{bmatrix} 0.3 & 0.7 \\ 0.4 & 0.6 \end{bmatrix} & \begin{bmatrix} 0.5 & 0.5 \\ 0.6 & 0.4 \end{bmatrix} & \begin{bmatrix} 0.7 & 0.3 \\ 0.2 & 0.8 \end{bmatrix} \\ \begin{bmatrix} 0.9 & 0.1 \\ 0.2 & 0.8 \end{bmatrix} & \begin{bmatrix} 0.7 & 0.3 \\ 0.4 & 0.6 \end{bmatrix} & \begin{bmatrix} 0.5 & 0.5 \\ 0.6 & 0.4 \end{bmatrix} & \begin{bmatrix} 0.7 & 0.3 \\ 0.2 & 0.8 \end{bmatrix} \\ \begin{bmatrix} 0.9 & 0.1 \\ 0.2 & 0.8 \end{bmatrix} & \begin{bmatrix} 0.7 & 0.3 \\ 0.4 & 0.6 \end{bmatrix} & \begin{bmatrix} 0.5 & 0.5 \\ 0.6 & 0.4 \end{bmatrix} & \begin{bmatrix} 0.7 & 0.3 \\ 0.2 & 0.8 \end{bmatrix} \\ \begin{bmatrix} 0.9 & 0.1 \\ 0.2 & 0.8 \end{bmatrix} & \begin{bmatrix} 0.7 & 0.3 \\ 0.4 & 0.6 \end{bmatrix} & \begin{bmatrix} 0.5 & 0.5 \\ 0.9 & 0.1 \end{bmatrix} & \begin{bmatrix} 0.2 & 0.8 \\ 0.4 & 0.6 \end{bmatrix} \end{array} \right\}$$

where the matrices are disposed from left to right and from top to bottom. The first matrix is M_{0000} and the last is M_{1111}.

Let us choose an automaton in the grid and name it x:

	1	
4	x	2
	3	

Assume that the five automata have the following probability vectors at a given moment (the indices identify the automata in the previous figure, the first position in the vector is the probability of being 0): $p_1 = (0.2, 0.8), p_2 = (0.6, 0.4), p_3 = (0.3, 0.7), p_4 = (0.9, 0.1), p_x = (0.1, 0.9)$. If the neighborhood configuration of automaton x were, for instance (0,1,0,0), the following matrix operation computes the next state vector for automaton x: $p_x \times M_{0100}$.

The probability of this situation is $p_1[0] \star p_2[1] \star p_3[0] \star p_4[0]$. We have to compute the equivalent probabilities for all possible neighborhood configurations and add the results, thus getting:
$p_x \times \left(\sum_{i,j,k,l \subset [0,1]} (p_1[i] \star p_2[j] \star p_3[k] \star p_4[l]) \times M_{ijkl} \right)$, *which can be expressed by means of the tensor product, where the dot operator represents the element by element matrix product:*

$$p_x \times \left(\sum \left(\prod_{m \in \{1,2,3,4\}}^{\tau} p_m \right) \bullet M \right).$$

2.1 A Configuration of a Probabilistic Cellular Automaton

A configuration C of a probabilistic cellular automaton is a time dependent mapping $C(t) : F \rightarrow Q$ that assigns a state to each automaton in the grid.

The probability that a probabilistic cellular automaton (A) is in a given configuration (C) at a given moment (t) will be denoted $p_{t,A}(C)$, where t and A will be omitted whenever they are obvious from the context.

Each automaton in the grid has a state vector that shows the probability for it being in each possible state. So, the event "the automaton is in configuration C" could be expressed as $\bigcap_{a \in G}$ "the automaton is in configuration C(a)".

If v_a is the state vector for automaton a, then $\pi_{C(a)}(v_a)$ represents the probability that automaton a is in state C(a):

$$p_{t,A}(C) = \prod_{a \in G} \pi_{C(a)}(v_a)$$

The sum of these probabilities over the set of all possible configurations must equal 1:

$$p_{t,A}(C) = \sum_{C \in set-of-possible-configurations} \left(\prod_{a \in G} \pi_{C(a)}(v_a) \right) = 1$$

This expression assumes that the set of possible configurations is ordered.

3   Bidimensional IL Systems

A bidimensional L System is an L System whose words are matrices of characters instead of linear strings. In order to clarify the notation, the following conventions will be followed: the context will always be written before the symbol changed by the production rule; the context will be determined by a function c that generates the horizontal and vertical displacements of the context symbols with respect to the current symbol.

Formally, a bidimensional $\langle K, 0 \rangle$IL System is defined as the five-fold $\langle \Sigma, P, g, \omega, c \rangle$ where Σ, P, g, ω are defined in the usual way and $c : [1, k] \cap \aleph \rightarrow \{-1, 0, +1\}$.

Example 2 *A bidimensional IL System with a von Neumann neighborhood is an extended $\langle 4, 0 \rangle$ system whose c function is defined as follows:* $c(1) = (0, +1)$, $c(2) = (+1, 0)$, $, c(3) = (0, -1)$, $c(1) = (-1, 0)$:

Graphically

	x_1	
x_4	x	x_2
	x_3	

If a Moore neighborhood is used we get an $\langle 8, 0 \rangle$ bidimensional IL System with the following c function: $c(1) = (-1, +1)$, $c(2) = (0, +1)$, $c(3) = (+1, +1)$, $c(4) = (+1, 0)$, $c(5) = (+1, -1)$, $c(6) = (0, -1)$, $c(7) = (-1, -1)$, $c(8) = (-1, 0)$:

Graphically

x_1	x_2	x_3
x_8	x	x_4
x_7	x_6	x_5

4   Probabilistic L Systems

We define a probabilistic L System as an L System in which each production rule has an associated probability with the restriction that the sum of the probabilities associated to all the rules applicable to a symbol at any time must be 1.

A deterministic L System (DL) can be seen as probabilistic with a probability of 1 associated to every rule.

In a DL System, a derivation is linear. In a probabilistic L System (S), it is a tree, with a probability associated to each branch and the sum of the probabilities associated to all the branches with the same origin being 1. This tree will be called $T_n(S)$, and n is the depth of the tree.

Formally, a probabilistic L System is an L System in which the rule set P has been replaced by a set of pairs $(R, p(R))$, where R is a derivation rule and $p(R)$ its probability, with the restriction that if $P' \subset P$ is the set of rules applicable to a symbol at a given context, $\sum_{R in P'} = 1$.

Example 3 *Assume the probabilistic $\langle 1 : 1 \rangle$ IL System $S_2 = \langle \sum_2, P_2, 0011 \rangle$, where:*

$$\sum_2 = \{0, 1, g\} \quad (g \text{ is the end marker})$$

$$P_2 = \left\{ \begin{array}{ll} (xsg ::= x, 1) & \forall x \in \sum_2, \forall s \in \{0, 1\} \\ (gsx ::= x, 1) & \forall x \in \sum_2, \forall s \in \{0, 1\} \\ (x0y ::= x, 0.3) & \forall x, y \in \{0, 1\} \\ (x0y ::= y, 0.7) & \forall x, y \in \{0, 1\} \\ (x1y ::= y, 0.3) & \forall x, y \in \{0, 1\} \\ (x1y ::= x, 0.7) & \forall x, y \in \{0, 1\} \end{array} \right\}$$

i.e. the first symbol in the string becomes its right neighbor, the last becomes its left neighbor, intermediate 0s become its left/right context with probability 0.3/0.7 and intermediate 1s do the same with probability 0.7/0.3.

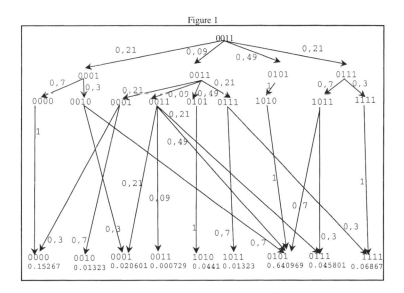

Figure 1

Figure 1 shows the first three derivations of system S_2 and indicates the probability of each branch.

Observe that the probability of reaching a node by a given path is the product of the probabilities of the branches that go from the axiom to that node along that path. The sum of the probabilities of all the nodes in the tree at a given derivation level is 1.

If a node at derivation level k may be reached by more than one path, the probability that the word is generated by a derivation of depth k is the sum of the probabilities of all the paths, computed as above. Formally, the probability that word x is generated by system S in n derivations is $p_{n,S}(x) = \sum_{\rho(D_i)=x \wedge D_i \in T_n(S)} \left(\prod_{b_j \in D_i} \left(\prod_{r_k \in b_j} p(r_k) \right) \right)$ where D_i is a path in tree $T_n(S)$; b_j is a branch in path D_i; r_k is a production rule applied at branch b_j; $\rho(D_i)$ is the result of derivation D_i.

Each path in the tree is a derivation. The expression could be written alternatively in Lindenmayer notation as follows:

$$p_{n,S}(x) = \sum_{\rho(D_i)=x \wedge D_i \in T_n(S)} \left(\prod_{(l,k) \in O_i} p\left(p_i\left((l,k)\right)\right) \right),$$

where $p\left(p_i\left((l,k)\right)\right)$ is the probability associated with the production rule $p_i\left((l,k)\right)$.

Let S be a probabilistic L System. Let n be a natural number ($n \in \aleph$). Let θ be a real number ($\theta \in \Re$).

The language generated by S_θ, is defined as:

$$L(S,\theta) = \{x | \exists \in \aleph : x \in L_n(S) \wedge p_{n,S}(x) \geq \theta\}.$$

In the previous example, the language for threshold 0 is:

$$L(S_2, 0) = \{0001, 0011, 0101, 0111, 0000, 0010, 1010, 1011, 1111\}.$$

5   Step-Equivalence Between Probabilistic L Systems and Probabilistic Cellular Automata

Let A be a probabilistic cellular automaton. Let S be a probabilistic bidimensional IL System.

Definition 1 *S is step-equivalent to A if and only if:*
$\forall t \in \aleph, \forall C$ *(configuration of A)* $\exists \theta \in \Re, w \in L(S)$ *such that*
$$p_{t,L}(w) = p_{t,A}(C).$$

Theorem 1 *Given a probabilistic cellular automaton $A = (G, G_0, N, M, Q)$, there is an equivalent probabilistic bidimensional IL System that is step-equivalent to the cellular automaton.*

Proof. (Constructive proof) Consider the bidimensional $\langle n : 0 \rangle$ IL System $S = \langle \Sigma, P, g, \omega, c \rangle$, where $\Sigma = Q \bigcup \{s_i\}$ and s_i expresses the axiom of the L System; g is a symbol not in Σ; P is the set of the pairs (*rules, probability*):

$\left(\overrightarrow{q}x ::= y, M_{\overrightarrow{q}}\,[x,y]\right)$; axiom ω is a matrix with the same dimensions as G and whose elements are all equal to s_i; and c and N refer to the same elements in their matrices. Rules with g use $M_{\overrightarrow{q}'}$, where $\overrightarrow{q}'$ is obtained from $\overrightarrow{q}$ with replacing g with the appropriate boundary symbol in the automaton.

It is easy to see that S is step equivalent to A .   □

Assume an automaton whose mean-field evolution follows the Lotka-Volterra equations for a predator (species Y, carnivorous) and a prey (species X, herbivorous), with a slight modification that accounts for the saturation of the herbivorous species:

$$\frac{dN_X(t)}{dt} = K_1 N_X(t)\left(1 - \frac{N_X(t)}{N_X^{sat}}\right) - K_2 N_Y(t) N_X(t),$$

$$\frac{dN_Y(t)}{dt} = -K_3 N_Y(t) + K_4 N_Y(t) N_X(t),$$

where N_X^{sat} is the saturation level of species X. The saturation term is necessary since the automaton cannot represent the unlimited growth of species X in the absence of individuals of the species Y. The territory in which the population dynamics takes place is a regular two-dimensional square lattice with periodic boundary conditions. Only nearest neighbors displacements are allowed. The solution to the inverse problem of finding the reactive rules that yield a specified set of mean-field equations was given by Boon et al. in their extensive review on reactive lattice-gas automata. The reactive rules are encoded into a reaction probability matrix, whose entries are the probability of obtaining an outgoing configuration $\mathbf{n}^{out} = \{n_Y^{out}, n_X^{out}\}$ from a given incoming $\mathbf{n}^{in} = \{n_Y^{in}, n_X^{in}\}$. In particular, one possible prescription leading to the previous expressions in the mean-field limit is:

$$p(\mathbf{n}^{in} \to \mathbf{n}^{out}) = h\ K_1\ n_X^{in}\ \delta\left(n_X^{out}, n_X^{in}+1\right)\delta\left(n_Y^{out}, n_Y^{in}\right)\left(1 - \delta\left(n_X^{in}, m\right)\right) +$$

$$h\left[K_2\ n_X^{in} n_Y^{in} + \frac{K_1}{N_X^{sat}}\frac{m}{m-1}n_X^{in}\left(n_X^{in} - 1\right) - K_1\ n_X^{in}\ \delta\left(n_X^{in}, m\right)\right]\delta\left(n_X^{out}, n_X^{in} - 1\right)\delta\left(n_Y^{out}, n_Y^{in}\right) +$$

$$h K_4 n_X^{in} n_Y^{in} \delta\left(n_X^{out}, n_X^{in}\right)\delta\left(n_Y^{out}, n_Y^{in} + 1\right)\left(1 - \delta\left(n_Y^{in}, m\right)\right) +$$

$$h\left[K_3\ n_Y^{in} - K_4\ n_X^{in} n_Y^{in}\delta\left(n_Y^{in}, m\right)\right]\delta\left(n_X^{out}, n_X^{in}\right)\delta\left(n_Y^{out}, n_Y^{in} - 1\right) \quad \text{for } \mathbf{n}^{in} \not\overrightarrow{q}\ \mathbf{n}^{out}$$

$$p(\mathbf{n}^{in} \to \mathbf{n}^{in}) = 1 - \sum_{\mathbf{n}^{out} \neq \mathbf{n}^{in}} p(\mathbf{n}^{in} \to \mathbf{n}^{out}) \tag{1}$$

where $\delta(n, n')$ is a Kronecker delta (an indicator equal to 1 if $n = n'$ and 0 otherwise), the inverse of h represents the reaction time-scale, and m is the maximum number of particles of a single species at a given node, which coincides with the number of channels associated to a node. In the present model $m = 4$, meaning that this is the maximum number of individuals of each species that may occupy a given node. The condition that $p(\mathbf{n}^{in} \to$

$\mathbf{n}^{out}$) be a probability (i.e. a non-negative number in the interval $[0, 1]$)
imposes restrictions on the possible values of the reaction constants K_i , h
and N_X^{sat} that can be used in the simulations. In particular, N_X^{sat} should be
smaller or equal to m (this upper limit corresponds to full occupation of an
automaton node).

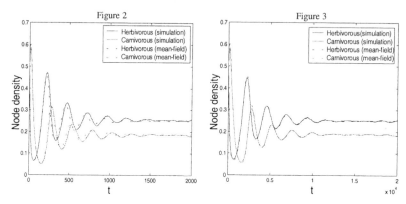

Figure 2 depicts the time-evolution of the automaton for $K_1 = K_2 =
K_3 = K_4 = 1$, $h = 0.0313$ (close to the maximum possible value of h). For
hese values of the reaction constants, there are configurations (in particular
those where $n_Y^{in} = m$) for which Eq. (1) yields negative values. These negative values are set to zero. This procedure does not alter the mean-field
behavior of the automaton in a significant manner, since the configurations
affected appear rather infrequently. The same plot also shows the dynamics
corresponding to the solution of the mean-field equations. The mean-field
equations in this automaton provide a very good approximation to the evolution of the species' node densities, even for the largest possible values of
h, even though, as the figure shows, the frequency of the damped oscilla-
tions predicted by the mean-field approximation is slightly lower than the
frequency of the actual simulated time-series. These small discrepancies can
once again be accounted for by the (limited) influence of correlations on the
dynamics. The observation that the influence of correlations is small in this
automaton is corroborated by the absence of spatial structure in the species'
populations. Figure 3 compares the results of simulations in an automaton
with the same characteristics as the previous one, except that the inverse
reaction time-scale is $h = 0.00313$. In this case, the mean-field approximation provides an excellent description of the global population dynamics in
the ecosystem.

5.1 The Step-Equivalent IL System

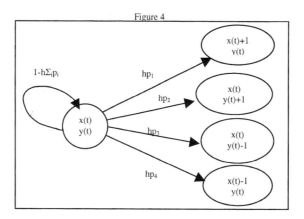

Figure 4

Each automaton in the grid contains several individuals of each species. Let us call l=x(t) and k=y(t) the number of individuals of species x and y at time t. The previous probability prescription can be represented by means of the state diagram in figure 4, where:

$$p_1 = \left\{ \begin{array}{ll} K_1 x(t) & x(t) \neq m \\ 0 & x(t) = m \end{array} \right\} , \quad p_2 = \left\{ \begin{array}{ll} K_4 x(t) y(t) & y(t) \neq m \\ 0 & y(t) = m \end{array} \right\} ,$$

$$p_3 = max \left(0, \left\{ \begin{array}{ll} K_3 y(t) & y(t) \neq m \\ K_3 y(t) - K_4 x(t) y(t) & y(t) = m \end{array} \right\} \right) ,$$

$$p_4 = max \left(0, \left\{ \begin{array}{ll} K_2 x(t) y(t) + \frac{K_1}{N_x^{sat}} \frac{m}{m-1} x(t) [x(t) - 1] & x(t) \neq m \\ K_2 x(t) y(t) + \frac{K_1}{N_x^{sat}} \frac{m}{m-1} x(t) [x(t) - 1] - K_1 x(t) & x(t) = m \end{array} \right\} \right) .$$

This probability distribution is shown in the following table, which represents the rules of the step-equivalent IL system:

$(l,k)_t$	$(l,k)_{t+1}$	$(l+1,k)_{t+1}$	$(l,k+1)_{t+1}$	$(l,k-1)_{t+1}$	$(l-1,k)_{t+1}$
(0,0)	(0,0),1	-	-	-	-
(0,1)	(0,1),1	-	-	-	-
(0,2)	(0,2),1	-	-	-	-
(0,3)	(0,3),1	-	-	-	-
(0,4)	(0,4),1	-	-	-	-
(1,0)	(1,0),1	-	-	-	-
(1,1)	$(1,1),1-\Sigma p_i$	$(2,1),hK_1$	$(1,2),hK_4$	$(1,0),hK_3$	$(0,1),hp_4$
(1,2)	$(1,2),1-\Sigma p_i$	$(2,1),hK_1$	$(1,3),2hK_4$	$(1,2),2hK_3$	$(0,2),hp_4$
(1,3)	$(1,3),1-\Sigma p_i$	$(2,3),hK_1$	$(1,4),3hK_4$	$(1,2),3hK_3$	$(0,3),hp_4$
(1,4)	$(1,4),1-\Sigma p_i$	$(2,4),hK_1$	-	$(1,3),4h(K_3-K_4)$	$(0,4),hp_4$
(2,0)	(2,0),1	-	-	-	-
(2,1)	$(2,1),1-\Sigma p_i$	$(3,1),2hK_1$	$(2,2),2hK_4$	$(2,0),hK_3$	$(1,1),hp_4$
(2,2)	$(2,2),1-\Sigma p_i$	$(3,2),2hK_1$	$(2,3),4hK_4$	$(2,1),2hK_3$	$(1,2),hp_4$
(2,3)	$(2,3),1-\Sigma p_i$	$(3,3),2hK_1$	$(2,4),6hK_4$	$(2,2),3hK_3$	$(1,3),hp_4$
(2,4)	$(2,4),1-\Sigma p_i$	$(3,4),2hK_1$	-	$(2,3),4h(K_3-2K_4)$	$(1,4),hp_4$
(3,0)	(3,0),1	-	-	-	-
(3,1)	$(3,1),1-\Sigma p_i$	$(4,1),3hK_1$	$(3,2),3hK_4$	$(3,0),hK_3$	$(2,1),hp_4$
(3,2)	$(3,2),1-\Sigma p_i$	$(4,2),3hK_1$	$(3,3),6hK_4$	$(3,1),2hK_3$	$(2,2),hp_4$
(3,3)	$(3,3),1-\Sigma p_i$	$(4,3),3hK_1$	$(3,4),9hK_4$	$(3,2),3hK_3$	$(2,3),hp_4$
(3,4)	$(3,4),1-\Sigma p_i$	$(4,4),3hK_1$	-	$(3,3),4h(K_3-3K_4)$	$(2,4),hp_4$
(4,0)	(4,0),1	-	-	-	-
(4,1)	$(4,1),1-\Sigma p_i$	-	$(4,2),4hK_4$	$(4,0),hK_3$	$(3,1),hp_4$
(4,2)	$(4,2),1-\Sigma p_i$	-	$(4,3),8hK_4$	$(4,1),2hK_3$	$(3,2),hp_4$
(4,3)	$(4,3),1-\Sigma p_i$	-	$(4,4),12hK_4$	$(4,2),3hK_3$	$(3,3),hp_4$
(4,4)	$(4,4),1-\Sigma p_i$	-	-	$(4,3),4h(K_3-4K_4)$	$(3,4),hp_4$

The alphabet of the step-equivalent Bidimensional IL System is the set of all possible configurations of the nodes (the number of individuals of species x and y in the node). The table above gives the rules. The axiom is a matrix of symbols s_i. The context is the current symbol itself.

References

[1] M. Alfonseca, *Teoría de Lenguajes, Gramáticas y Autómatas*. Promosoft, Madrid, 1997.

[2] J.P. Boon, D. Dab, R. Kapral and A. Lawniczak, Lattice gas automata for reactive systems. *Physics Reports*, 273 (1996), 55–147.

[3] G.T. Herman and G. Rozenberg, *Developmental Systems and Languages*. North Holland/American Elsevier, Amsterdam, 1975.

[4] G. Rozenberg and A. Salomaa (eds.), *Lindenmayer Systems: Impacts on Theoretical Computer Science, Computer Graphics, and Developmental Biology*. Springer, Berlin, 1992.

On Iterated Sequential Transducers

Henning Bordihn

Faculty of Informatics
Otto-von-Guericke University
Magdeburg, Germany
bordihn@iws.cs.uni-magdeburg.de

Henning Fernau

Wilhelm-Schickard Institute of Informatics
University of Tübingen
Germany
fernau@informatik.uni-tuebingen.de

Markus Holzer

I.R.O. Department
University of Montréal
Québec, Canada
holzer@iro.umontreal.ca

Abstract.   We continue to explore the relationships between Lindenmayer Systems and iterated sequential transducers introduced by Manca, Martín-Vide, and Păun in [10]. We investigate nondeterministic as well as deterministic transducers. The latter case was stated as an open question by Manca, Martín-Vide, and Păun.

1   Introduction and Definitions

Iterated transducers are a natural extension of Lindenmayer Systems, as already observed by Wood in [18]. Independently of this, there has been steady interest in this topic in the formal language community, as can be seen by studying [1, 2, 11, 12, 13, 16]. Recently, this field of research has been revived by the carving paradigm for computing [9]. In this paper, we continue to explore the relationships between Lindenmayer Systems and iterated sequential transducers introduced by Manca, Martín-Vide, and Păun in [10]. We investigate both nondeterministic and deterministic transducers. The latter case was stated as an open question in [10].

We assume the reader to be familiar with the basic notions of formal languages, as contained in [15]. In general, we have the following conventions: $\subseteq$ denotes inclusion, while $\subset$ denotes strict inclusion. The empty word is denoted by λ. For $x \in V^*$, where V is some alphabet, and for $a \in V$, $|x|_a$ denotes the number of occurrences of the letter a in x.

The families of languages generated by regular, context-free, context-sensitive, general type-0 Chomsky grammars, D0L, 0L, F0L, ED0L, E0L, and ET0L systems are denoted by REG, CF, CS, RE, D0L, 0L, F0L, ED0L, E0L, and ET0L, respectively. Details about these families can be found in [6, 15]. Recall that an ED(1,0)L system is given by a quadruple $G = (V, \Sigma, P, \omega)$, where V and Σ are the total alphabet and the terminal alphabet, respectively, $\Sigma \subseteq V$, $\omega \in V^+$ is the axiom, and P is a mapping from $(V \cup \{\lambda\}) \times V$ into V^*. We write $(\alpha, a) \rightarrow w$ instead of $P(\alpha, a) = w$ and call it production in P. A word x directly yields the word y, in symbols $x \Rightarrow y$, if and only if $x = a_1 a_2 \ldots a_k$, $y = w_1 w_2 \ldots w_k$, $k \geq 1$, $a_i \in V$, $w_i \in V^*$, and $(a_{i-1}, a_i) \rightarrow w_i$ is a production in P, $1 \leq i \leq k$. Here, we set $a_0 = \lambda$. Let $\overset{*}{\Rightarrow}$ be the reflexive transitive closure of the relation $\Rightarrow$. By definition, G generates the language $L(G) = \{w \in \Sigma^* \mid \omega \overset{*}{\Rightarrow} w\}$. Intuitively, an ED(1,0)L system is a parallel rewriting system with one-sided context. We denote the corresponding family of languages by ED(1,0)L. Similarly, ED2L is the family of languages generated by parallel rewriting systems with two-sided context. Vitányi has shown in [17, Section 3.2] that ED2L equals the family of recursively enumerable languages RE.

An *iterated (finite state) sequential transducer (IFT)* [10] is a construct $\gamma = (K, V, s_0, a_0, F, P)$, where K, V are disjoint alphabets (the set of *states* and the *alphabet* of γ), $s_0 \in K$ (the *initial state*), $a_0 \in V$ (the *starting symbol*), $F \subseteq K$ (the set of *final states*), and P is a finite set of *transition rules* of the form $sa \rightarrow xs'$, for $s, s' \in K$, $a \in V$, $x \in V^*$ (in state s, the device reads the symbol a, passes to state s', and produces the string x). For $s, s' \in K$ and $u, v, x \in V^*$, $a \in V$, we define:

$$usav \vdash uxs'v \quad \text{if and only if} \quad sa \rightarrow xs' \in P.$$

This is a *direct transition step* with respect to γ. We denote by $\vdash^*$ the reflexive transitive closure of the relation $\vdash$.

Then, for $w, w' \in V^*$, we define:

$$w \Rightarrow w' \quad \text{if and only if} \quad s_0 w \vdash^* w's, \quad \text{for some } s \in K.$$

We say that w *derives* w'; note that this means that w' is obtained by translating the string w, starting from the initial state of γ and ending in any state of γ, not necessarily a final one. We denote by $\overset{*}{\Rightarrow}$ the reflexive transitive closure of $\Rightarrow$. If $s_0 w \vdash^* w' s$, for some $s \in F$, i.e. a derivation stops in a final state, we write $w \overset{f}{\Rightarrow} w'$. The *language generated* by γ is:

$$L(\gamma) = \{\, w \in V^* \mid a_0 \overset{*}{\Rightarrow} w' \overset{f}{\Rightarrow} w, \text{ for some } w' \in V^* \,\}.$$

In other words, we iteratively translate the strings obtained by starting from a_0, without caring about the states we reach at the end of each translation; we necessarily stop in a final state, only after the last step.

The IFT's, as defined above, are *nondeterministic*. If for each pair $(s, a) \in K \times V$, there is at most one transition rule $sa \to xs'$ in P, then we say that γ is *deterministic*.

For $n \geq 1$, let IFT_n denote the family of languages of the form $L(\gamma)$, where γ is a nondeterministic IFT with at most n states; similarly, DIFT_n is the family of languages generatable by DIFT's with at most n states.

Let $\mathrm{IFT} = \bigcup_{n \geq 1} \mathrm{IFT}_n$ and $\mathrm{DIFT} = \bigcup_{n \geq 1} \mathrm{DIFT}_n$.

2 Nondeterministic Language Families

In [10] it was shown that the hierarchy of nondeterministic finite state sequential transducers induced by the number of states collapses, and that the fourth level characterizes the recursively enumerable languages. Moreover, the following chains of inclusions were exhibited in [6, 10]:

$$\mathrm{IFT}_1 = 0\mathrm{L} \subset \mathrm{E0L} \subseteq \mathrm{IFT}_2 \subseteq \mathrm{IFT}_3 \subseteq \mathrm{IFT}_4 = \mathrm{IFT} = \mathrm{RE}$$

and $\mathrm{ET0L} \subseteq \mathrm{IFT}_3$. The next theorem corrects and shows the relationships of the first two levels of the IFT hierarchy, hence solving a problem posed after [10, Theorem 3].

Theorem 1 $0\mathrm{L} \subset \mathrm{IFT}_1 = \mathrm{F0L} \subset \mathrm{E0L} \subset \mathrm{IFT}_2$.

Proof. The strictness of the trivial inclusion $0\mathrm{L} \subset \mathrm{F0L}$ follows from [6, Theorem 2.1]. Since the finite axiom set $\{\omega_1, \ldots, \omega_k\}$ of an F0L system can be generated by rules of the form $s_0 a_0 \to \omega_i s_0$, the inclusion $\mathrm{F0L} \subseteq \mathrm{IFT}_1$ is clear. On the other hand, since the mappings defined by one-state finite state transducers are exactly the substitutions, $\mathrm{IFT}_1 \subseteq \mathrm{F0L}$ is obvious. The strict inclusion $\mathrm{F0L} \subset \mathrm{E0L}$ follows from [6, Theorem 7.1], together with the fact that E0L is closed under finite union.

The inclusion relation $\mathrm{E0L} \subseteq \mathrm{IFT}_2$ has been shown in [10, Lemma 3]. Finally, consider the IFT $\gamma = (\{s_0, s_1\}, \{a, b, c, a_0\}, s_0, a_0, \{s_0\}, P)$, with the set of transition rules:

$$P = \{s_0 a_0 \to cb s_1, s_0 c \to c s_1, s_0 c \to s_0\} \cup \{s_1 b \to bb s_1\}$$
$$\cup \{s_0 b \to ab s_0, s_0 b \to ba s_0, s_0 b \to b s_0, s_0 a \to a s_0, s_0 a \to aa s_0\}.$$

In the following, $L(\gamma) = \{w \in \{a,b\}^* \mid |w|_b = 2^n \text{ and } n \geq 0\}$ is shown. Hence, $L(\gamma)$ is non-E0L according to [3, Example 3]. The first step $s_0 a_0 \to cb s_1$ introduces the word cb. Each subsequent translation $s_0 cb^n \vdash^* cb^{2n} s_1$ starting with the rule $s_0 c \to c s_1$ doubles the number of b's. After removing the leading letter c with the rule $s_0 c \to s_0$ or when starting with a word without a leading c, the iteration continues by introducing a's on arbitrary positions, while the number of b's remains unchanged.   □

Theorem 2 *For every $L \in \mathrm{RE}$, there is an IFT γ with at most three states and a regular set R, such that $L = L(\gamma) \cap R$.*

Proof.   The proof parallels the construction $\mathrm{RE} = \mathrm{IFT}_4$ given in [10]. Let $L \subseteq \Sigma^*$ be a recursively enumerable language. Then:

$$L = \bigcup_{a \in \Sigma} (\delta_a^r(L) \cdot a) \cup (\{\lambda\} \cap L).$$

Since L is in RE, the right derivative $\delta_a^r(L) = \{w \mid wa \in L\}$ is also in RE due to the well-known closure properties of RE. Thus, $\delta_a^r(L)$ is generated by a grammar $G_a = (N_a, \Sigma, S_a, P_a)$ in Geffert normal form [4], i.e. $N_a = \{S_a, A_a, B_a, C_a\}$ and the rules are of the forms $S_a \to \alpha$, for $\alpha \in (N_a \cup \Sigma)^+$, and $A_a B_a C_a \to \lambda$. Assume $N_a \cap N_b = \emptyset$, for $a,b \in \Sigma$ with $a \neq b$. We consider the IFT $\gamma = (\{s_0, s_A, s_B\}, N \cup \Sigma \cup \{a_0\}, s_0, a_0, \{s_0\}, P)$, where $N = \bigcup_{a \in \Sigma} N_a$ and:

$$\begin{aligned} P = \; & \{s_0 a_0 \to S_a a s_0 \mid a \in \Sigma\} \cup \{s_0 X \to X s_0 \mid X \in N \cup \Sigma\} \\ & \cup \{s_0 S_a \to \alpha s_0 \mid a \in \Sigma \text{ and } S_a \to \alpha \in P_a\} \\ & \cup \{s_0 A_a \to s_A, s_A B_a \to s_B, s_B C_a \to s_0 \mid a \in \Sigma\}. \end{aligned}$$

We add $s_0 a_0 \to s_0$ if $\lambda \in L$. The first step $s_0 a_0 \to S_a a s_0$ introduces the axiom of G_a or the empty word by $s_0 a_0 \to s_0$. Each subsequent translation $s_0 w \vdash^* w' s_0$ corresponds to an equivalent derivation step sequence $w \overset{*}{\Rightarrow} w'$ in G_a. Note that derivation steps from different grammars G_a and G_b, for $a \neq b$ cannot be mixed due to the distinct nonterminals. Moreover, the presence of the rightmost symbol (introduced in the first step) ensures that γ does not reach the end of the string in state s_A or s_B. Consequently, $L(\gamma)$ is the language of all sentential forms induced by the grammars G_a, for $a \in \Sigma$. Thus, $L = L(\gamma) \cap \Sigma^*$.   □

Note that if L is an undecidable language, then the constructed language of all sentential forms in the previous proof is also undecidable. Since we have seen that the latter language belongs to IFT_3, we have:

Corollary 1 *The family* IFT$_3$ *contains non-recursive languages.*

Due to [10, Lemma 7], ET0L is included in IFT$_3$. The previous corollary implies that the inclusion is strict, as already claimed in [10], because ET0L languages are recursive.

Corollary 2 ET0L $\subset$ IFT$_3$.

The following theorem answers an open question posed in the report version of [10] after Theorem 6, since the operation "intersection with regular languages" is a sequential transducer mapping,[1] and it can be seen as a sort of generalization of that theorem, since morphisms are sequential transducer mappings. We state our theorem without proof, because the construction is quite similar to the well-known triple construction.

Theorem 3 *Let* $n \geq 1$. *For* $L \in IFT_n$ *and a sequential transducer* γ, *we have* $\gamma(L) \in IFT_{n+1}$.

3 Deterministic Language Families

Here, we consider *deterministic* IFT's. This case was left almost completely open in [10]. Firstly, we show an analogue to Theorem 1, using the deterministic variants of E0L and IFT$_2$. This solves an open problem stated in the report version of [10], where, with respect to DIFT$_2$, only DIFT$_2 \setminus$ CF $\neq \emptyset$ and DIFT$_2 \setminus$ 0L $\neq \emptyset$ was shown.

Lemma 1 ED0L $\subset$ DIFT$_2$.

Proof. Firstly, we show the inclusion. Consider a given ED0L system $G = (V, \Sigma, P, \omega)$. Without loss of generality, we can assume that $\omega \notin \Sigma^*$. Construct the deterministic IFT $\gamma = (\{s_0, s_1\}, V \cup \{a_0\}, s_0, a_0, \{s_0\}, P')$, where a_0 is a new symbol not contained in $V \supseteq \Sigma$ and:

$$
\begin{aligned}
P' \;=\; &\{s_0 a_0 \rightarrow \omega s_1\} \\
&\cup \{\, s_0 a \rightarrow x s_0 \mid a \rightarrow x \in P \text{ and } x \in \Sigma^* \,\} \\
&\cup \{\, s_0 a \rightarrow x s_1 \mid a \rightarrow x \in P \text{ and } x \notin \Sigma^* \,\} \\
&\cup \{\, s_1 a \rightarrow x s_1 \mid a \rightarrow x \in P \,\}.
\end{aligned}
$$

The first step, $s_0 a_0 \rightarrow \omega s_1$, introduces the axiom of G. Each subsequent translation $s_0 w \vdash^* w' s_i$, with $i = 0$ if $w' \in \Sigma^*$ and $i = 1$ otherwise, corresponds precisely an equivalent derivation step $w \Rightarrow w'$ in G. Thus,

---

[1]If, in the definition of an IFT, we distinguish between an input and an output alphabet, we call those devices *(finite state) sequential transducers, FT* for short. If γ is an FT and L a language over the input alphabet of γ, then the FT image of L under γ is defined as $\gamma(L) = \{\, w \mid u \Rightarrow^f w, u \in L \,\}$.

the final state s_0 is reached whenever a terminal string in the sense of G is derived. Consequently, $L(G) = L(\gamma)$.

Secondly, we prove the strictness of the inclusion. We construct the deterministic IFT $\gamma = (\{s_0, s_1\}, \{a, b, a_0\}, s_0, a_0, \{s_0, s_1\}, P'')$, where:

$$P'' = \{s_0 a_0 \rightarrow aas_0, s_0 a \rightarrow bs_1, s_1 a \rightarrow as_0, s_0 b \rightarrow as_0\}.$$

It is easily verified that γ generates the language $\{a^2, ab, ba\}$, which is a non-ED0L language [6, Exercise 4.9]. □

With the lemma given above and the fact that DIFT_1 equals the family of D0L languages, we state without proof:

Corollary 3 $\text{D0L} = \text{DIFT}_1 \subset \text{DIFT}_2$ *and* $\text{D0L} = \text{DIFT}_1 \subset \text{IFT}_1$.

In the following, we shall prove that any recursively enumerable language with and end-marker, i.e. $L\{\#\}$ with $L \in \text{RE}$, can be generated by some deterministic IFT. From the equality $\text{RE} = \text{ED2L}$, Vitányi concludes in [17, Theorem 3.47]:

Lemma 2 *If* $L \subseteq \Sigma^*$ *is a recursively enumerable language, then* $L\{\#\} \in \text{ED}(1,0)\text{L}$, *where* $\#$ *is a new symbol, i.e.* $\# \notin \Sigma$.

Lemma 2 is also true when taking DIFT's instead of $\text{ED}(1,0)\text{L}$ systems. More precisely, we find:

Theorem 4 $\text{ED}(1,0)\text{L} \subseteq \text{DIFT}$.

Proof. Consider an $\text{ED}(1,0)\text{L}$ system $G = (V, \Sigma, P, \omega)$ generating L. We shall construct a DIFT γ for L. Let $K = \{(\alpha), [\alpha] \mid \alpha \in (V \cup \{\lambda\})\}$ be the set of states of γ, (λ) its initial state, and $F = \{(\alpha) \mid \alpha \in (V \cup \{\lambda\})\}$ its set of final states. The alphabet of γ equals $V \cup \{a_0\}$, where a_0 is a new letter. The transition rules are the following:

$$
\begin{array}{lll}
(\lambda)a_0 \rightarrow \omega(\lambda) & \text{for} & \omega \in \Sigma^*, \\
(\lambda)a_0 \rightarrow \omega[\lambda] & \text{for} & \omega \notin \Sigma^*, \\
(\alpha)a \rightarrow w(a) & \text{for} & a \in V, (\alpha, a) \rightarrow w \in P, \text{ and } w \in \Sigma^*, \\
(\alpha)a \rightarrow w[a] & \text{for} & a \in V, (\alpha, a) \rightarrow w \in P, \text{ and } w \notin \Sigma^*, \\
[\alpha]a \rightarrow w[a] & \text{for} & a \in V, (\alpha, a) \rightarrow w \in P.
\end{array}
$$

Obviously, γ is deterministic and it simulates the derivation of G. □

Therefore, [17, Theorem 3.42] yields:

Corollary 4 *The family* DIFT *contains non-recursive languages.*

Clearly, this implies that almost all DIFT_n classes contain non-recursive languages [2] Due to [17, Theorem 3.46], we can state:

---

[2] It would be of interest to determine the smallest n such that DIFT_n contains a non-recursive language.

Corollary 5 *The closure of* DIFT *under letter-to-letter homomorphisms is equal to* RE.

Finally, we mention that the proofs of the closure properties for IFT's given in the report version of [10] are not transferable to the deterministic case, but we are able to give a result concerning the closure with respect to intersections with regular sets.

Lemma 3 *Let* $n \geq 1$. *For* $L \in DIFT_n$ *and a regular language* R *which is accepted by a deterministic finite automaton with* k *states and* m *accepting states, we have* $L \cap R \in \mathrm{DIFT}_{\ell+m}$, *where* $\ell = \max\{n, k\}$.

Proof. Consider a DIFT $\gamma = (K, V, s_0, a_0, F, P)$ generating L and a deterministic finite automaton $\tau = (Z, X, z_0, Q, \delta)$ accepting R, where $Z = \{z_0, z_1, \ldots, z_{k-1}\}$ is the set of states, z_0 the initial state, $Q \subseteq Z$ the set of accepting states, X the input alphabet and $\delta : Z \times X \to Z$ the transition function. Moreover, let $K = \{s_0, s_1, \ldots, s_{n-1}\}$ and $\ell = \max\{n, k\}$. Let $K' = \{y_0, y_1, \ldots, y_{\ell-1}\} \cup \{y_i^f \mid z_i \in Q\}$. Furthermore, set $V' = \{a' \mid a \in V\}$ and consider the morphism h defined by $h(a) = a'$ for $a \in V$; we simply write x' instead of $h(x)$ for strings $x \in V^*$. Now, construct the DIFT:

$$
\begin{aligned}
\gamma' &= (K', V \cup V' \cup \{\Lambda, \#\}, y_0, a_0, \{y_i^f \mid z_i \in Q\}, P'), \quad \text{where:} \\
P' &= \{y_i a \to x' y_j \mid s_i a \to x s_j, x \neq \lambda, \text{ and } s_j \notin F\} \\
&\quad \cup \{y_i a \to x' \# y_j \mid s_i a \to x s_j, x \neq \lambda, \text{ and } s_j \in F\} \\
&\quad \cup \{y_i a \to \Lambda y_j \mid s_i a \to s_j \text{ and } s_j \notin F\} \\
&\quad \cup \{y_i a \to \Lambda \# y_j \mid s_i a \to s_j \text{ and } s_j \in F\} \\
&\quad \cup \{y_i a' \to a y_j \mid \delta(z_i, a) = z_j\} \\
&\quad \cup \{y_i \# \to y_i^f \mid z_i \in Q\} \cup \{y_i \# \to y_i \mid z_i \notin Q\} \\
&\quad \cup \{y_i^f \# \to y_i^f \mid z_i \in Q\} \\
&\quad \cup \{y_i^f a' \to y_j \mid \delta(z_i, a) = z_j\} \\
&\quad \cup \{y_i \Lambda \to y_i \mid z_i \in Z\} \cup \{y_i^f \Lambda \to y_i \mid z_i \in Z\}.
\end{aligned}
$$

γ' is deterministic and has $\max\{k, n\} + m$ states. γ' works as follows. It simulates a derivation step $w_1 \Rightarrow w_2$ of γ essentially by $w_1 \Rightarrow w_2'$, but a symbol Λ is written whenever γ performs an erasing step $s_i a \to s_j$, and an additional symbol $\#$ is written whenever γ is led into a final state. In those phases, the states $y_i \in K'$ are interpreted as renamed states of γ, and the last symbol of the output is $\#$ if and only if the corresponding output belongs to $L(\gamma)$. In the next step, γ' rewrites any symbol a' by a and erases all occurrences of Λ and $\#$. Meanwhile, it simulates the automaton τ, interpreting the states $y_i \in K'$ as states of τ. Whenever a letter $\#$ is read and, at the same time, γ' is led to a state corresponding to an accepting state of τ, then γ' enters a final state y_i^f instead of y_i. Thus, at the end of such a phase, γ' is in a final state if and only if a word in $L(\gamma) \cap R$ is generated. This proves our assertion. $\square$

4   Extended Iterated Sequential Transducers

Iterated sequential transducers are a sort of "pure" rewriting mechanism. Thus, "extended versions" such as Lindenmayer languages can also be meaningfully considered. Let γ be an IFT with input/output alphabet V and let $\Delta \subseteq V$ be a "terminal alphabet." Then $L(\gamma, \Delta) = L(\gamma) \cap \Delta^*$ is an *extended language* generated by γ. The corresponding language families are denoted by EIFT_n and EDIFT_n, respectively. Some easy consequences of our above findings are:

Theorem 5   *1. $\mathrm{DIFT}_n \subseteq \mathrm{EDIFT}_n \subseteq \mathrm{DIFT}_{n+1}$ for all $n \geq 1$ and*

   2. $\mathrm{IFT}_n \subseteq \mathrm{EIFT}_n \subseteq \mathrm{IFT}_{n+1}$ for all $n \geq 1$.

Looking more carefully, we also find $\mathrm{ED0L} = \mathrm{EDIFT}_1$, $\mathrm{E0L} = \mathrm{EIFT}_1$ and $\mathrm{RE} = \mathrm{EIFT}_3$ because of Theorem 2. Finally, we state:

Theorem 6 $\mathrm{ET0L} \subseteq \mathrm{EIFT}_2$.

Proof.   Let L be an ET0L language. By using [15, Theorems V.1.3 and V.1.4], we can assume that L is generated by an ET0L system $G = (V, \Sigma, \{P_1, P_2\}, \omega)$ with $\omega \notin \Sigma^*$. Let a_0 and c be new symbols, i.e. $a_0, c \notin V$. Define $\gamma = (\{s_0, s_1\}, V \cup \{a_0, c\}, s_0, a_0, \{s_0, s_1\}, P)$, where:

$$
\begin{aligned}
P \;=\; & \{s_0 a_0 \to c\omega s_0\} \\
& \cup \{s_0 c \to c s_0, s_0 c \to s_0\} \cup \{\, s_0 a \to x s_0 \mid a \to x \in P_1 \,\} \\
& \cup \{s_0 c \to c s_1, s_0 c \to s_1\} \cup \{\, s_1 a \to x s_1 \mid a \to x \in P_2 \,\}.
\end{aligned}
$$

It is easy to see that $L = L(\gamma) \cap \Sigma^*$ holds. $\square$

5   Conclusions and Further Research Proposals

In this paper, we obtained further results about the relationships between iterated transducers and Lindenmayer systems. Similarly, the relationships with models that are obviously akin such as restart automata (see e.g. [7]), clog automata [5] and "sequentialized versions" of programmed 0L systems [14] are of interest, particularly as far as the descriptional complexity is concerned (similar to the number of states as studied in the present note for IFT's). Furthermore, (E)T0L systems with regular context conditions (which are specific to each table and which determine the applicability of a table) can be easily simulated by IFT's and, to our knowledge, have not been investigated before, although they seem to be quite natural and interesting in their own right.

Besides the power of iterated "propagating" (i.e. λ-free) IFT's and DIFT's, the main open problem concerning iterated transducer languages is whether the hierarchy $DIFT_1 \subset DIFT_2 \subseteq \ldots$ is infinite. We conjecture that the hierarchy is infinite, but it is very hard to give a satisfying proof. Many derivation steps might not contribute to the candidate language, because the transducer is not in a final state after the present sentential form has been scanned.

References

[1] P.R.J. Asveld, On controlled iterated gsm mappings and related operations, *Revue Roumaine des Mathématiques Pures et Appliquées*, 25 (1980), 136–145.

[2] J.M. Autebert and J. Gabarró, Iterated gsm's and co-cfl, *Acta Informatica*, 26 (1989), 749–769.

[3] A. Ehrenfeucht and G. Rozenberg, The number of occurrences of letters versus their distribution in some E0L languages, *Information and Control*, 26 (1974), 256–271.

[4] V. Geffert, Normal forms for phrase structure grammars, *RAIRO Informatique Théorique et Applications*, 25 (1991), 473–496.

[5] L. Haines, Representation theorems for context-sensitive languages, *Notices of the American Mathematical Society*, 16.3 (1969), 527.

[6] G.T. Herman and G. Rozenberg, *Developmental Systems and Languages*. North-Holland, Amsterdam, 1975.

[7] P. Jančar et al., On restarting automata with rewriting. In Gh. Păun and A. Salomaa (eds.), *New Trends in Formal Languages*, Springer, Berlin, 1997, 119–136.

[8] J. van Leeuwen, The membership problem for ET0L-languages is polynomially complete, *Information Processing Letters*, 3 (1975), 138–143.

[9] V. Manca, C. Martín-Vide and Gh. Păun, New computing paradigms suggested by DNA computing: computing by carving. In L. Kari, H. Rubin and D.H. Wood (eds.), *Preliminary Proceedings of the 4th DIMACS Workshop on DNA Based Computers*. University of Pennysylvania, Philadelphia, Pa., 1998, 41–56.

[10] V. Manca, C. Martín-Vide and Gh. Păun, Iterated gsm-mappings: a collapsing hierarchy. In J. Karhumäki, H. Maurer, Gh. Păun and G. Rozenberg (eds.), *Jewels are Forever*. Springer, Berlin, 1999, 182–193.

[11] Gh. Păun, On the iteration of gsm mappings, *Revue Roumaine des Mathématiques Pures et Appliquées*, 23 (1978), 921–937.

[12] Gh. Păun, Classes of iterated gsm's suggested by suspicious communication questions, *Revue Roumaine de Linguistique, Cahiers de Linguistique Théorique et Appliquée*, XXIV.2 (1987), 139–144.

[13] Gh. Păun, The complexity of language translation by gsm's, *Revue Roumaine de Linguistique, Cahiers de Linguistique Théorique et Appliquée*, XXV.1 (1988), 49–58.

[14] K.S. Rajasethupathy and R.K.Shyamasundar, Programmed 0L-systems, *Information Sciences*, 20 (1980), 137–150.

[15] G. Rozenberg and A. Salomaa, *The Mathematical Theory of L Systems*. Academic Press, New York, 1980.

[16] H. Takahashi, The maximum invariant set of an automaton system, *Information and Control*, 32 (1976), 307–354.

[17] P.M.B. Vitányi, Lindenmayer Systems: Structure, Languages, and Growth Functions, Technical Report 96, Mathematisch Centrum, Amsterdam, 1980.

[18] D. Wood, Iterated a-NGSM maps and Γ systems, *Information and Control*, 32 (1976), 1–26.

Distributed Real-Time Automata

Cătălin Dima

Department of Fundamentals of Computer Science
University of Bucharest
Romania
cdima@funinf.cs.unibuc.ro

Abstract. We introduce a class of automata with real-time constraints, automata which can be seen as tuples of real-time automata working on their own tape and synchronized on their transitions. Though the general class is equivalent to timed automata and hence is not closed under language complementation, the subclass consisting of automata with a so-called *stuttering-free* condition is shown to be closed under language complementation.

1 Introduction

Automata theory remains one of the most fruitful and influential domains in theoretical computer science. It is used in such diverse fields as linguistics, semantics and formal verification and it has influenced such distant disciplines as logic and control theory.

In the past decade much interest has been shown in adapting automata theory to the problems of real-time systems, and particular emphasis has been put on generalizing finite automata by taking time into account.

A few types of automata have proved to be useful, but unfortunately their properties are rather complementary. Nevertheless, no class of automata can claim to be the most natural and expressive extension of finite automata. Timed automata of [1] have an undecidable universality problem; event-clock automata [2, 7], though complementable, have rather complicated algebraic properties and are not closed under substitution; and real-time automata [4] cannot support semantics of real-time concurrent programming languages.

In this paper, we build upon an idea that arose in [6]: a tuple of real-time automata working on their own tape but synchronized on their transitions might still be closed under complementation. This idea has its roots in the semantics of real-time synchronous programming languages, which (when working on variables with finite domains) can be given in the so-called Simple Duration Calculus [5].

From the very beginning it is apparent that the automata built this way are too powerful - they are equivalent to the timed automata of [1] and the complementation procedure from [6] is therefore lost. However, we find that by adding the so-called *stuttering freeness* condition automata become determinizable and hence complementable. It is the same condition that on real-time automata required some properties of the Kleene algebra of sets of intervals; see [6].

The paper is organized as follows. In the next section we recall the definitions and properties of signals and real-time automata. The third section contains the definitions of distributed real-time automata and the observation that stuttering implies nonclosure under complementation. The fourth section contains the constructions for determinization and complementation of the automata being discussed. We end by formalizing the problem of constructing the semantics of real-time synchronous programs from the semantics of each component.

2   Preliminaries

Let us first fix some notations: $\mathbb{R}_+$ denotes the set of nonnegative numbers while $Intv_{\mathbb{Q}_+}$ denotes the set of intervals with nonnegative rational bounds. We denote by π_i the usual projection of a Cartesian product on the i-th component $\pi_i : \prod_{j=1}^n A_j \longrightarrow A_i$, $\pi_i(a_1, \ldots, a_n) = a_i$ and $Diag(A)$ denotes the diagonal relation on A, i.e. $Diag(A) = \{(a, a) \mid a \in A\}$.

For a function $f : \mathbb{R}_+ \longrightarrow A$ and for some $\alpha > 0$, if there exist some $a, b \in A$, $a \neq b$ and $\epsilon > 0$ such that $f(x) = a$ for all $x \in (\alpha - \epsilon, \alpha)$ and $f(x) = b$ for all $x \in (\alpha, \alpha + \epsilon)$, then we call a the *left limit* of f at α and b the *right limit* of f at α and denote them $f(\alpha - 0)$ and $f(\alpha + 0)$, respectively.

We say that f *has a discontinuity at* α iff $f(\alpha - 0) \neq f(\alpha + 0)$. The discontinuity is *left* iff we also have that $f(\alpha) = f(\alpha + 0)$.

Definition 1 *A signal over an alphabet Σ is a function $\sigma : [0, \alpha) \longrightarrow \Sigma$ whose domain is an initial interval of the positive numbers and which has* *finitely many discontinuities, all being left discontinuities.*

Given n-sets of symbols $\Sigma_1, \ldots, \Sigma_n$ (not necessarily pairwise disjoint), *an n-signal is a signal over $\Sigma_1 \times \ldots \times \Sigma_n$.*

We assume throughout this paper that the alphabets are finite. We denote the domain of σ as $dom(\sigma)$ and the endpoint of this domain as $endp(\sigma)$. The set of all signals over Σ is denoted $Sig(\Sigma)$ and subsets of it are called *real-time languages*, in analogy with sets of words over Σ. Also the set of n-signals is denoted $Sig(\Sigma_1, \ldots, \Sigma_n)$. Note that for each $i \in [n]$, the projection $\pi_i : \Sigma_1 \times \ldots \times \Sigma_n \longrightarrow \Sigma_i$ extends to a function that is also denoted π_i from $Sig(\Sigma_1, \ldots, \Sigma_n)$ to $Sig(\Sigma)$ and defined as $endp(\pi_i(\sigma)) = endp(\sigma)$ and $\pi_i(\sigma)(t) = \pi_i(\sigma(t))$ for all $t \in dom(\sigma)$.

Now let $\sigma \in Sig(\Sigma)$. For each discontinuity α of σ we define the *duration of the symbol after* α as the difference $d(\sigma, \alpha) = \alpha - \beta$, where β is the next discontinuity after α, if it exists, or $endo(\sigma)$ otherwise. Similarly, define $d(\sigma, \alpha)$ for $\alpha = 0$ too. Then we extend this definition for n-signals: the *duration of the i-symbol after* α in an n-signal σ is $d_i(\sigma, \alpha) = d(\pi_i(\sigma), \alpha)$, if α is a discontinuity of $\pi_i(\sigma)$ and undefined otherwise.

$Sig(\Sigma)$ can be endowed with a noncommutative monoidal structure by defining a concatenation operation: given two signals σ_1 and σ_2, their concatenation can be defined as the signal σ with domain $[0, endp(\sigma_1) + endp(\sigma_2))$ and whose value is:

$$\sigma(t) = \begin{cases} \sigma_1(t) & \text{iff } t \in [0, endp(\sigma_1)), \\ \sigma_2(t - endp(\sigma_1)) & \text{iff } t \in [endp(\sigma_1), endp(\sigma_1) + endp(\sigma_2)). \end{cases}$$

We denote the concatenation of σ_1 with σ_2 as $\sigma_1; \sigma_2$. The unit of concatenation is the *unique* signal with empty domain $\sigma_\epsilon : [0, 0) \longrightarrow \Sigma$.

Then $\mathcal{P}(Sig(\Sigma))$, the powerset of $Sig(\Sigma)$, can be given a structure that is similar to $\mathcal{P}(\Sigma^*)$, namely a *Kleene algebra structure* [3] by defining the star of a real-time language $L \subseteq Sig(\Sigma)$ as $L^* = \bigcup_{n \in \mathbb{N}} L^n$, where $L^0 = \{\sigma_\epsilon\}$ and $L^{n+1} = L^n; L$ ($n \in \mathbb{N}$). Here, concatenation on languages is the natural extension of concatenation on words.

Definition 2 *A real-time automaton (RTA for short) is a tuple $\mathcal{A} = (Q, \Sigma, \lambda, \iota, \delta, S, F)$ where Q is the (finite) set of* states, $\delta \subseteq Q \times Q$ *is the* transition relation, $S, F \subseteq Q$ *are the sets of* initial, *resp.* final *states*, $\lambda : Q \longrightarrow \Sigma$ *is the* state labeling function *and* $\iota : Q \longrightarrow Intv_{\mathbb{Q}_+}$ *is the* interval labeling function.

RTA works over signals: a *run* of length n is a sequence of locations $(q_i)_{i \in [n]}$ connected by δ, i.e. $(q_{i-1}, q_i) \in \delta, \forall i \in [n]$. A run is called *accepting* iff it starts in S and ends in F. The run *accepts* a signal σ iff the signal can be split into n parts such that the state within the i-th part equals $\lambda(q_i)$

while the length of this i-th part is in the interval $\iota(q_i)$. Formally there exist $e_i \in \iota(q_i)$ such that $dom(\sigma) = [0, \sum_{i=1}^{n} e_i)$ and $\sigma(t) = \lambda(q_i)$ for all $t \in [\sum_{j=1}^{i-1} e_j, \sum_{j=1}^{i} e_j)$ and all $i \in [n]$.

This definition may also be extended to any runs and signals: we say that a run is *associated* to a signal iff the above properties hold for the respective run and signal.

Clearly the splitting points must contain all the time points where state changes occur but there might be more splitting points than discontinuities, as two consecutive locations in the run might have the same state label. This is a form of nondeterminism specific to automata with real-time constraints.

Definition 3 *A RTA $\mathcal{A}$ is language deterministic iff each signal in $L(\mathcal{A})$ is associated to a unique run. $\mathcal{A}$ is stuttering-free iff it does not have a transition (q, r) with $\lambda(q) = \lambda(r)$. $\mathcal{A}$ is state-deterministic iff initial locations have disjoint labels and transitions starting in the same locations have disjoint labels too, i.e. whenever $r \neq s$ and either $r, s \in S$ or $(q, r), (q, s) \in \delta$ then $\lambda(r) \neq \lambda(s)$ or $\iota(r) \cap \iota(s) = \emptyset$. $\mathcal{A}$ is simply called deterministic whenever it is both state-deterministic and stuttering-free.*

Hence, for stuttering-free RTA, the splitting points are unique for each signal. Notice also that there might be state-deterministic RTA that are not language deterministic due to stuttering steps. The important property that stuttering-free RTA have is that the usual determinization procedure of finite automata can be adapted for them:

Theorem 1 *The class of languages which are accepted by stuttering-free real-time automata is closed under complementation.*

This theorem is a corollary of the results proven in [6].

3   Distributed Real-Time Automata

Definition 4 *An n-distributed real-time automaton (n-DRTA) is a tuple:*

$$\mathcal{A} = (Q, \Sigma_1, \ldots, \Sigma_n, \delta, \lambda_1, \ldots, \lambda_n, \iota_1, \ldots, \iota_n, S, F),$$

where $n \geq 1$, Q is the set of states, *$\lambda_i : Q \longrightarrow \Sigma_i$ are the* state labeling functions, *$\iota_i : Q \longrightarrow Intv_{\mathbb{Q}_+}$ are the* interval labeling functions, *$\delta \subseteq Q \times \mathcal{P}([n]) \times Q$ is the* transition relation *and $S, F \subseteq Q$ are the sets of* initial, *resp.* final, *states. The transitions $(q, K, r) \in \delta$ also obey the following condition: if $\lambda_i(q) \neq \lambda_i(r)$ then $i \in K$.*

The second component of each transition is called the *resetting component*. Its use will be apparent in defining the language of these automata.

A *run* is some alternating sequence $(q_1, K_1, q_2, \ldots, K_{m-1}, q_m)$ consisting of states $q_i \in Q$ and sets of indices $K_i \subseteq [n]$ with the property that $(q_j, K_j, q_{j+1}) \in \delta$ for all $j \in [m-1]$. An *accepting run*, therefore, is a run that begins in an initial state and ends in a final state. We denote the set of runs of $\mathcal{A}$ by Runs($\mathcal{A}$) and the set of accepting runs of $\mathcal{A}$ by ARuns($\mathcal{A}$).

Intuitively, in n-DRTA there are n real-time automata which work in parallel but which are synchronized on certain transitions. The resetting component is used by each RTA to determine where its current state has to be changed. These ideas are formalized as follows: the *i-th underlying RTA* of the above n-DRTA is the RTA $\mathcal{A}_i = (Q, \Sigma_i, \delta_i, \lambda_i, \iota_i, S, F)$ where:

$$\delta_i = \{(q, r) \in Q^2 \mid \exists (q = q_1, K_1, q_2, \ldots, K_{m-1}, q_m = r) \in \text{Runs}(\mathcal{A}) \text{ and}$$
$$\exists j_0 \in [m-1] \text{ s.t. } i \in K_{j_0} \text{ and } i \notin K_j, \forall j \in [m-1], j \neq j_0\}. \qquad (1)$$

Then, given a run $\rho = (q_1, K_1, q_2, \ldots, K_{m-1}, q_m)$ we may define the *i-th projection* of this run as the set of runs in $\mathcal{A}_i$ induced by ρ:

$$pr_i(\rho) = \{(q_{i_j})_{j \in [k]} \mid k \leq m, \ i_j \in [m], \ i_1 = 1, i_k = m \text{ and}$$
$$(q_{i_j}, q_{i_{j+1}}) \in \delta_i \ \forall j \in [k-1]\}.$$

Note that there might be more than one run in $pr_i(\rho)$ because definition 1 implies that a single transition in δ induces a whole range of transitions in δ_i. We also call the runs in $pr_i(\rho)$ *unidimensional*.

Then, given an n-signal $\sigma \in Sig(\Sigma_1, \ldots, \Sigma_n)$, we say it is *accepted* by the accepting run $\rho = (q_1, K_1, q_2, \ldots, K_{m-1}, q_m)$ (and hence *accepted* by the n-DRTA) iff for each $i \in [n]$ the signal is accepted by some unidimensional run in the i-th projection of ρ. The *language accepted* by $\mathcal{A}$ is the set of n-signals accepted by $\mathcal{A}$. We say two n-DRTA are *equivalent* iff they have the same language. We will denote by $DRTA$ the class of all n-DRTA, for all $n \in \mathbb{N}$.

We should point out here that "some" may be replaced by "any" in the definition of acceptance of a signal.

Proposition 1 *The language emptiness problem is decidable for $DRTA$.*

This follows since n-DRTA are special cases of timed automata with n-clocks which have a decidable emptiness problem [1].

Definition 5 *A n-DRTA $\mathcal{A} = (Q, \Sigma_1, \ldots, \Sigma_n, \delta, \lambda_1, \ldots, \lambda_n, \iota_1, \ldots, \iota_n, S, F)$ is called language deterministic iff each n-signal $\sigma \in L(\mathcal{A})$ is associated to a unique accepting run.*

$\mathcal{A}$ is called state deterministic iff the following conditions are met:

- *for each pair of transitions $(q, K, r), (q, K, s) \in \delta$ with the property that $\lambda_i(r) = \lambda_i(s)$ for all $i \in [n]$, there is some $j \in [n]$ such that $\iota_j(r) \neq \iota_j(s)$;*

- *for each pair of initial states $r, s \in S$ with the property that $\lambda_i(r) = \lambda_i(s)$ for all $i \in [n]$, there is some $j \in [n]$ such that $\iota_j(r) \neq \iota_j(s)$.*

The automaton is called stuttering free iff, for each transition $(q, K, r) \in \delta$, K is the set of indices i such that $\lambda_i(q) \neq \lambda_i(r)$.

If $\mathcal{A}$ is both stuttering-free and state deterministic then it is simply called deterministic.

Finally, the automaton is called complete iff for each n-signal (not necessarily in the language of $\mathcal{A}$) there exists a unique associated run that starts in an initial state.

Hence, in a stuttering-free n-DRTA the transition relation could just be given as a *subset of Q^2* with the property that if $(q, r) \in \delta$ then there exists some $i \in [n]$ such that $\lambda_i(q) \neq \lambda_i(r)$. We will use this definition of the transition relation for these automata. This also has the advantage that runs can be described then simply as (finite) sequences of states $(q_i)_{i \in [m]}$.

It is clear that a deterministic n-DRTA is language deterministic. It is also easy to find some language nondeterministic automata that are still deterministic or some state deterministic automata that are not language deterministic. All these notions are important to prove the closure under complementation: for the usual complementation construction, a language deterministic automaton is needed which is also complete and its set of final states must then be complemented.

Note that if the n-DRTA is stuttering-free then all its projections are stuttering-free too.

3.1   Comments on the Stuttering-Free Condition

The stuttering-free condition looks rather strong and it can easily be seen that it does not allow $DRTA$ to be closed under concatenation of languages. However, it is essential in the determinization procedure: without the stuttering condition, n-DRTA have essentially the same expressive power as timed automata of [1], hence they have an undecidable universality problem.

It is mentioned in [1] that the following language[1] is accepted by some timed automaton, while its complement cannot be accepted:

$$L = \{\sigma \in Sig(\{a, b\}) \mid \exists t \in dom(\sigma) \text{ such that } \sigma(t - 0) = a, \sigma(t + 0) = b,$$
$$\sigma((t + 1) - 0) = a \text{ and } \sigma((t + 1) + 0) = b\}.$$

---

[1]Actually, this is a modification of the example in [1], which was stated for the timed words semantics of timed automata.

The language consists of signals in which there exist two (possibly non consecutive) discontinuities where the signal jumps from a to b which are separated by a unit interval.

Consider the language $\overline{L}$ which consists of 2-signals over $\{a, b\} \times \{c\}$ and whose first projection is L. We can easily find a 2-DRTA for this language:

$\mathcal{A} = ([6], \{a, b\}, \{c\}, \delta, \lambda_1, \lambda_2, \iota_1, \iota_2, \{1\}, \{6\})$ where:

$\lambda_1(1) = \lambda_1(3) = \lambda_1(5) = a, \quad \lambda_1(2) = \lambda_1(4) = \lambda_1(6) = b,$
$\lambda_2(i) = c$ for all $i \in [6]$,

$\iota_1(i) = (0, \infty)$ for all $i \in [6]$,
$\iota_2(1) = \iota_2(2) = \iota_2(5) = \iota_2(6) = (0, \infty), \quad \iota_2(3) = \iota_2(4) = [1, 1],$

$$\begin{aligned}
\delta \;=\; & \{(1, \{1\}, 2), (2, \{1\}, 1), (1, \{1, 2\}, 4), (2, \{1, 2\}, 3), (3, \{1\}, 4), \\
& (4, \{1\}, 3), (3, \{1, 2\}, 6), (4, \{1, 2\}, 5), (5, \{1\}, 6), (6, \{1\}, 5)\}.
\end{aligned}$$

Hence, 2-DRTA cannot be closed under complementation, otherwise some timed automata would accept the complement of L.

4   Complementation of Stuttering-Free n-DRTA

The aim of the determinization construction is for each signal to have a single run that accepts the signal. For stuttering-free n-DRTA this means that at each discontinuity there must be only one choice to make between the different transitions enabled at that point.

Let's call a transition (q, r) *enabled* at some discontinuity α in an n-signal σ iff the right limit of each projection of the signal is consistent with the labels of r. It is clear that even for deterministic n-DRTA at each discontinuity of σ more transitions might be enabled. But the length of the next i-symbol uniquely determines the transition which is to be taken. Therefore, in the classical determinization procedure we need to take into account not only sets of states which have the same state label, but also those which share a specific interval label that is not available to any other (similarly state-labeled) set of states. This idea is formalized in the sequel.

Start with a n-DRTA $\mathcal{A} = (Q, \Sigma_1, \ldots, \Sigma_n, \delta, \lambda_1, \ldots, \lambda_n, \iota_1, \ldots, \iota_n, S, F)$ which is stuttering free. Construct the set of subsets of Q consisting of identically labeled states:

$$Id(Q) = \{T \subseteq Q \mid \forall i \in [n] \, \exists a_i \in \Sigma_i \text{ such that } \forall q \in T, \lambda_i(q) = a_i\}.$$

Also for each $T \in Id(Q)$ and $T' \subseteq T$ denote $I_i(T, T')$ the intersection of all intervals which label T' together with the complements of the intervals which label $T \setminus T'$:

$$I_i(T, T') = \bigcap_{q \in T'} \iota_i(q) \cap \bigcap_{q \in T \setminus T'} (\mathbb{R} \setminus \iota_i(q)).$$

As usual, the intersection of an empty family of intervals is $\mathbb{R}$.

Note that the collection of all sets $I_i(T, T')$, where T is fixed but T' ranges over all subsets of T, forms a *partition* or $\mathbb{R}$. This property is essential in the determinization construction and it also assures that the deterministic automaton is *complete*.

On the other hand, it is clear that $I_i(T, T')$ may not be an interval, but rather a finite union of intervals. So $int(I_i(T, T'))$ denotes the pairwise disjoint and *minimal* (in cardinality) set of intervals of which $I_i(T, T')$ is composed and $p_i(T, T')$ denotes the cardinal of $int(I_i(T, T'))$. Consider also some enumeration of the intervals in $int(I_i(T, T'))$:

$$int(I_i(T, T')) = \{J_1^i, \ldots, J_{p_i(T,T')}^i\}.$$

The states of the deterministic n-DRTA which is equivalent to $\mathcal{A}$ are $(n+2)$-uples $(T, T', k_1, \ldots, k_n)$, where $T \in Id(Q)$, $T' \subseteq T$ and $k_i \in [p_i(T, T')]$ for all $i \in [n]$. Denote this set of tuples $\overline{Q}$. The state labeling functions are then clear:

$$\overline{\lambda}_i : \overline{Q} \longrightarrow \Sigma_i, \qquad \overline{\lambda}_i((T, T', k_1, \ldots, k_n)) = \overline{\lambda}_i(q) \text{ for some } q \in T. \quad (2)$$

The interval labeling functions are defined using $int(I_i(T, T'))$:

$$\iota_i(T, T', k_1, \ldots, k_n) = J_{k_i}^i \quad \forall i \in [n].$$

The set of initial states $\overline{S}$ consists of all the tuples $(S, S', k_1, \ldots, k_n)$ in which the set of initial states S is paired with all its *nonempty subsets* S' and some tuple of indices $k_i \in [p_i(S, S')]$. Similarly, the set of final states $\overline{F}$ consists of all the tuples $(U, U', k_1, \ldots, k_n)$, where U' has a nonempty intersection with the set of final states F (and therefore is itself nonempty) and $k_i \in [p_i(U, U')]$.

To define the transitions of $\overline{\mathcal{A}}$ we need another notation: for each $(T, T', k_1, \ldots, k_n) \in \overline{Q}$ denote by $R(T, T', k_1, \ldots, k_n)$ the element of $Id(Q)$ which collects the destinations of all transitions that start from T' and lead to identically labeled states:

$$R(T, T', k_1, \ldots, k_n) = \{r \in Q \mid \overline{\lambda}_i(r) = a_i \text{ for some fixed } a_i \in \Sigma_i \text{ and }$$
$$\exists q \in T' \text{ s.t. } (q, r) \in \delta\}.$$

Then $\overline{\mathcal{A}}$ will have as its transitions all the tuples of the kind:

$$((T, T', k_1, \ldots, k_n), (R, R', l_1, \ldots, l_n)),$$

where $R = R(T, T', k_1, \ldots, k_n)$, $R' \subseteq R$ and $l_i \in [p_i(T, T')]$. Hence:

Theorem 2 *Stuttering-free n-DRTA and deterministic n-DRTA have the same expressive power.*

Yet another important outcome of the above construction is that the n-DRTA obtained is *complete*. The n-DRTA which accepts $Sig(\Sigma_1, \ldots, \Sigma_n) \setminus L(\mathcal{A})$ will be, then, the n-DRTA whose components are the same as for $\overline{\mathcal{A}}$ with the exception of the set of final states which is $\overline{Q} \setminus \overline{F}$. Hence:

Theorem 3 *The class of languages which are accepted by n-DRTA is closed under complementation.*

5 Constructing the Semantics of Real-Time Synchronous Programs

As noted in [4], RTA may be used to provide semantics for sequential real-time programs, i.e. programs in which real-time constraints are put on the delays between the executions of instructions. We can then define the semantics of an n-component synchronous real-time program as a suitable n-DRTA whose projections are the n RTAs that are the semantics of each component but also whose transition relation models the synchronizations between the components. This idea is formalized as follows:

Definition 6 *Given n stuttering-free RTA $\mathcal{A}_i = (Q_i, \Sigma_i, \lambda_i, \iota_i, \delta_i, S_i, F_i)$, a synchronization pattern on $(\mathcal{A}_i)_{i\in[n]}$ is a set of tuples $\delta \subseteq (\prod_{i=1}^n Q^i)^2 \setminus Diag(\prod_{i=1}^n Q^i)$ with the following property:*

$$((q_1, \ldots, q_n), (r_1, \ldots, r_n)) \in \delta \Rightarrow (q_i, r_i) \in \delta_i \cup Diag(Q_i).$$

A synchronization pattern δ is just a special case of stuttering-free n-DRTA:

$$\mathcal{A} = \left(\prod_{i=1}^n Q_i, \Sigma_1, \ldots, \Sigma_n, \lambda_1, \ldots, \lambda_n, \iota_1, \ldots, \iota_n, \prod_{i=1}^n S_i, \prod_{i=1}^n F_i\right).$$

Therefore, the accepting condition and language of a synchronization pattern are particular cases of the respective definitions for n-DRTA.

However, the reverse relationship is unclear because it is unclear whether each n-DRTA is equivalent to some synchronization pattern of its underlying RTAs. The choice of a synchronization pattern consisting of all $2n$-uples of the form $((q, \ldots, q), (r, \ldots, r))$ is wrong since it is possible that $(q, r) \notin \delta_i$ because it is also possible that $i \notin K$, for any transition (q, K, r) of the given n-DRTA. Hence, we conjecture that synchronization patterns are less expressive than n-DRTA.

References

[1] R. Alur and D.L. Dill, A theory of timed automata, *Theoretical Computer Science*, 126 (1994), 183–235.

[2] R. Alur, L. Fix and T.A. Henzinger, A determinizable class of timed automata. In *Computer-Aided Verification*. Springer, Berlin, 1994, 1–13.

[3] J.H. Conway, *Regular Algebra and Finite Machines*. Chapman and Hall, London, 1971.

[4] C. Dima, Automata and regular expressions for real-time languages. In *Proceedings of the AFL'99 workshop*, Vasszeczeny, Hungary, 1999.

[5] C. Dima, Simple Duration Calculus Semantics of a Real-Time Synchronous Programming Language, 1999, unpublished ms.

[6] C. Dima, Real-time automata and the Kleene algebra of sets of real numbers. In *Proceedings of STACS'2000*. Springer, Berlin, 2000.

[7] T.A. Henzinger, J.F. Raskin and P.Y. Schobbens, The regular real-time languages. In *Proceedings of the 25th ICALP*. Springer, Berlin, 1998.

[8] J.E. Hopcroft and J.D. Ullman, *Introduction to Automata Theory, Languages and Computation*. Addison-Wesley, Reading, Mass., 1992.

On Commutative Directable Nondeterministic Automata[1]

Balázs Imreh

Department of Informatics
József Attila University
Szeged, Hungary
imreh@inf.u-szeged.hu

Masami Ito

Department of Mathematics
Faculty of Science
Kyoto Sangyo University
Japan
ito@ksuvx0.kyoto-su.ac.jp

Magnus Steinby

Department of Mathematics
University of Turku
Finland
steinby@utu.fi

Abstract. In [8] an input word w of a nondeterministic automaton (nda) $\mathcal{A}$ was called:

(1) D1-directing if the set of states aw in which $\mathcal{A}$ may be after reading w is the same singleton set $\{b\}$ for all initial states a;

[1]This work has been supported by the Hungarian-Finnish S & T Co-operation Programme for 1997-1999, Grant SF-10/97, the Japanese Ministry of Education, Mombusho International Scientific Research Program, Joint Research 10044098, the Hungarian National Foundation for Science Research, Grant T030143, and the Ministry of Culture and Education of Hungary, Grant FKFP 0704/1997.

(2) D2-directing if the set aw is the same for every initial state a;

(3) D3-directing if some state b appears in every set aw.

Here we consider these notions for commutative nda. Commutativity is a very strong assumption which virtually eliminates the distinction between general nda and complete nda (cnda). Moreover, the sets of Di-directing words of a given nda are essentially of the same form in all six cases considered. We also give bounds for the maximal lengths of minimum-length Di-directing words of an n-state nda or cnda ($i = 1, 2, 3$).

1   Introduction

An input word of an automaton $\mathcal{A}$ is *directing* if it takes $\mathcal{A}$ from every state to the same fixed state. An automaton is called *directable* if it has a directing word. Directability has been studied extensively from several points of view and for various types of automata (cf. [3], [7], [8] for some references). In particular, the directability of nondeterministic automata (nda) has also received some attention. In [8] an input word w of an nda $\mathcal{A}$ was said to be:

(1) D1-*directing* if the set of states in which $\mathcal{A}$ may be after reading w consists of the same single state b regardless of the starting state, *i.e.* if $aw = \{b\}$ for all states a,

(2) D2-*directing* if the set aw is the same for all states a, and

(3) D3-*directing* if there is a state b which appears in every set aw.

Similar notions have been considered by Goralčik et al. [4] in connection with a game with binary relations on a finite set. Moreover, D1-directable complete nda (cnda) were explicitly studied by Burkhard [1].

Here we consider D1-, D2- and D3-directing words of commutative nda and commutative cnda. Commutativity turns out to be a very strong property in this context. In particular, the considerable differences between general nda and cnda observed in [8] are mostly eliminated. A D1- or D3-directing word of a commutative nda cannot contain any incomplete letters, that is to say, letters for which there are no transitions from some states. Also, incomplete letters may appear only in relatively short minimal D2-directing words, which do not affect the bounds we are interested in.

Section 2 introduces some basic notions. In Section 3 we consider the commutative equivalence of words, the commutative subword relation, and commutative closures and cones. A commutative cone is a set which with any word w also contains every word in which w is a commutative subword. It follows from a well-known theorem by König [9] that any commutative cone is finitely generated and, hence, a regular language. This fact is used in Section 4, where we describe the sets of Di-directing words of a given nda or cnda ($i = 1, 2, 3$); in all six cases these sets are commutative cones. This also applies to the sets of directing words of ordinary commutative automata, and altogether we get just two different families of languages as opposed to the five families obtained in [6] without assuming commutativity.

In Section 5, we give upper bounds for the length of a minimum-length Di-directing word of an n-state commutative nda or cnda ($n \geq 1, i = 1, 2, 3$). For deterministic commutative automata, the general lower bound $(n - 1)^2$ can be replaced by the exact linear bound $n - 1$ (*cf.* [7], [10]), and here too commutativity has a considerable effect. In particular, the bounds for Di-directing words of complete nda and general nda are the same. The exact bound $2^n - n - 1$ for Burkhard's D1-directing words [1] can be replaced by the exact linear bound $n - 1$. A considerable improvement also takes place for for D2-directing words. For D3-directing words the bounds given in [8] can be lowered considerably for general nda. However, it seems that commutativity has not yet been fully utilized for D2- and D3-directability.

2 Preliminaries

In what follows, X is always a finite nonempty alphabet. We denote by X^* the set of all finite words over X and by $\lg(w)$ the length of a word w. The symbol ε represents the empty word. The set of nonempty words over X is denoted by X^+. The number of the occurrences of a given letter $x \in X$ in a word $w \in X^*$ is denoted by $\lg_x(w)$.

A *deterministic finite automaton* with input alphabet X is a system (A, X, δ), where A is a finite nonempty set of *states* and $\delta : A \times X \rightarrow A$ is the *transition function*. The transition function is extended to $A \times X^*$ as usual. An automaton (A, X, δ) may also be viewed as the finite algebra $\mathcal{A} = (A, X)$, where each $x \in X$ is realised as the unary operation $x^{\mathcal{A}} : A \rightarrow A$, $a \mapsto \delta(a, x)$. For any $a \in A$ and $w \in X^*$, we also denote $\delta(a, w)$ by $aw^{\mathcal{A}}$ or aw. An automaton $\mathcal{A} = (A, X)$ is *commutative* if $axy = ayx$ for all $a \in A$ and $x, y \in X$. The class of commutative automata is denoted by **Com**.

A word $w \in X^*$ is a *directing word* of an automaton $\mathcal{A} = (A, X)$ if $aw = bw$ for all $a, b \in A$, and $\mathcal{A}$ is called *directable* if it has a directing word. The set of all directing words of an automaton $\mathcal{A}$ is denoted by DW$(\mathcal{A})$ and the class of all directable automata by **Dir**.

A *recognizer* over X is a system $\mathbf{A} = (A, X, \delta, a_0, F)$, where (A, X, δ) is an automaton, $a_0(\in A)$ is the *initial state*, and $F(\subseteq A)$ is the set of *final states*. The *language recognized* by $\mathbf{A}$ is the set $L(\mathbf{A}) = \{w \in X^* : \delta(a_0, w) \in F\}$. A language L is *recognizable* (or *regular*) if $L = L(\mathbf{A})$ for some recognizer $\mathbf{A}$. The set of all recognizable languages over X is denoted by Rec(X).

We define a *nondeterministic automaton* (an *nda* for short) as a generalized automaton $\mathcal{A} = (A, X)$, where each letter $x \in X$ is realised as a binary relation $x^{\mathcal{A}}(\subseteq A \times A)$ on A. For any $a \in A$ and $x \in X$, $ax^{\mathcal{A}} = \{b \in A : (a, b) \in x^{\mathcal{A}}\}$ is the set of states which $\mathcal{A}$ may assume when it receives the input x in state a. For any $C \subseteq A$ and $x \in X$, we set $Cx^{\mathcal{A}} = \bigcup\{cx^{\mathcal{A}} : c \in C\}$. The set $Cw^{\mathcal{A}}$ of states reachable from some state in $C \subseteq A$ by reading the input word w can now be defined as follows:

(1) $C\varepsilon^{\mathcal{A}} = C$;

(2) $Cw^{\mathcal{A}} = (Cx^{\mathcal{A}})v^{\mathcal{A}}$ if $w = xv$ for some $x \in X$ and $v \in X^*$.

For any $w = x_1 x_2 \ldots x_k$, where $k \geq 0$ and $x_i \in X$, we may view $w^{\mathcal{A}}$ as the relational product $x_1^{\mathcal{A}} x_2^{\mathcal{A}} \ldots x_k^{\mathcal{A}}$. If there is no danger of confusion, we write simply aw and Cw for $aw^{\mathcal{A}}$ and $Cw^{\mathcal{A}}$, respectively.

A *complete nda* (a *cnda* for short) is an nda $\mathcal{A} = (A, X)$ such that $ax \neq \emptyset$ for all $a \in A$ and $x \in X$. An nda $\mathcal{A} = (A, X)$ is *commutative* if $x^{\mathcal{A}} y^{\mathcal{A}} = y^{\mathcal{A}} x^{\mathcal{A}}$ for all $x, y \in X$. The classes of all commutative nda and commutative cnda are denoted by **COM** and **cCOM**, respectively. Since any commutative automaton can be regarded as a commutative cnda, we have **Com** $\subseteq$ **cCOM** $\subseteq$ **COM**.

An nda $\mathcal{B} = (B, Y)$ is called the *Y-reduct* of an nda $\mathcal{A} = (A, X)$ if $B = A$, $Y \subseteq X$ and $y^{\mathcal{B}} = y^{\mathcal{A}}$ for every $y \in Y$.

Let us recall the three notions of directability studied in [8]. For any nda $\mathcal{A} = (A, X)$ and any $w \in X^*$, we consider the following three conditions:

(D1) $(\exists c \in A)(\forall a \in A)(aw = \{c\})$;

(D2) $(\forall a, b \in A)(aw = bw)$;

(D3) $(\exists c \in A)(\forall a \in A)(c \in aw)$.

If w satisfies (Di), then it is a Di-*directing word* of $\mathcal{A}$ ($i = 1, 2, 3$). For each $i = 1, 2, 3$, the set of Di-directing words of $\mathcal{A}$ is denoted by $D_i(\mathcal{A})$, and $\mathcal{A}$ is Di-*directable* if it has a Di-directing word. The class of Di-directable nda is denoted by **Dir**(i) and the class of Di-directable cnda by **CDir**(i).

3   Commutative Equivalence, Closures and Cones

A word $u \in X^*$ is a *commutative subword* of a word $v \in X^*$, and we express this by writing $u \leq_c v$, if $\lg_x(u) \leq \lg_x(v)$ for every $x \in X$. Similarly, the words u and v are said to be *commutatively equivalent*, $u \equiv_c v$ in symbols, if $\lg_x(u) = \lg_x(v)$ for every $x \in X$. The following facts are obvious.

Lemma 1 *Let* $\mathcal{A} = (A, X)$ *be a commutative nda and let* $u, v \in X^*$:

(a) *If* $u \equiv_c v$, *then* $au = av$ *and* $Cu = Cv$ *for all* $a \in A$ *and* $C \subseteq A$.

(b) *If* $u \leq_c v$, *then* $Av \subseteq Au$.

(c) *If* $u \leq_c v$ *and* $au = bu$ *for some* $a, b \in A$, *then also* $av = bv$.

The *commutative closure* of a language $L \subseteq X^*$ is the language:

$$c(L) = \{v \in X^* : u \equiv_c v \text{ for some } u \in L\},$$

and the *commutative cone* generated by L in X^* is defined as the language:

$$[L)_c = \{v \in X^* : u \leq_c v \text{ for some } u \in L\}.$$

For $L = \{w\}$, we write simply $c(w)$ and $[w)_c$ for $c(L)$ and $[L)_c$, respectively. A language $L \subseteq X^*$ is called *commutative* if $c(L) = L$. The following facts are easily verified.

Lemma 2 *The mappings $L \mapsto c(L)$ and $L \mapsto [L)_c$ are algebraic closure operators on X^*. Moreover, for any $L \subseteq X^*$:*

(a) $c(L) = \bigcup \{c(w) : w \in L\}$ *and* $[L)_c = \bigcup \{[w)_c : w \in L\}$,

(b) $c(L) \subseteq [L)_c = c(X^*L) = c(LX^*) = c(X^*LX^*)$,

(c) $c([L)_c) = [L)_c$, *and*

(d) $[c(L))_c = [L)_c$.

Furthermore, for any words $u, v \in X^*$, $[u)_c = [v)_c$ iff $u \equiv_c v$ iff $c(u) = c(v)$.

The (*internal*) *shuffle* $K \diamond L$ of two languages $K, L \subseteq X^*$ is the set of all words $u_0 v_1 u_1 v_2 \ldots u_{m-1} v_m$, where $m \geq 1$, $u_i, v_j \in X^*$, $u_0 u_1 \ldots u_{m-1} \in K$ and $v_1 v_2 \ldots v_m \in L$ (cf. [2], for example). It is easy to see that commutative cones can be expressed in terms of the shuffle operation as follows.

For any $L \subseteq X^*$, $[L)_c = X^* \diamond c(L) \diamond X^* = X^* \diamond c(L) = c(L) \diamond X^*$. If the language L is commutative, then $[L)_c = X^* \diamond L \diamond X^* = X^* \diamond L = L \diamond X^*$.

Clearly, $\leq_c$ is a quasi-order on X^* and $\equiv_c$ is the corresponding equivalence: for any $u, v \in X^*$, $u \equiv_c v$ iff $u \leq_c v$ and $v \leq_c u$. The proper part $<_c$ of $\leq_c$ is defined so that for any $u, v \in X^*$, $u <_c v$ iff $u \leq_c v$ but not $v \leq_c u$. A word w is *minimal* in $L \subseteq X^*$ if $w \in L$ and there is no $u \in L$ such that $u <_c w$. Let us denote the set of minimal words of a language L by $m(L)$.

The *Parikh-vector* $\Psi(w)$ of a word $w \in X^*$ is the X-indexed family $(\lg_x(w) : x \in X)$ of natural numbers. If $|X| = k$ and the letters of X are given a fixed order, we may regard $\Psi(w)$ as an element of N^k. If $(N^k, \leq)$ is the lattice of all k-tuples of natural numbers with the usual componentwise order, then for any $u, v \in X^*$, $u \leq_c v$ iff $\Psi(u) \leq \Psi(v)$, and similarly, $u \equiv_c v$ iff $\Psi(u) = \Psi(v)$. Hence, the quasi-order $\leq_c$ is the order-kernel and the equivalence $\equiv_c$ is the kernel of the Parikh-mapping $\Psi : X^* \to N^k$. These observations yield the following proposition and its corollary. Analogous results for the piecewise subword order were proved by Haines [5].

Proposition 1 *Every commutaive cone is finitely generated. In particular, for every language $L \subseteq X^*$:*

(a) $m(L)$ *is finite and* $[L)_c = [m(L))_c$,

(b) $[L)_c = [K)_c$ *for a finite commutative language K, and*

(c) $[L)_c = [I)_c$ *for a finite set I of pairwise $\equiv_c$-unrelated words.*

Proof. Consider any $L \subseteq X^*$. It is clear that $[L]_c = [m(L)]_c$. As shown by König [9], any antichain in the lattice $(N^k, \leq)$ is finite, and hence the set of minimal elements of any subset $S \subseteq N^k$ is finite. Since each $\equiv_c$-class is finite, $m(L)$ is finite for every language L. Hence (a) holds. Now (b) follows by Lemma 2 (d). The set I required for (c) is obtained by selecting exactly one representative from each $\equiv_c$-class intersecting with $m(L)$. $\square$

Corollary 1 *Every commutative cone is a regular language.*

4   Languages of Directing Words

For any class **K** of automata, let $\mathcal{L}_D(\mathbf{K}) = \{\mathcal{L}_D(\mathbf{K}, X)\}_X$ be the family of languages of directing words of automata in **K**. That is to say, for each alphabet X, $\mathcal{L}_D(\mathbf{K}, X)$ is the set of languages $DW(\mathcal{A})$ with $\mathcal{A} = (A, X)$ in **K**. Note that $\emptyset \in \mathcal{L}_D(\mathbf{K}, X)$ iff **K** contains a non-directable automaton with input alphabet X. Similarly, for any class **K** of nda and each $i = 1, 2, 3$, let $\mathcal{L}_{Di}(\mathbf{K}) = \{\mathcal{L}_{Di}(\mathbf{K}, X)\}_X$ be the family of languages such that for each X, $\mathcal{L}_{Di}(\mathbf{K}, X)$ consists of the languages $D_i(\mathcal{A})$, where $\mathcal{A} = (A, X)$ is in **K**.

First we describe the families $\mathcal{L}_D(\mathbf{Com})$ and $\mathcal{L}_{Di}(\mathbf{cCOM})$ $(i = 1, 2, 3)$.

Lemma 3 *For any commutative automaton $\mathcal{A}$, the set $DW(\mathcal{A})$ is a commutative cone.*

Proof. If $\mathcal{A} \in \mathbf{Com}$ is not directable, then $DW(\mathcal{A}) = \emptyset = [\emptyset]_c$. On the other hand, if $u \in DW(\mathcal{A})$ and $u \leq_c v$, then $v \in DW(\mathcal{A})$ by Lemma 1 (c).
                                                                                    $\square$

Lemma 4 *For any commutative cone $[L)_c \subseteq X^*$, there is a commutative automaton $\mathcal{A} = (A, X)$ such that $DW(\mathcal{A}) = [L)_c$.*

Proof. If $[L)_c = \emptyset$, then $DW(\mathcal{A}) = [L)_c$ for any non-directable commutative automaton $\mathcal{A} = (A, X)$. Let $[L)_c \neq \emptyset$. By Corollary 1, $[L)_c \in \text{Rec}(X)$. Let $\mathbf{A} = (A, X, \delta, a_0, F)$ be a minimal recognizer of $[L)_c$. It is easy to see that the automaton $\mathcal{A} = (A, X, \delta)$ is commutative. We claim that $DW(\mathcal{A}) = [L)_c$. If $w \in DW(\mathcal{A})$, then $wu \in DW(\mathcal{A})$ for every $u \in X^*$. Since $\mathbf{A}$ is reduced, this means that there is exactly one final state, say $F = \{b\}$, and that $bx = b$ for every $x \in X$. Therefore, $DW(\mathcal{A}) \subseteq L(\mathbf{A})$ since $Aw = \{b\}$ must hold for all $w \in DW(\mathcal{A})$. On the other hand, if $w \in L(\mathbf{A})$ and $a \in A$ is any state, then $a_0 w = b$ and $a_0 v = a$ for some word $v \in X^*$. By commutativity, $aw = a_0 vw = a_0 wv = bv = b$, and, hence, $L(\mathbf{A}) \subseteq DW(\mathcal{A})$ also holds. $\square$

These two lemmata and Proposition 1 yield the following result.

Proposition 2 $\mathcal{L}_D(\mathbf{Com})$ *consists exactly of the commutative cones. Hence, for any alphabet X and any language $L \subseteq X^*$, $L \in \mathcal{L}_D(\mathbf{Com}, X)$ iff $L = [K)_c$ for some finite set $K \subseteq X^*$.*

Let us consider the sets of Di-directing words of commutative cnda.

Lemma 5 *For any commutative cnda $\mathcal{A} = (A, X)$, all of the languages* $D_1(\mathcal{A})$, $D_2(\mathcal{A})$ *and* $D_3(\mathcal{A})$ *are commutative cones.*

Proof. Let $u, v \in X^*$ be words such that $u \leq_c v$. If $u \in D_1(\mathcal{A})$, then there is a state b such that $au = \{b\}$ for every $a \in A$. By the completeness of $\mathcal{A}$ and Lemma 1 (b) this implies that for every $a \in A$, $\emptyset \neq av \subseteq Av \subseteq Au = \{b\}$, and thus $v \in D_1(\mathcal{A})$. Hence, $D_1(\mathcal{A})$ is a commutative cone. Also $D_2(\mathcal{A})$ is a commutative cone since $u \in D_2(\mathcal{A})$ implies $v \in D_2(\mathcal{A})$ by Lemma 1 (c).

Finally, let $u \in D_3(\mathcal{A})$ and let b be a state such that $b \in au$ for every $a \in A$. There is a word $w \in X^*$ such that $wu \equiv_c v$. Since $\mathcal{A}$ is complete, $aw \neq \emptyset$ for every $a \in A$, and, hence, $b \in (aw)u = av$, and therefore $v \in D_3(\mathcal{A})$. □

Any ordinary automaton $\mathcal{A}$ can be regarded as a cnda, so $D_i(\mathcal{A}) = DW(\mathcal{A})$ for all $i = 1, 2, 3$. It therefore follows from Lemma 4 that every commutative cone is for each $i = 1, 2, 3$ of the form $D_i(\mathcal{A})$, where $\mathcal{A}$ is a cnda. Hence, we reach the following conclusion.

Proposition 3 $\mathcal{L}_{D1}(\mathbf{cCOM}) = \mathcal{L}_{D2}(\mathbf{cCOM}) = \mathcal{L}_{D3}(\mathbf{cCOM}) = \mathcal{L}_D(\mathbf{Com})$.

Let us now turn to general commutative nda. As regards sets of D2-directing words, the above description is still valid.

Proposition 4 *For any commutative nda $\mathcal{A}$, $D_2(\mathcal{A})$ is a commutative cone, and hence* $\mathcal{L}_{D2}(\mathbf{COM}) = \mathcal{L}_D(\mathbf{Com})$.

Proof. Since $\mathcal{L}_{D2}(\mathbf{COM}) \supseteq \mathcal{L}_{D2}(\mathbf{cCOM}) = \mathcal{L}_D(\mathbf{Com})$, it suffices to verify that $D_2(\mathcal{A})$ is a commutative cone for any commutative nda $\mathcal{A} = (A, X)$. But this is obvious: if $u \in D_2(\mathcal{A})$ and $u \leq_c v$, then $v \equiv_c uw$ for some $w \in X^*$, and hence $av = auw = buw = bw$ for all $a, b \subset A$. □

For any nda $\mathcal{A} = (A, X)$, we call an input letter $x \in X$ *complete* if $ax \neq \emptyset$ for every state $a \in A$, and *incomplete* if $ax = \emptyset$ for some $a \in A$. We denote the sets of these letters by X_c and X_i, respectively. The partition $X = X_c \cup X_i$ depends naturally on the nda considered, although our notation does not show this. The following facts are easy to verify.

Lemma 6 *Let $\mathcal{A} = (A, X)$ be a commutative nda. If $w \in X^*$ contains an incomplete letter, then $aw = \emptyset$ for some $a \in A$. If $w \in D_2(\mathcal{A})$ is a D2-directing word with an incomplete letter, then $aw = \emptyset$ for all $a \in A$.*

Moreover, the following facts can be noted.

Proposition 5 *Let $\mathcal{A}$ be a commutative nda. If w is a D2-directing word of $\mathcal{A}$ containing at least one incomplete letter, then any word obtained from w by erasing from it any number of complete letters is also a D2-directing word. In particular, if $\mathcal{A}$ has a D2-directing word containing an incomplete letter, then it has a D2-directing word consisting of incomplete letters only.*

Proof. Let $w \in D_2(\mathcal{A})$ contain an incomplete letter and let v be any commutative subword of w which contains complete letters only. Then there are words $u \in X_i^+$ and $v' \in X_c^*$ such that $w \equiv_c uv'v$. By Lemma 6, $auv'v = aw = \emptyset$ for every $a \in A$. Since v contains complete letters only, this must mean that $auv' = \emptyset$ for every $a \in A$, and, hence, $uv' \in D_2(\mathcal{A})$. $\qquad\square$

By Lemma 6, D1- and D3-directing words of a commutative nda $\mathcal{A} = (A, X)$ cannot contain any incomplete letters and, hence, $D_1(\mathcal{A})$ and $D_3(\mathcal{A})$ are not complete commutative cones in X^* if there are incomplete letters.

Proposition 6 $\mathcal{L}_{D1}(\mathbf{COM}) = \mathcal{L}_{D3}(\mathbf{COM})$, *and for every alphabet* X, $\mathcal{L}_{D1}(\mathbf{COM}, X)$ *consists of all* $[L]_c \subseteq Y^*$, *where* $\emptyset \subset Y \subseteq X$.

Proof. Let $\mathcal{A} = (A, X)$ be a commutative nda. It follows from Lemma 6 that if $X_c = \emptyset$, then $D_1(\mathcal{A}) = D_3(\mathcal{A}) = \emptyset$ and that otherwise $D_1(\mathcal{A}) = D_1(\mathcal{B})$ and $D_3(\mathcal{A}) = D_3(\mathcal{B})$ for the X_c-reduct $\mathcal{B} = (A, X_c)$ of $\mathcal{A}$. On the other hand, $D_1(\mathcal{B})$ and $D_3(\mathcal{B})$ are commutative cones in X_c^* by Lemma 5. $\qquad\square$

If $[L]_c \subseteq Y^*$ is a commutative cone in Y^* for a subalphabet Y of X, then by Lemma 4 there is a commutative automaton $\mathcal{A} = (A, Y)$ such that $L = DW(\mathcal{A})$. By setting $y^{\mathcal{B}} = y^{\mathcal{A}}$ for each $y \in Y$, and $x^{\mathcal{B}} = \emptyset$ for each $x \in X \setminus Y$, we get a commutative nda $\mathcal{B} = (A, X)$ such that $D_1(\mathcal{B}) = D_3(\mathcal{B}) = L$.

5   Minimum-Length Directing Words

Let $1 \le i \le 3$. For any Di-directable nda $\mathcal{A}$, let $d_i(\mathcal{A}) = \min\{\lg(w) : w \in D_i(\mathcal{A})\}$. If $\mathbf{K}$ is a class of nda, we set for each $n \ge 1$:

$$d_{\mathbf{K},i}(n) = \max\{d_i(\mathcal{A}) : \mathcal{A} = (A, X) \in \mathbf{K} \cap \mathbf{Dir}(i), |A| = n\},$$
$$cd_{\mathbf{K},i}(n) = \max\{d_i(\mathcal{A}) : \mathcal{A} = (A, X) \in \mathbf{K} \cap \mathbf{CDir}(i), |A| = n\}.$$

Here X ranges over all finite alphabets. If $\mathbf{K}$ is the class of all nda, then $d_{\mathbf{K},i}(n)$ and $cd_{\mathbf{K},i}(n)$ are the functions $d_i(n)$ and $cd_i(n)$ introduced in [8]. We consider the functions $d_{\mathbf{COM},i}(n)$ and $cd_{\mathbf{COM},i}(n) = d_{c\mathbf{COM},i}(n)$.

Proposition 7 $d_{\mathbf{COM},1}(n) = cd_{\mathbf{COM},1}(n) = n - 1$, *for all* $n \ge 1$.

Proof. The equality $d_{\mathbf{COM},1}(n) = cd_{\mathbf{COM},1}(n)$ follows from Proposition 6: for every D1-directable nda $\mathcal{A}$, there is a D1-directable cnda which has the same state set and the same D1-directing words as $\mathcal{A}$. Moreover, $d_1(\mathcal{A}) = n-1$ for the n-state commutative cnda $\mathcal{A} = (\{1, \ldots, n\}, \{x\})$ defined so that $ax^{\mathcal{A}} = \{\min\{a + 1, n\}\}$ for every $a \in \{1, \ldots, n\}$.

Consider now any n-state D1-directable commutative cnda $\mathcal{A} = (A, X)$. If $x_1 \ldots x_k$ is a minimum-length D1-directing word of $\mathcal{A}$, then by Lemma 1:

$$A \supseteq Ax_1 \supseteq \ldots \supseteq Ax_1 \ldots x_{k-1} \supseteq Aw,$$

and all inclusions must now be proper since $|Aw| = 1$ and $\mathcal{A}$ is complete. Hence, $k \le n - 1$. $\qquad\square$

Proposition 8 *For every $n \geq 2$:*

$$(n-1)^2 + 1 \leq \mathrm{cd}_{\mathbf{COM},2}(n) = \mathrm{d}_{\mathbf{COM},2}(n) \leq 2^n - 2.$$

Of course, $\mathrm{cd}_{\mathbf{COM},2}(1) = \mathrm{d}_{\mathbf{COM},2}(1) = 0.$

Proof. The case $n = 1$ is trivial, so let $n \geq 2$. The shortest D2-directing word of the 2-state cnda $\mathcal{A} = (\{1,2\}, \{x\})$ defined so that $1x^{\mathcal{A}} = \{2\}$ and $2x^{\mathcal{A}} = \{1,2\}$ is xx. For any $n \geq 3$, the claimed lower bound $(n-1)^2 + 1$ is attained by the commutative cnda $\mathcal{A}_n = (\{1,\ldots,n\}, \{x\})$ defined so that $1x^{\mathcal{A}_n} = \{2,3\}$, $ix^{\mathcal{A}_n} = \{i+1\}$ for $1 < i < n$, and $nx^{\mathcal{A}_n} = \{1\}$.

Consider now any n-state $(n \geq 2)$ commutative cnda $\mathcal{A} = (A, X)$. It is clear that $\mathrm{D}_2(\mathcal{A}) = \mathrm{DW}(\mathcal{B})$ if $\mathcal{B} = (B, X, \delta)$ is the automaton such that B is the set of all nonempty subsets of A and $\delta(C, x) = Cx^{\mathcal{A}}$ for all $C \in B$ and $x \in X$. Moreover, $\mathcal{B}$ is commutative and therefore it has a directing word of length $\leq |B| - 1$ (*cf.* [7] or [10]) and, hence, also $\mathrm{d}_2(\mathcal{A}) = 2^n - 2$.

Let us now drop the assumption that $\mathcal{A}$ is complete. If $\mathcal{A}$ has a minimum-length D2-directing word consisting of complete letters only, then $\mathrm{d}_2(\mathcal{A}) = \mathrm{d}_2(\mathcal{B}) \leq 2^n - 2$, where $\mathcal{B}$ is the X_c-reduct of $\mathcal{A}$. If $w = x_1 \ldots x_k$ is a minimum-length D2-directing word with an incomplete letter, then:

$$A \supseteq Ax_1 \supseteq \ldots \supseteq Ax_1 \ldots x_{k-1} \supseteq Aw = \emptyset.$$

Moreover, all inclusions are proper. Indeed, if $w = ux_iv$ and $Au = Aux_i$ for some i, $1 \leq i < k$, then $Auv = Aw = \emptyset$ would imply $uv \in \mathrm{D}_2(\mathcal{A})$. Hence $k \leq n-1 \leq 2^n - 2$. This also proves $\mathrm{cd}_{\mathbf{COM},2}(n) = \mathrm{d}_{\mathbf{COM},2}(n)$ since we have seen that no minimum-length D2-directing word containing an incomplete letter can be of length $\geq (n-1)^2 + 1$. □

That in the commutative case there is no difference between cnda and nda with respect to D3-directability means that the upper bound for $\mathrm{d}_3(n)$ given in [8] can be lowered considerably when only commutative nda are considered.

Proposition 9 *For every $n \geq 2$:*

$$n^2 - 3n + 3 \leq \mathrm{cd}_{\mathbf{COM},3}(n) = \mathrm{d}_{\mathbf{COM},3}(n) \leq \frac{1}{2}n(n-1)(n-2) + 1.$$

Of course, $\mathrm{cd}_{\mathbf{COM},3}(1) = \mathrm{d}_{\mathbf{COM},3}(1) = 0.$

Proof. The equality $\mathrm{cd}_{\mathbf{COM},3}(n) = \mathrm{d}_{\mathbf{COM},3}(n)$ follows again from the fact that a D3-directing word of a commutative nda cannot contain any incomplete letters. It is also easy to see that for $n = 1$ and $n = 2$ both the lower and upper bounds are accurate. For any $n \geq 3$:

$$\mathrm{d}_3(\mathcal{A}_n) = (n-2)^2 + (n-1) = n^2 - 3n + 3,$$

when $\mathcal{A}_n$ is the cnda defined in the proof of Proposition 8. As shown in [8], the upper bound is valid for the class of all cnda. □

References

[1] H.V. Burkhard, Zum Längenproblem homogener Experimente an determinierten und nicht-deterministischen Automaten, *Elektronische Informationsverarbeitung und Kybernetik, EIK*, 12 (1976), 301–306.

[2] S. Eilenberg, *Automata, Languages, and Machines*. Academic Press, New York, 1974.

[3] W. Göhring, Minimal initializing word: a contribution to Černý's conjecture, *Automata, Languages and Combinatorics*, 2 (1977), 209–226.

[4] P. Goralčik et al., A game of composing binary relations, *RAIRO Informatique Théorique/Theoretical Informatics*, 16 (1982), 365–369.

[5] L.H. Haines, On free monoids partially ordered by embedding, *Journal of Combinatorial Theory*, 6 (1969), 94–98.

[6] B. Imreh and M. Ito, On some special classes of regular languages, to appear.

[7] B. Imreh and M. Steinby, Some remarks on directable automata, *Acta Cybernetica*, 12 (1995), 23–35.

[8] B. Imreh and M. Steinby, Directable nondeterministic automata, *Acta Cybernetica*, 14 (1999), 105–115.

[9] D. König, *Theorie der endlichen und unendlichen Graphen*. Akademische Verlagsgesellschaft, Leipzig, 1936.

[10] I. Rystsov, Reset words for commutative and solvable automata, *Theoretical Computer Science*, 172 (1977), 273–279.

Testing Non-Deterministic X-Machines

Florentin Ipate

Department of Computer Science for Economics
Romanian-American University
Bucharest, Romania
`fipate@ifsoft.ro`

Mike Holcombe

Department of Computer Science
University of Sheffield
United Kingdom
`m.holcombe@dcs.shef.ac.uk`

Abstract. The paper presents a method for generating test sets from non-deterministic generalised stream X-machines. X-machines are generalisations of finite automata, that use processing functions or relations to label state transitions instead of mere symbols. The method is proved to detect all faults of the implementation provided that the processing functions or relations are implemented correctly and the system meets certain initial requirements, called "design for test conditions".

1 Introduction

The subject of software quality, of delivering a correct implementation of the correct system, has occupied the attention of many software engineers and generated a substantial literature. Two current areas of emphasis in the development of higher quality software are the use of formal methods for the specification and verification of software and the development of sophisticated methods of software and system testing. In general, these areas have been regarded as mutually exclusive and testing issues are very seldom mentioned by those within the formal methods community.

However, in recent years there has been some interest in trying to use the information in a formal specification as a basis for test set generation. The specification, being a formal description, can be used by a tool to generate test inputs. Laycock's [19] case study shows how to generate test cases using Ostrand and Balcer [20] category-partition method, but based on a Z specification. Chow [6], Fujiwara et al. [10] generate test cases from a finite state machine and the test produced is proved to find all faults of a system providing that it can be modelled as a finite state machine. Chow's suggestion was to separate the control of the program from its data structure and to represent the former as a finite state machine. Thus, the control structure could be tested using existing finite state machine methods. It appears that a model that integrates these two aspects -control and data- of a software system and their testing is needed; such a model is the X-machine [8], [12].

In its essence an X-machine is like a finite state machine but with one important difference. A basic data set, X, is identified together with a set of basic processing relations or functions, which operate on X. Each arrow in the finite state machine diagram is then labeled by a relation or function from, the sequences of state transitions in the machine determine the processing of the data set and thus the function or relation computed. The data set X can contain information about the internal memory of a system as well as different sorts of output behaviour so it is possible to model very general systems in a transparent way. Introduced by Eilenberg [8] in 1974, X-machined are proposed by Holcombe [11] as a basis for a possible specification language and since then a number of further investigations have demonstrated that the model is intuitive and easy to use as well as general enough to cater for a wide range of applications (Holcombe and Ipate [12], Fairtlough et al. [9]). Furthermore Bălănescu et al. [2] define the communicating X-machine model for the specification of communicating systems, such as distributed systems.

A number of important classes of X-machines have been identified and studied (Ipate [14], Holcombe and Ipate [12]). Typically, the classes are defined by restrictions on the underlying data set X and the set of basic processing relations or functions Φ of the machines. In particular, stream X-machines (SX machines) and generalised stream X-machines (GSX-machines) have been found to be extremely useful in practice and most of the theory developed so far has concentrated on these particular classes. As suggested by their name, SX-machines and GSX-machines are those in which the input and the output sets are streams of symbols. The input stream is processed in a straightforward manner, producing, in turn, a stream of outputs and a regularly updated internal memory state. In the case of SX-machines, each basic relation or function processes exactly an input symbol and produces exactly an output symbol. GSX-machines are slightly more general: here, a basic relation or function processes an input symbol but can produce a sequence of outputs of any length. Such X-machines are used by Bălănescu et al. [5], where the processing relations are defined by generative grammars and the translation power of such mechanisms is investigated for different types of grammars.

Ipate and Holcombe [15], [17] present a method for testing systems specified as deterministic SX-machines. The method, called deterministic stream X-machine (DSXM) testing, is proved to detect all faults of the implementation provided that two major requirements are met. Firstly, the system has to satisfy some "design for test" conditions. Basically these conditions require that the specified system has to have detectable behaviour under all conditions. Without such conditions, the system may be very difficult, or indeed impossible, to test. Secondly, the method assumes that the implementations of the basic processing functions Φ are correct, thus the system implementation uses the same set of basic processing functions as the specification. In practice this will be done with a separate testing process, depending on the nature of Φ.

However, there is a practical need for testing non-deterministic models; non-determinism and concurrency are two important features of formal specification languages for communicating software, in particular communication protocols. Moreover, a system of communicating deterministic X-machines may have non-deterministic behaviour (Bălănescu et al. [2]). In this paper we present a method for generating test sets from non-deterministic GSX-machines, called non-deterministic generalised stream X-machine (NGSXM) testing. The method is a generalisation of the DSXM method and is based on the theoretical results of Ipate and Holcombe [18].

Before we go any further, we will introduce the notations used in this paper. When considering sequences of inputs or outputs, we will use A^* to denote the set of finite sequences with members in A. ϵ will denote the empty sequence and $A^+ = A^* \setminus \{\epsilon\}$. For $a, b \in A^*$, ab will denote the concatenation of sequences a and b and a^n will be defined by $a^0 = \epsilon$ and $a^n = a^{n-1}a$ for $n \geq 1$. For a sequence $a \in A^+$, $length(a)$ will denote the number of elements of a, $length(\epsilon) = 0$. For $U, V \subseteq A^*$, $UV = \{ab \mid a \in U, b \in V\}$; U^n will be defined by $U^0 = \{\epsilon\}$ and $U^n = U^{n-1}U$ for $n \geq 1$.

For a relation $f : A \to B$, $dom f = \{a \in A \mid f(a) \neq \emptyset\}$. If $U \subseteq A$ then $f(U) = \cup_{a \in U} f(a)$.

2   Automata Theory: Basics and Notations

This section defines the finite automaton and introduces a few basic concepts and results of the automata theory.

A finite automaton A is a tuple (Σ, Q, F, I, T), where Σ is a finite set called the input alphabet, Q is the finite set of states, F is the (partial) transition function, $F : Q \to 2^Q$ is usually described as a transition diagram, and $I, T \subseteq Q$ are the sets of initial and terminal states respectively. If $q, q' \in Q$, $\sigma \in \Sigma$, and $q' \in F(q, \sigma)$, we say that $\sigma : q \to q'$ is an arc from q to q' and write $\sigma : q \to q'$. If $q, q' \in Q$ are such that there exist $q_0, ..., q_n \in Q$ with $q_0 = q$ and $q_n = q'$ so that $\sigma_1 : q_0 \to q1$, $\sigma_2 : q_1 \to q_2, ..., \sigma_n : q_{n-1} \to q_n$, we say that we have a path $p = \sigma_1 ... \sigma_n$ from q to q' and write $p : q \to q'$. If $n = 0$ then $\epsilon : q \to q$ is the empty path. The language accepted by the automaton, L, is defined by: $L = \{s \in \Sigma^* \mid \exists q_0 \in I, q \in T$ so that $s : q_0 \to q$ is a path in A$\}$. An automaton is called deterministic if there is one initial state, i.e. $I = \{q_0\}$ F maps each state/input pair into at most one single state, i.e. $F : Q \times \Sigma \to Q$. In the sequel we will refer to deterministic automata with all states terminal (i.e. $T = Q$), denoted (Σ, Q, F, q_0). Let $A = (\Sigma, Q, F, q_0)$ be a deterministic automaton and let $q \in Q$. Then the language accepted by A in q, L_q, is defined by: $L_q = \{s \in \Sigma^* \mid s : q_0 \to q$ is a path in $A\}$ The concepts of minimal automaton, transition cover and characterisation set of a minimal automaton are defined next. Let $A = (\Sigma, Q, F, q_0)$ and $A' = (\Sigma, Q', F', q_0')$ be two deterministic automata and let $X \subseteq \Sigma^*$. Then two states $q \in Q$ and $q' \in Q'$ are called X-equivalent if $L_q \cap X = L_q' \cap X$. If q, q' are not X-equivalent then they are called X-distinguishable. A and A' are called X-equivalent if q_0 and q_0' are X-equivalent. Otherwise, A and A' are called X-distinguishable. A deterministic automata $A = (\Sigma, Q, F, q_0)$ is called minimal if the following are true. $\forall q \in Q \exists s \in \Sigma^*$ such that $s : q_0 \to q$ is a path in A; $\forall q, q' \in Q$ two distinct states of A $\exists X \subseteq \Sigma^*$ such that q and q' are X-distinguishable. Given a deterministic automaton A, there exists a minimal automaton A' that accepts the same language as A. Furthermore, A' is unique up to a renaming of the state set. A' is called the minimal automaton of A.

Definition 1 *Let $A = (\Sigma, Q, F, q_0)$ be a minimal automaton. Then a set of input sequences $W \subseteq \Sigma^*$ is called a characterisation set of A if any two distinct states of A are $W-$distinguishable.*

Definition 2 *Let $A = (\Sigma, Q, F, q_0)$ be a minimal automaton. Then a set of input sequences $S \subseteq \Sigma^*$ is called a state cover of A if $\forall q \in Q$ $\exists s \in S$ such that $s : q_0 \to q$ is a path in A.*

Note that the minimality of the automaton ensures the existence of a characterisation set and of a transition cover.

3 Generalised Stream X-Machines and the Theoretical Basis of NDXM Testing

This section introduces the GSX-machine, the notation used and introduces the theoretical concepts used in the presentation of the NDXM testing method.

Definition 3 *A generalised stream X-machine (GSX-machine) Z is a tuple $(\Sigma, \Gamma, Q, M, \Phi, F, I, T, m_0)$ as follows:*

- *Σ and Γ are finite sets called the input and output alphabet, respectively.*

- *M is a (possibly) infinite set called memory.*

- *Q is the finite set of states.*

- *Φ is the type of Z, a set of basic non-empty relations that the machine can use, of the form $\varphi : M \times \Sigma \longleftrightarrow \Gamma^* \times M$.*

- *F is the "next state" (partial) function, $F : Q \times \Phi \to 2^Q$. F is usually described as a state-transition diagram.*

- *I and T are the sets of initial and terminal states respectively. $I \subseteq Q, T \subseteq Q$.*

- *m_0 is the initial memory value $m_0 \in M$.*

Thus, GSX-machines are X-machines for which the basic processing relations have the form $\varphi : M \times \Sigma \longleftrightarrow \Gamma^* \times M$ i.e. each such relation will read an input symbol, discard it and produce a sequence of output symbols while (possibly) changing the value of the memory. If the sequence of outputs is always of length one, i.e. the processing relations have the form $\varphi : M \times \Sigma \longleftrightarrow \Gamma \times M$ then the machine is called a stream X-machine (SX-machine).

It is sometimes helpful to think of an X-machine as an automaton with the arcs labeled by relations from the type Φ. The automaton $A = (\Phi, Q, F, I, T)$ over the alphabet Φ is called the associated automaton of Z.

Definition 4 *If $q, q' \in Q, \varphi \in \Phi, \varphi : q \to q'$ is called an arc of Z if $\varphi : q \to q'$ is an arc of A. Similarly, for $q, q' \in Q, p \in \Phi^*, p : q \to q'$ is called a path of Z if $p : q \to q'$ is a path of A. Each path $p = \varphi_1 \ldots \varphi_n$ with $n \geq 0$, gives rise to a relation (the path relation): $|p| : M \times \Sigma* \longleftrightarrow \Gamma^* \times M$ where $(m, s)|p|(g, m')$ if and only if there exist $m_1, \ldots, m_{n+1} \in M$ with $\forall i = 1 \ldots n(m_i, \sigma_i)\varphi_i(g_i, m_{i+1})$, where $m_1 = m, m_{n+1} = m', s = \sigma_1 \sigma_2 \ldots \sigma_n$ and $g = g_1 g_2 \ldots g_n$. The relation corresponding to the empty path δ is $(m, \epsilon)|\delta|(\epsilon, m)$.*

A machine computation takes the form of a traversal of a path in the state space and the application, in turn, of the path labels (which represent basic processing relations) to the initial memory value. The correspondence between the input sequence applied to the machine and the output produced give rise to the relation computed by the machine, as defined next. In general, an GSX-machine is non-deterministic, in the sense that the application of an input sequence can produce more than one single output sequence.

Definition 5 *Given a GSX-machine Z, we define the relation $f : \Sigma^* \longleftrightarrow \Gamma^*$ by: sfg if and only if $\exists q_0 \in I, q \in T, m' \in M$ and a path $p : q_0 \to q$ with $(m_0, s)|p|(g, m')$. We say that Z computes f.*

We now prepare the ground for the presentation of the NDXM method and identify certain conditions that a GSX-machine will have to meet in order to be successfully tested using this method. Since the NDXM method is primarily aimed at testing non-deterministic GSX-machines, we will make no assumptions with regard to the determinism of the machine specifications. However, we will assume that the associated automaton of the GSX-machine specification is deterministic. Nevertheless, this requirement will not restrict the generality of our discussion in any way, as shown by the following result (Ipate and Holcombe [18]): given a (non-deterministic) GSX-machine Z, there exists a GS X-machine Z' with the same type and initial memory as Z so that Z' computes the same relation as Z and the associated automaton of Z' is deterministic.

Furthermore, even though in practice Φ is usually a set of (partial) functions, we will consider the more general case when Φ is a set of relations.

When dealing with specifications of real systems, it is natural to assume that the application of any input symbol in any state and memory value is specified. Such a machine will be said to be completely specified.

Definition 6 *A GSX-machine Z is called completely specified if $\forall q \in Q, \sigma \in \Sigma, m \in M \ \exists \varphi \in \Phi$ such that $F(q, \varphi) = \emptyset$ and $\varphi(m, \sigma) = \emptyset$.*

An arbitrary X-machine can be easily transformed into one that is completely specified by adding an extra state that "traps" all the unspecified transitions, the technique is similar to that used for finite state machines (Cohen [7]). Since the GSX-machine model is used in this paper for testing purposes, naturally we would like to have as much information about the outputs produced as possible. For this reason, the NDXM testing method assumes that all the states of the machine specification are terminal (i.e. $T = Q$). This means that the output produced by the machine can be viewed in any of its states, even though the machine is allowed to terminate its computation only in certain states. Of course, the intermediary outputs can be filtered out once the system has been tested. Clearly, if Z is completely specified and all its states are terminal then $dom f = \Gamma^*$. If a GSX-machine is to be successfully tested using the NDXM method, then the behaviour of the machine will have to be detectable under all conditions. This is ensured by two requirements, completeness and output-distinguishability, that the set of processing relations Φ will have to meet. These two requirements will be formalised next.

Definition 7 *Let* $V \subseteq M$ *and* $U = \{U_\varphi | \varphi \in \Phi\}$ *a family of non-empty subsets of* $\Sigma, U_\varphi \subseteq \Sigma$, *indexed by* Φ. *Then* Φ *is called closed w.r.t.* (V, U) *if the following is true:*

- $m_0 \in V$.

- $\forall \varphi \in \Phi, m \in V, \sigma \in U_\varphi, \gamma \in \Gamma$ *if* $(m, \sigma)\varphi(g, m')$ *then* $m' \in V$.

Definition 8 *Let* $V \subseteq M$ *and* $U = \{U_\varphi | \varphi \in \Phi\}$ *with* $U_\Phi \subseteq \Sigma$ *so that* Φ *is closed w.r.t.* (V, U). *Then* Φ *is called complete w.r.t* (V, U) *if:* $\forall \varphi \in \Phi, m \in V, \exists \sigma \in U_\varphi$ *such that* $\varphi(m, \sigma) \neq \emptyset$. *If* $V = M$ *and* $\forall \varphi \in \Phi U_\varphi = \Sigma$ *then* Φ *is called complete.*

In other words, any basic relation φ will be able to process all memory values of V using inputs in U_φ. This guarantees that any path of the associated automaton can be exercised from the initial state and initial memory value.

Definition 9 *Let* $V \subseteq M$ *and* $U = \{U_\varphi | \varphi \in \Phi\}$ *with* $U_\varphi \subseteq \Sigma$ *so that* Φ *is closed w.r.t.* (V, U). *Then* Φ *is called output-distinguishable w.r.t.* (V, U) *if:* $\forall \varphi_1, \varphi_2 \in \Phi, m, m'_1, m'_2 \in V, \sigma \in U_{\varphi_1} \cap U_{\varphi_1}, g \in \Gamma^*$, *if* $(m, \sigma)\varphi_1(g, m'_1)$ *and* $(m, \sigma)\varphi_2(g, m'_2)$ *then* $\varphi_1 = \varphi_2$ *and* $m'_1 = m'_2$. *If* $V = M$ *and* $\forall \varphi \in \Phi U_\varphi = \Sigma$ *then* Φ *is called output-distinguishable.*

What this is saying is that the memory/input pair processed and the output produced will uniquely determine the processing relation applied and the next memory value. Note that if Φ is a set of partial functions rather than relations then the output-distinguishability condition has the following, simpler, form: $\forall \varphi_1, \varphi_2 \in \Phi, m, m'_1, m'_2 \in V, \sigma \in U_{\varphi_1} \cap U_{\varphi_1}, g \in \Gamma^*$ if $\varphi_1(m, \sigma) = (g, m'_1)$ and $\varphi_2(m, \sigma) = (g, m'_2)$ then $\varphi_1 = \varphi_2$.

These two conditions are required of the specification machine and they will be referred to as design for test conditions. Without them, it is going to be difficult to test a system properly: there may be hidden behavioural faults in the implementation which cannot be exposed. Although these conditions might appear to be quite restrictive, they can be easily introduced into a specification by simply extending the definitions of the Φ relations in a suitable manner, introducing extra input and output symbols. A very simple algorithm is given in [18].

The idea of the NDXM method is to prove that the two GSX-machines (one representing the specification and the other the implementation) have identical behaviour by showing that their associated automata have identical behaviour. In this way, a finite state machine testing method (e.g. Chow's [6] W-method) can be used. Now, using such a method, we can construct a set of sequences of elements from Φ^* that will establish whether the two associated automata accept the same language. However, this is not really very convenient, as we really want a set of input sequences from Σ^*. We thus need to convert sequences from Φ^* into sequences from Σ^*. We do this by using a test function as discussed next. The scope of a test function is to test whether a certain path exists or not in Z using appropriate input symbols (hence the name).

Definition 10 Let $Z = (\Sigma, \Gamma, Q, M, \Phi, F, q_0, m_0)$ be a GSX-machine and let $V \subseteq M$ and $U = \{U_\varphi \mid \varphi \in \Phi\}$ with $U_\varphi \subseteq \Sigma$ so that Φ is complete w.r.t. (V, U). Then a function $t : \Gamma^* \to \Sigma^*$ is called a test function of Z w.r.t. U if:

- $t(\epsilon) = \epsilon$.

- For any $\varphi_1, \ldots \varphi_n \in \Phi$ with $n > 0$, $t(\varphi_1, \ldots \varphi_n)$ satisfies the following requirements (in what follows, for $i \in \{1, \ldots n\}$, p_i denotes the path $\varphi_1 \ldots \varphi_i$):

  - If p_n is a path in Z starting in q_0 then $t(\varphi_1, \ldots \varphi_n) = \sigma_1 \ldots \sigma_n$, where $\sigma_1 \in U_{\varphi_1} \ldots \sigma_n \in U_{\varphi_n}$ and $(m_0, \sigma_1 \ldots \sigma_n) \in dom|p_n|$.
  - Otherwise, $t(\varphi_1, \ldots \varphi_n) = \sigma_1 \ldots \sigma_{k+1}$, where $\sigma_1 \in U_{\varphi_1} \ldots \sigma_{k+1} \in U_{\varphi_{k+1}}$ and $(m_0, \sigma_1 \ldots \sigma_{k+1}) \in dom|p_{k+1}|$, where $k \in \{0, \ldots, n-1\}$ is the largest number for which p_k is a path in Z starting in q_0.

Note that since Φ is complete w.r.t. (V, U) there always exists $t(p)$ that meets the above requirements. Also note that a test function w.r.t. U is not uniquely determined, and in general many different possible test functions exist. If $\forall \varphi \in \Phi U_\varphi = \Sigma$ then t is simply called a test function of Z. For two family of sets $U = \{U_\varphi \mid \varphi \in \Phi\}$, $U' = \{U'_\varphi \mid \varphi \in \Phi\}$, with $\forall \varphi \in \Phi, U_\varphi \subseteq U'_\varphi$ it is obvious that if t is a test function of Z w.r.t. U then t is a test function of Z w.r.t. U'. In particular, for any family of sets U as above, any test function of Z w.r.t. U is also a test function of Z.

The following lemma provides a simple and intuitive way of generating the values of a test function.

Lemma 1 Let $Z = (\Sigma, \Gamma, Q, M, \Phi, F, q_0, m_0)$ be a GSX-machine and let $V \subseteq M$ and $U = \{U_\varphi \mid \varphi \in \Phi\}, U_\varphi \subseteq \Sigma$ so that Φ is complete w.r.t. (V, U). Then there exists a test function of Z w.r.t. U so that, for any $\varphi_1 \ldots \varphi_n$ with $n > 0$, $t(\varphi_1 \ldots \varphi_n)$ satisfies the following - in what follows, for $i \in \{0, \ldots n\}, p_i$ denotes the path $\varphi_1 \ldots \varphi_n$:

- *If $p_{n-1} = \varphi_1 \ldots \varphi_{n-1}$ is a path in Z starting in q_0 then $t(\varphi_1 \ldots \varphi_n) = t(\varphi_1 \ldots \varphi_{n-1})\sigma_n$, where $\sigma_n \in U_{\varphi_n}$, so that $(m_0, t(\varphi_1 \ldots \varphi_n)) \in dom|p_n|$.*

- *Otherwise, $t(\varphi_1 \ldots \varphi_n) = t(\varphi_1 \ldots \varphi_{n-1})$.*

4   The NGSXM Testing Method

Once all of this mechanism is in place we can generate a test set mechanically. This is the basis of the NGSXM testing method, which is described in detail next.

4.1   Pre-requisites

The method works under the following conditions:

- The system specification is a (non-deterministic) GSX-machine that is completely specified.

- There exist $V \subseteq M$ and $U = \{U_\varphi \mid \varphi \in \Phi\}$, with $U_\varphi \subseteq \Sigma$ so that the set of basic relations Φ is complete w.r.t. (V, U) and output-distinguishable w.r.t. (V, U). As already discussed, this design for test conditions can be easily enforced by extending the input and output alphabets.

  Obviously, the extra functionality resulted from this extension can be filtered out once the system has passed testing.

- The associated automaton of the specification has to be deterministic and minimal. This can be easily arranged by the designer: standard techniques from finite state machine theory are available (Cohen[7], Eilenberg [8]).

- The implementation can be modelled as a GSX-machine machine with the same set of basic processing relations Φ. In practice, this means that the basic processing relations have to be shown to be correct and this is done with a separate testing process depending on the nature and complexity of $\varphi's$, as discussed in the first section of the paper.

- The difference between the (unknown) number of states of the implementation and the (known) number of states of the specification has to be estimated - we will denote this by k. In practice this is not usually large, for especially sensitive applications one can make very pessimistic assumptions about k at the cost of a large test set.

4.2   Generation of the Test Set

The generation of the test set consists of the following two steps:

- First, we generate X, a k-test set of the associated automaton of the specification, using the W-method (Chow [6]), thus: $X = S(\Sigma^{k+1} \cup \ldots \Sigma \cup \{\epsilon\})W$, where:

  - W is a characterisation set of the associated automaton of the specification.

  - S is a state cover of the associated automaton of the specification.

  - k is the estimated difference between the number of states of the implementation and the number of states of the specification.

- The set X is converted into sequences of inputs through a test function w.r.t. U, t, thus $Y = t(X)$ will be a k-test set of the specification.

4.3   Application of the Test Set and Evaluation of the Results

In order to test non-deterministic implementations, one usually makes a so-called complete-testing assumption: it is possible, by applying a given input sequence s to a given implementation a finite number of times, to exercise all the paths of the implementation that can be traversed by s. Without such an assumption, no test suites can guarantee full fault coverage for non-deterministic implementations. Obviously, the quality of testing increases with the number of repetitions of test sequence application; in actual testing, this number is limited by practical and economical considerations. If the complete-testing assumption is in place then the theoretical results of Ipate and Holcombe [18] show that Y is a test set of the system, i.e. if Y is applied to the implementation and the results coincide with those in the specification then the implementation and the specification will be guaranteed to have identical behaviour.

5   Conclusions

The NDXM method guarantees that if the implementation passes all the tests in the test set then all the faults of the implementation are detected modulo the correct implementation of the basic relations. In practice, the correctness of the basic relations will endure through a separate testing process, depending on their nature.

The method is reductionist in the sense that it reduces the problem of proving that the implementation is correct to proving that the basic processing relations are correctly implemented. At first sight, it might appear that what the method really does is shifting the burden onto the testing of Φ's.

This is not so. What is proposed here is a gradual testing process in which at each level the system is assumed to be made of reliable components. In likely applications of the method, this will be successively applied to the hierarchy of GSX-machines that are created when the basic functions are considered at each level. Thus, testing a specific function will involve considering it as the computation defined by a simpler GSX-machine and so on. Ultimately, at the bottom level, the basic relations are usually quite simple and can be tested using suitable alternative methods - for example, category partition testing (Ostrand and Balcer [20]) or a variant - or even assumed to be fault free if they are routines or objects from a library.

References

[1] T. Bălănescu, Generalized stream X-machines with output delimited types. Submitted to *Formal Aspects of Computer Science*, 1999.

[2] T. Bălănescu et al., Communicating stream X-machines are no more than X-machines, *Journal of Universal Computer Science*, 5.9 (1999), 494–507.

[3] T. Bălănescu, H. Georgescu and M. Gheorghe, Grammar systems with counting derivation and dynamical priorities. In Gh. Păun and A. Salomaa (eds.), *New Trends in Formal Languages: Control, Cooperation and Combinatorics*. Springer, Berlin, 1997, 151–166.

[4] T. Bălănescu and M. Gheorghe, A set of string functions and its power, *Analele Universității București. Seria Matematică-Informatică*, XCV.1 (1996), 9–14.

[5] T. Bălănescu, M. Gheorghe and H. Georgescu, Stream X-machines with underlying distributed grammars. Submitted to *Informatica*, 1999.

[6] T.S. Chow, Testing software design modelled by finite state machines, *IEEE Transactions on Software Engineering*, 4.3 (1978), 178–187.

[7] D.I.A. Cohen, *Introduction to Computer Theory*. John Wiley, New York, 1991.

[8] S. Eilenberg, *Automata, Languages and Machines*. Academic Press, New York, 1974.

[9] M. Fairtlough et al., Using an X-machine to model a video cassette recorder, *Current Issues in Electronic Modelling*, 3 (1995), 141–161.

[10] K. Fujiwara et al., Test selection based on finite state models, *IEEE Transactions on Software Engineering*, 17.6 (1991), 591–603.

[11] M. Holcombe, X-machines as a basis for dynamic system specification, *Software Engineering Journal*, 3.2 (1988), 69–76.

[12] M. Holcombe and F. Ipate, *Correct Systems: Building a Business Process Solution*. Springer, Berlin, 1998.

[13] M. Holcombe, F. Ipate and A. Grondoudis, Complete functional testing of safety-critical systems. In *Proceedings of the Second IFAC Workshop on Safety and Reliability in Emerging Control Technologies*, Daytona Beach, Fl., 1995.

[14] F. Ipate, Theory of X-Machines and Applications in Specification and Testing, PhD dissertation, University of Sheffield, 1995.

[15] F. Ipate and M. Holcombe, An integration testing method that is proved to find all faults, *International Journal of Computer Mathematics*, 63 (1997), 159–178.

[16] F. Ipate and M. Holcombe, A method for refining and testing generalised machine specifications, *International Journal of Computer Mathematics*, 68 (1998), 197–219.

[17] F. Ipate and M. Holcombe, Specification and testing using generalised machines: a presentation and a case study, *Journal of Software Testing, Verification and Reliability*, 8 (1998), 61–81.

[18] F. Ipate and M. Holcombe, Generating test sequences from non-deterministic stream X-machines. Submitted to *Formal Aspects of Computer Science*.

[19] G.T. Laycock, Formal specification and testing, *Journal of Software Testing, Verification and Reliability*, 2 (1992), 7–23.

[20] T.J. Ostrand and M.J. Balcer, The category-partition method for specifying and generating functional tests, *Communications of the ACM*, 31.6 (1989), 667–686.

Note on Minimal Automata and Uniform Communication Protocols[1]

Galina Jiráskóvá

Mathematical Institute
Slovak Academy of Sciences
Košice, Slovakia
jiraskova@duro.upjs.sk

Abstract. In this paper, we describe regular languages with an essential difference between their nondeterministic message complexity and the size of their minimal nondeterministic finite automata. This solves an open problem posed by Hromkovič [2]. We also define a two-way message complexity and we show that the two-way message complexity of a regular language L provides a lower bound on the size of the minimal two-way deterministic finite automaton for L. We find specific regular languages with an exponential gap between these two complexity measures and we also do the same for the nondeterministic case.

1   Introduction

The communication complexity of two-party protocols is well-established as a successful method for proving lower bounds on several fundamental complexity measures of sequential and parallel computations. In this paper,

---

[1]Supported by grant 2/7007/20.

we study how to use communication protocols to prove lower bounds on the size of minimal finite automata.

It is well-known that the one-way communication complexity of a regular language L provides a direct lower bound on the logarithm of the number of states of the minimal finite automaton recognizing L [1, 2]. In order to establish a closer relation between communication complexity and finite automata, Hromkovič and Schnitger [4]a introduced a uniform model of two-party communication protocols. They defined the message complexity of a regular language L as the number of distinct messages used by the optimal one-way uniform communication protocol recognizing L and they showed that the message complexity of L provides a lower bound on the size of the minimal finite automaton for L. This relation was extended to the nondeterministic case by Hromkovič [2].

This, then, is the method for proving lower bounds on the size of minimal deterministic and nondeterministic finite automata and until now it has been the best method known for this purpose. It has been shown to be very successful in the deterministic case, in which the message complexity of a regular language L is exactly equal to the size of the minimal deterministic finite automaton for L [4, 2].

The aim of this paper is to show that this method has weaknesses in the cases of nondeterministic finite automata and of two-way (deterministic and nondeterministic) finite automata. We give specific regular languages for which the difference between their nondeterministic message complexity and the size of their minimal nondeterministic automata is exponential. Further, using two-way uniform protocols we define the two-way message complexity of a regular language L. We show that it provides a lower bound on the size of the minimal two-way finite automaton for L but that the difference between these two complexity measures may be exponential. The same holds for the nondeterministic case, too.

The paper is organized as follows. Section 2 contains the definitions of a nondeterministic and two-way message complexity. In Section 3, we give specific regular languages with an exponential difference between their nondeterministic message complexity and the size of their minimal nondeterministic finite automata. In Section 4, we study the relation between the two-way message complexity and two-way finite automata.

2   Definitions

To define one-way uniform nondeterministic protocols we follow [2]. Informally, a *one-way uniform nondeterministic protocol* P over an alphabet Σ accepting a regular language $L \subseteq \Sigma^*$ can be described as follows. The first computer (C_I) receives the first part x of an input $xy \in \Sigma^*$ and the second one (C_{II}) receives the rest y. The first computer looks at its input x and nondeterministically sends binary messages to the second one. The second computer must then decide whether the input xy is in L or not (see [2] for details). The *message complexity of the protocol P* is the number of distinct messages that can be sent by the first computer. The *protocol P accepts a regular language L over the alphabet* Σ if for all $x, y \in \Sigma^*$ there is an accepting computation of P on xy if and only if $xy \in L$. The *nondeterministic message complexity of L* is the message complexity of the best one-way uniform nondeterministic protocol accepting L.

A *two-way uniform deterministic protocol P* accepting a regular language L over an alphabet Σ can be informally described as follows. The first computer (C_I) receives the first part x of an input $xy \in \Sigma^*$ and the second one (C_{II}) receives the rest y. Then they can communicate, i.e. exchange binary messages, until one of them knows whether the input xy is in L or not. The messages exchanged are not stored by the computers, i.e. the next message sent by a computer is a function of its input and the preceding message received from the other computer. The *two-way message complexity of the protocol P* is the number of distinct messages exchanged between the two computers. The *two-way message complexity of the language L* is the two-way message complexity of the best two-way uniform deterministic protocol P accepting L. Now, let us formalize these informal definitions of two-way uniform protocols.

Definition 1 *Let* Σ *be an alphabet and let* $L \subseteq \Sigma^*$. *A two-way uniform protocol over* Σ *is a pair* $P = \langle \Phi, \varphi \rangle$, *where:*

$$\Phi, \varphi : \Sigma^* \times \{0,1\}^* \to \{0,1\}^* \cup \{\overline{0}, \overline{1}\}$$

are functions which have the prefix freeness property (i.e. $\Phi(x, c)$ *is not a proper prefix of* $\Phi(x', c)$; *the same for* φ).

A computation *of P on an input* $xy \in \Sigma^*$ *is a string* $c = c_1 \$ c_2 \$ \ldots$ $\$ c_k \$ c_{k+1}$, *where* $k \geq 0$, $c_1, \ldots, c_k \in \{0,1\}^*, c_{k+1} \in \{\overline{0}, \overline{1}\}$ *and such that:*

(i) $c_1 = \Phi(x, \lambda)$,

(ii) *if* l *is odd, then* $c_{l+1} = \varphi(y, c_l)$,

(iii) *if* l *is even, then* $c_{l+1} = \Phi(x, c_l)$.

A computation $c = c_1 \$ c_2 \$ \ldots \$ c_k \$ c_{k+1}$ *is called* accepting (rejecting) *if* $c_{k+1} = \overline{1}$ $(\overline{0})$. *We say that a* protocol P accepts *a language* L *if for all* $x, y \in \Sigma^*$ *the computation of P on the word* xy *is accepting iff* $xy \in L$.

The *two-way message complexity of the protocol P is:*

$$2mc(P) = |\{\Phi(x, c) \mid x \in \Sigma^*, c \in \{0,1\}^*\} \cup \{\varphi(y, c) \mid y \in \Sigma^*, c \in \{0,1\}^*\}|,$$

i.e. the number of distinct messages used by the protocol P.

The two-way message complexity of a language L *is:*

$$2mc(L) = \min\{2mc(P) \mid P \text{ is a two-way uniform protocol accepting } L\}.$$

Definition 2 *Let* $M = (Q, \Sigma, \delta, q_0, F)$ *be a (nondeterministic, two-way) finite automaton recognizing a regular language L over the alphabet Σ. The size of the automaton M is the number of its states (i.e. $|Q|$). A (nondeterministic, two-way) finite automaton M is called* minimal *for L if it recognizes L with the minimal number of states.*

3   Nondeterministic Message Complexity Versus Nondeterministic Finite Automata

Hromkovič [2] showed that the nondeterministic message complexity of a regular language L provides a lower bound on the size of the minimal nondeterministic finite automaton for L.

Klauck and Schnitger [5] showed that there are regular languages with an essential difference between the nondeterministic message complexity and the size of the minimal nondeterministic finite automaton.

In this section, we give specific regular languages whose nondeterministic message complexity is much smaller then the size of their minimal nondeterministic finite automata. For an even integer k, let:

$$A_k = \{xy \in \{0,1\}^* \mid |x| = |y| = k, x \neq y \text{ or } x = ww\}$$

be the regular language over the alphabet $\{0,1\}$ that contains words of length $2k$ with different halves or having equal halfs of the first half of the word.

Theorem 1 *The nondeterministic message complexity of A_k is $O(k^2)$.*

Proof. It is sufficient to show that there exists a one-way uniform nondeterministic protocol P_k over the alphabet $\{0,1\}$ accepting A_k with the message complexity $O(k^2)$. Let us informally describe it. The computation of the protocol P_k on a word $xy \in \{0,1\}^*$ is as follows.

(i) If $x = \lambda$ then C_I submits the message "I have no bit". In this case, C_{II} accepts (rejects) if $y \in A_k$ ($y \notin A_k$).

(ii) If $|x| > 2k$ then C_I submits the message "I have more than 2k bits" and C_{II} rejects the input xy.

(iii) If $1 \leq |x| < k$ then C_I nondeterministically sends the messages $(|x|, i, x_i), i = 1, \ldots, |x|$. Because computer C_{II} has the information about the length of x and the $i - th$ bit of x, it accepts if $|y| + |x| = 2k$ and the corresponding bit of y is different from x_i, or $|y| + |x| = 2k$ and the last k bits of y can be written in the form ww. Otherwise, it rejects.

(iv) If $k \leq |x| \leq 2k$, then C_I nondeterministically sends the messages $(|x|, i, x_i), i = 1, \ldots, k$ and the message whether the first k bits of x can be written in the form ww or not. With this information C_{II} can again accept or reject the input xy.

It is not difficult to see that P_k is a one-way uniform nondeterministic protocol over the alphabet $\{0, 1\}$ accepting A_k with the message complexity $O(k^2)$. □

Theorem 2 *Any nondeterministic finite automaton recognizing the language A_k has at least $2^{k/6}$ states.*

Proof. Let $M = (Q, \{0, 1\}, \delta, q_0, F)$ be a nondeterministic finite automaton recognizing the language A_k. For any two different words $u, v \in \{0, 1\}^{k/2}$ the words $uuuu$ and $vvvv$ belong to A_k and so there are accepting computations of the automaton M on these words. Let $q_0, q_1, q_2, \ldots, q_{2k}$ be an accepting computation of M on $uuuu$ and $q_0, p_1, p_2, \ldots, p_{2k}$ be an accepting computation of M on $vvvv$ ($q_i, p_i \in Q$). We claim that the 3-tuple $(q_{k/2}, q_k, q_{3k/2})$ is different from the 3-tuple $(p_{k/2}, p_k, p_{3k/2})$. If we assume the contrary, that $q_{k/2} = p_{k/2}, q_k = p_k$, and $q_{3k/2} = p_{3k/2}$, then $q_0, q_1, \ldots, q_{k/2} = p_{k/2}, p_{k/2+1}, \ldots, p_k = q_k, q_{k+1}, \ldots, q_{3k/2} = p_{3k/2}, p_{3k/2+1}, \ldots, p_{2k}$ is an accepting computation of M on the word $uvuv$ which does not belong to the language A_k, a contradiction. So, we have proved that $(q_{k/2}, q_k, q_{3k/2}) \neq (p_{k/2}, p_k, p_{3k/2})$. This implies that $|Q|^3$ is at least $2^{k/2}$ (the number of all words of length $k/2$) and so the number of states of M is at least $2^{k/6}$. □

Corollary 1 *For any language A_k considered above, there is an exponential difference between the nondeterministic message complexity of A_k and the size of the minimal nondeterministic finite automaton for A_k.*

4 Two-Way Message Complexity Versus Two-Way Finite Automata

In this section, we show that the two-way message complexity of a regular language L provides a lower bound on the size of the minimal two-way deterministic automaton for L. We also find specific regular languages with the exponential difference between these two complexity measures. These results hold in the nondeterministic case, too.

Theorem 3 *For any regular language L over an alphabet Σ, the two-way message complexity of L is not greater than the size of the minimal two-way deterministic automaton for L.*

Proof. Let $M = (Q, \Sigma, \delta, q_0, F)$ be a minimal $2dfa$ for L. It is sufficient to prove that there is a two-way uniform deterministic protocol P accepting L with at most $|Q|$ messages. The computation of the protocol P on an input $xy \in \Sigma^*$ can be informally described as follows. The first computer looks at its input x and simulates the work of M on x until the last symbol of x is read by M and the step to the right has to be made by M. In this case, the first computer sends a message coding the state of M to the second computer. Then the second computer can simulate M on its input y until the first symbol of y is read by M and the step to the left has to be made by M. In this case, the second computer sends the message that codes the state of M to the first computer. Further, the communication between these two computers proceeds in the same way.It is not difficult to see that P is a two-way uniform deterministic protocol accepting L with the message complexity $|Q|$.   □

In the following part of this section we give examples of regular languages with the essential difference between their two-way message complexity and the size of their minimal two-way deterministic automata.

Lemma 1 *For an integer k, let $B_k = \{1^k\}$ be the unary language that contains the only word (of length k). Any two-way deterministic finite automaton for B_k has at least k states.*

Proof. Assume to the contrary that $M = (Q, \{1\}, \delta, q_0, F)$ is a $2dfa$ for B_k that has fewer than k states. Since $1^k \in B_k$, the automaton M accepts the word 1^k. Let us consider the sequence of states which M enters when computing on 1^k. Denote by q_i $(i = 1, 2, \ldots, k)$ the state that M enters on the move which takes M to the $i + 1$st cell for the first time (before reaching this cell M may move its head back and forth on cells 1 through i many times). Since M has less than k states, there exist $i < j$ such that $q_i = q_j$. But then the computation of M on the word $1^i 1^{j-i} 1^{j-i} 1^{k-j}$ of the length more than k is also accepting, which is a contradiction.   □

Lemma 2 *Let $C_k = \{1^{k^k - 1}\}$ be the unary language that contains the only word of length $k^k - 1$. There is a two-way uniform deterministic protocol P_k accepting C_k and using $O(k^2)$ messages.*

Proof. The protocol P_k over the alphabet $\{1\}$ can be informally described as follows. Let $xy \in \Sigma^*$ be an input word. If $|x| \geq k^k$ or $|y| \geq k^k$ a constant number of messages is sufficient for the computers C_I and C_{II} to reject the input. In the other case (i.e. $|x| < k^k$ and $|y| < k^k$) both computers can unambiguously write the length of their inputs in the form:

$C_I \cdot |x| = a_0 + a_1 k + a_2 k^2 + \ldots + a_{k-1} k^{k-1}$,

$C_{II} : |y| = b_0 + b_1 k + b_2 k^2 + \ldots + b_{k-1} k^{k-1}$, where $a_i, b_i \in \{0, 1, \ldots, k-1\}$. Note that $|x| + |y| = k^k - 1$ iff $a_i + b_i = k - 1$ for all $i = 0, 1, \ldots, k - 1$. So, C_I will successively send a subscript i and the value of the coefficient a_i (starting with a_0) and C_{II} will check whether $a_i + b_i = k - 1$. $O(k^2)$ messages are sufficient for them to accept or reject the input xy (the number of the coefficients a_i is k and their values are from the set $\{0, 1, 2, \ldots, k-1\}$). $\qquad \square$

By the two lemmata above, for the language $C_k = \{1^{k^k-1}\}$ it holds, that there is a two-way uniform deterministic protocol accepting C_k with the message complexity $O(k^2)$, while any two-way deterministic finite automaton recognizing $C_k = B_{k^k-1}$ has at least $k^k - 1$ states. So, there is an exponential difference between the two-way message complexity of C_k and the size of the minimal two-way deterministic automaton for C_k.

The two-way nondeterministic message complexity of regular languages can be defined simply by allowing Φ and φ in the definition 2.1 to be relations on $\Sigma^* \times \{0,1\}^* \times (\{0,1\}^* \cup \{\bar{0}, \bar{1}\})$, as opposed to functions. Theorem 3 and Lemma 1 can be proved for the nondeterministic case, too. So, we get that the two-way nondeterministic message complexity of a regular language L provides a lower bound on the size of the minimal two-way nondeterministic automaton for L. But for the language C_k the difference between these two complexity measures is exponential.

The languages considered in this section were over the alphabet $\{1\}$. Similar considerations can be made for the regular languages $\{(01)^{k^k-1}\}$ over the alphabet $\{0, 1\}$ or the regular languages $\{(abc)^{k^k-1}\}$ over the alphabet $\{a, b, c\}$.

Acknowledgement

I am grateful to Prof. Juraj Hromkovič for his comments concerning this work.

References

[1] J. Hromkovič, Relation between Chomsky hierarchy and communication complexity hierarchy, *Acta Mathematica Univ. Com.*, 48-49 (1986), 311–317.

[2] J. Hromkovič, *Communication Complexity and Parallel Computing.* Springer, Berlin, 1997.

[3] J. Hromkovič and G. Schnitger, Determinismus versus Las Vegas, 1995, unpublished ms.

[4] J. Hromkovič and G. Schnitger, On the power of the number of advice bits in nondeterministic computations. In *Proceedings of the 28th ACM STOC*, 1996, 551–560.

[5] H. Klauck and G. Schnitger, Nondeterministic finite automata versus nondeterministic uniform communication complexity, University of Frankfurt, 1996, unpublished ms.

On Universal Finite Automata and a-Transducers

Manfred Kudlek

Computer Science Department
University of Hamburg
Germany
kudlek@informatik.uni-hamburg.de

Abstract. We show that universal finite automata and a-transducers do not exist, and that for bounded finite automata there are universal finite automata, but not with the same bound.

1 Introduction and Definitions

Some traditional wisdom states that universal finite automata and universal a-transducers do not exist, but there is no published proof. This will be given here. However, if the number of states and the size of the alphabet is bounded, there exists a universal deterministic finite automaton for this class. But this finite automaton does not belong to the same class. We also discuss the impossibility of a universal pushdown automaton. All basic definitions of finite automata, a-transducers, and generalized sequential machines can be found in [1, 2, 3, 4, 5].

1.1 Encoding Finite Automata

Special NFA (DFA) should be encoded by a deterministic a-transducer (DAT).

An input $w \in V^*$ can be encoded by a DAT with the encoding $x_i \in V = \{x_0, \cdots, x_{k-1}\}$ in a binary way : $e(x_i) = bin_k(i)$, and extending this to V^* by $e(\lambda) = \lambda$ and $e(wx) = e(w)e(x)$. Note that $|e(x_i)| = \lceil log_2(k) \rceil$.

The encoding can be done by a DAT $A = (V, \{0, 1\}, \{q\}, \{q\}, \rho, \sigma)$ with $\rho(q, x) = q$ and $\sigma(q, x) = bin_k(i)$. The encoding needs $\lceil log_2(k) \rceil \cdot |w|$ space and time, where $k = |V|$.

Note also that both the input and output of a DAT can be encoded in this way.

Since k is not a constant, a second coding of $0, 1$ is necessary, e.g. by $c(0) = 00$, $c(1) = 01$, leaving sequences 10 and 11 for deliminators. E.g. $ce(x_i) = c(bin_k(i))10$.

The entire encoding ce, and also the decoding $(ce)^{-1} = e^{-1}c^{-1}$ can still be achieved by a DAT.

The following lemma holds:

Lemma 1 *For any DFA $A = (Q, V, \{q_0\}, F, \delta)$, there can be constructed another such DFA $A' = (Q', \{0, 1\}, \{q_0'\}, F', \delta')$ with $L(A') = e(L(A))$ and $L(A) = e^{-1}(L(A'))$.*

Proof. Replace any transition (q, x, q') in A, with $e(x) = bin_k(j) = b_0(x) \cdots b_{k-1}(x)$, by a sequence of transitions $(q, b_0(x), r_{1,x}), \cdots, (r_{j,x}, b_j(x), r_{j+1,x}), \cdots, (r_{k-1,x}, b_{k-1}(x), q')$.

The transitions for $r_{j,x}$ and other inputs $b \neq b_j(x)$ are either undefined or just lead to a sink s.

Therefore, it is obvious that $w \in L(A) \Leftrightarrow e(w) \in L(A') = e(L(A))$, and also that $u \in L(A') \Leftrightarrow e^{-1}(u) \in L(A) = e^{-1}(L(A'))$.   □

A similar construction can be done for a DAT.

2   Universal Finite Automata

The first theorem states that a Universal deterministic finite automaton cannot exist.

Theorem 1 *There is no universal DFA.*

Proof. A universal DFA must use an encoding of a special DFA A (which is assumed to work on $\{0, 1\}$) and the input w of A, e.g. $f(A)Dw$ with the encoding $f(A) \in \{0, 1\}^*$ and the special delimiter D.

Now assume that the universal DFA U has N states. Assume that the special DFA A to be simulated has $M > N$ states, and that all M states of A are reached by an input w for A. Such special A and w can easily be constructed. For example, take $V = \{x_0, \cdots, x_{M-1}\}$ and $L = \{x_0 \cdots x_{M-1}\}^*$. Then $A = (Q, V, \{q_0\}, \{q_0\}, \delta)$ with $\delta(q_i, x_i) = q_{i+1}$ for $0 \leq i < M - 1$ and $\delta(q_{M-1}, x_{M-1}) = q_0$, and $L(A) = L$. Take also $w = x_0 \cdots x_{M-1}$.

Then, construct the corresponding DFA A' with $L(A') = ce(L(A))$ and take the input $w' = ce(w)$. Obviously, $|Q'| = \lceil log_2(M) \rceil \cdot M$.

For universal U to simulate A, it must reach states corresponding to states q of A after it has read $f(A)$. But since $N < M$, and because all M states of A are reached by input w, U must reach identical states $z_i = z_j$ for some corresponding states $q_i \neq q_j$ (U is deterministic, too). This is a contradiction, because q_i, q_j are inequivalent, i.e. $\delta(q_i, 0) \neq \delta(q_j, 0) \vee \delta(q_i, 1) \neq \delta(q_j, 1)$, but U continues as if $q_i = q_j$. $\qquad \square$

The next theorem states the same for nondeterministic finite automata.

Theorem 2 *Neither is there a universal NFA that simulates all special NFA.*

Proof. Assume that there exists such a U. Then an equivalent universal DFA U' can be constructed that simulates all NFA, and also all DFA. This is a contradiction. $\qquad \square$

3   Universal a-Transducers

Lemma 2 *For each DAT (NAT) T there exists an equivalent literal DAT (NAT) T' with transitions of the forms $(q, x, y, r), (q, x, \lambda, r), (q, \lambda, y, r)$ only.*

Proof. Consider any transition $t = (q, x_1 \cdots x_m, y_1 \cdots y_n, r)$ of T. Replace it with a sequence of transitions in the following way :

$m \leq n$:
$(q, x_1, y_1, s_{t,1}), \cdots, (s_{t,m-1}, x_m, y_m, s_{t,m}), (s_{t,m}, \lambda, y_{t,m+1}, s_{t,m+1}), \cdots,$
$(s_{t,n-1}, \lambda, y_n, r)$
(This also covers the case $m = 0$.)

$m > n$:
$(q, x_1, y_1, s_{t,1}), \cdots, (s_{t,n-1}, x_n, y_n, s_{t,n}), (s_{t,n}, x_{n+1}, \lambda, s_{t,n+1}), \cdots,$
$(s_{t,m-1}, x_m, \lambda, r)$
(This also covers the case $n = 0$.) $\qquad \square$

The next lemma states a one-to-one relation between a-transducers and finite automata.

Lemma 3 *There is a one-to-one relation between DFA (NFA) and DAT (DFA).*

Proof. Let $T = (Q, X, Y, S, F, \rho, \sigma)$ be any DAT (NAT) with an accepted language $L(T)$ and a transduced set $M(T)$.

Define $\phi(T) = (Q, X, Y, S, F, \rho, \sigma')$ by $\sigma'(q, x) = \lambda$. Then $L(\phi(T)) = L(T)$ and $M(\phi(T)) = \{\lambda\}$.

Define $\psi(T) = (Q, X, S, F, \rho)$. Then it is obvious that $\psi(T)$ is a DFA (NFA) that accepts $L(T)$.

On the other hand, it is also obvious that for each DFA (NFA) $A = (Q, X, S, F, \delta)$ a DAT (NAT) $\chi(A) = (Q, X, Y, S, F, \delta, \sigma)$ can be defined by $\sigma(q, x) = \lambda$.

Thus, there is a one-to-one relation between DAT (NAT) $\phi(T)$ and DFA (NFA) A. $\Box$

This relation can be used to show that universal a-transducers cannot exist.

Theorem 3 *There is no universal DAT U that simulates all special DAT.*

Proof. Assume that there exists a universal DAT U that simulates all special DAT T. Then U also simulates all DAT $\phi(T)$.

Construct $\phi(U)$. If it receives the same inputs as U, it can simulate any special DAT $\phi(T)$ (with output $\{\lambda\}$ only).

Now, construct $\psi(U)$. $\psi(U)$ is a DFA. By construction, $\psi(U)$ can simulate any special DFA, since there is a one-to-one relation between DAT $\phi(T)$ and DFA A. But this is a contradiction. $\Box$

The same can be shown in an analogous way for NAT.

Theorem 4 *There is no universal NAT U that simulates all special NAT T.*

Proof. As for DAT, the assumption that a universal NAT exists is a contradiction, because it implies that a universal NFA also exists. $\Box$

4   Universal Bounded Finite Automata

This section shows that universal finite automata can be constructed for fixed sets of states and symbols. But these universal automata have more states or symbols.

Theorem 5 *Let S be a finite set of states, and V a finite set of symbols, with $|S| = n$ and $|V| = m$. Then there exists a universal DFA U that simulates all special DFA $A = (Q, X, \{q_0\}, F, \delta)$ with $Q \subseteq S$ and $X \subseteq V$.*

Proof. Let $V = \{x_0, \cdots, x_{m-1}\}$. Then each such special A can be encoded by $c(A) = \delta(q_0, x_0) \cdots \delta(q_0 x_{m-1}) \cdots \delta(q_{n-1}, x_0) \cdots \delta(q_{n-1}, x_{m-1}) D f(q_0) \cdots f(q_{m-1})$ with a special delimiter D, and $f(q_i) = E$ ($f(q_i) = G$) denoting $q_i \in F$ ($q_i \notin F$), respectively, and with $\delta(q_i, x_j) = *$ if this transition doesn't exist in A.

The universal DFA consists of an initial part in the form of a tree to the 'leaves' of which the special $DFA's$ are attached.

U will first read $c(A)$, and reach the initial state of A. Then it starts reading the input for A, working exactly like A.

The size (number of states) of this universal DFA is bounded by:

$$\frac{(n \cdot m)^{n+2} - 1}{n \cdot m - 1} \cdot 2 \cdot (2^n - 1) \cdot (n \cdot m) \leq (2 \cdot n \cdot m)^{n+1} \cdot (n \cdot m)^2. \qquad \square$$

Theorem 6 *There also exists a universal NFA U that simulates all special NFA $A = (Q, X, I, F, \delta)$ with $Q \subseteq S, X \subseteq V$.*

Proof. This is shown in an analogous way. Just encode A by:

$$\begin{aligned} c(A) = {}& d(q_0, x_0) \cdots d(q_0, x_{m-1}) \cdots d(q_{n-1}, x_0) \cdots d(q_{n-1}, x_{m-1})D \\ & s(q_0) \cdots s(q_{m-1})Df(q_0) \cdots f(q_{m-1}), \end{aligned}$$

where $d(q_i, x_j) = h(q_0, q_i, x_j) \cdots h(q_{m-1}, q_i, x_j)$, $h(q_k, q_i, x_j) = E(h(q_k, q_i, x_j) = G)$ if $q_k \in \delta(q_i, x_j)$ $(q_k \notin \delta(q_i, x_j))$, respectively, and $s(q_k) = E(s(q_k) = G)$ if $q_k \in I$ $(q_k \notin I)$, respectively. $\qquad \square$

5 Conclusion and Outlook

It is generally accepted that there are no such things as finite automata or a-transducers. Likewise, it is also generally accepted that universal pushdown automata and universal transducers with pushdown do not exist. Nevertheless, until now there was no formal proof for these hypotheses. Special pushdown automata should be encoded (and decoded) by deterministic transducers with pushdown. Unfortunately, the method presented for finite automata and a-transducers does not work in the case of pushdown automata. Kolmogorov complexity methods may have to be used to state that a universal pushdown automaton has higher Kolmogorov complexity than any special pushdown automaton.

References

[1] S. Eilenberg, *Automata, Languages, and Machines*. Academic Press, New York, 1974.

[2] M.A. Harrison, *Introduction to Formal Language Theory*. Addison-Wesley, Reading, Mass., 1978.

[3] J.E. Hopcroft and J.D. Ullman, *Introduction to Automata Theory, Languages and Computation*. Addison-Wesley, Reading, Mass., 1979.

[4] G. Rozenberg and A. Salomaa (eds.), *Handbook of Formal Languages*. Springer, Berlin, 1997.

[5] A. Salomaa, *Theory of Automata*. Pergamon, Oxford, 1969.

Electronic Dictionaries and Acyclic Finite-State Automata: A State of the Art

Denis Maurel

LI, Computer Laboratory
University of Tours
France
`maurel@univ-tours.fr`

Abstract. This paper presents: first, a brief history of the applications of the finite state automata theory to the lexicon, with the implementation of electronic dictionaries; and, second, three algorithms: to build a transducer that gives the rank of a word in alphabetic order, to build directly a minimal automaton and to reduce the size of memory representation.

1 Motivation

Today, the finite state automata theory is used in many areas of natural language processing, as is pointed out in [14]. I present here a brief history of its application to the lexicon, with the implementation of electronic dictionaries. This specific field restricts research to acyclic automata. Some algorithms make specific reference to this property.

I begin this paper with a brief history (Section 2) and then I discuss three algorithms: the first (Section 3) is used to build a transducer that gives the rank of a word in alphabetic order [12]; the second (Section 4) is used to directly build a minimal automaton [7]; and the third (Section 5) is used to reduce the size of memory representation [16, 5, 12].

2   A Brief History

An electronic dictionary can be seen simply as a list of words and associated codes. Natural language processing needs very large coverage dictionaries with more and more codes. In a French dictionary of simple inflected forms, there is a total of 900 000 entries and in a German dictionary of compounds, there is a total of ten million entries. The first electronic dictionary just gave the syntactic category of a word (e.g. [1]), but today, we also want to know syntactic or semantic features. So, one encounters two problems: first, the access time to data and, second, the very large size of these data.

In 1989, Maurice Gross and Dominique Perrin proposed to represent electronic dictionaries by finite state automata (FSA) [3]. A priori, a lexicographic tree is convenient for a list, but it falls short of our expectations. First, the access time to data is not optimal because of backward; second, the size of this tree is proportional to the number of characters in the whole dictionary. To meet our needs, for the first point, we use a deterministic finite state automata (DFA) [11] and, for the second point, we use a minimal one (MFA) [10]. For instance, see Figure 1.

Sure, MFA is the best representation of a word list, but how to represent an electronic dictionary (i.e. a list of words and associated codes)? We can use a Moore machine, i.e. an FSA with an output function to associate final states with codes. This is the choice of Intex software [15]. But, if one wants to minimize a Moore machine, one can't merge two final states with different codes and one is less size efficiency. Mehryar Mohri has proposed using subsequential transducers (i.e. DFA with two output functions) to associate transitions and states with outputs that are concatenated to obtain the code of a recognized word [9]. He has given a general algorithm to minimize a subsequential transducer, with the outputs as far to the left as possible [8]. In his doctoral thesis, Dominique Revuz gave a nice algorithm to associate a word with its place in alphabetic order, using an MFA (the H-Transducer [12], see Section 3).

But what algorithm can we use to build an MFA? The classic way to do it is to build an DFA and to minimize it. Provided the list is sorted, Dominique Revuz builds an DFA that is almost minimized [12] and a linear algorithm to minimize acyclic DFAs [13]. In 1998, J. Daciuk gave an algorithm to incrementally build an MFA from a sorted list [2]. But Stoyan Mihov proposed, at the same time, another algorithm [6] (see Section 4); he has been followed by Bruce Watson [17] shortly after.

There is still another problem to solve: how to represent an MFA in

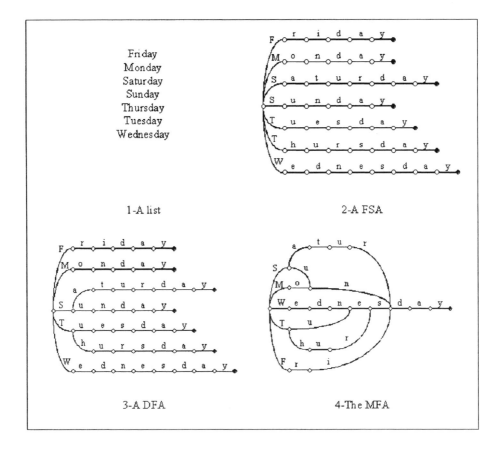

Figure 1: An example.

memory? George Anton Kiraz presents some classical responses [4]. R.E. Tarjan and A.C. Yao gave a nice solution with an array whose number of entries is almost the same as the number of transition [16, 5, 12] (see Section 5).

3 The H-Transducers [12]

3.1 Definitions

Dominique Revuz attempted to associate a word with its place in alphabetic order. Let (A, Q, q_0, F, δ) be an MFA representing an alphabetical list of n words. The first word in this list received the number 0 and the last one, the number $n - 1$. First, he gave two definitions:

Definition 1 *The cardinal number of a state q is the sum of the cardinal numbers of states pointed out by an output transition of q. We add 1 to this* *number if q is a final state:*

$$\forall q \in Q \setminus F,\ Card(q) = \Sigma_{\{q'/\exists a \in A, \delta(q,a)=q'\}} Card(q'),$$

$$\forall q \in F,\ Card(q) = 1 + \Sigma_{\{q'/\exists a \in A, \delta(q,a)=q'\}} Card(q').$$

Definition 2 *The cardinal number of a transition (q, a, q') is the sum of cardinal numbers of states pointed out by an output transition of q labeled with a letter a' lexicographically less than a. We add 1 to this number if q is a final state:*

$$\forall q \in Q \setminus F,\ Card((q,a,q')) = \Sigma_{\{q''/\exists a' \in A, a' < a, \delta(q,a')=q''\}} Card(q''),$$

$$\forall q \in F, Card((q,a,q')) = 1 + \Sigma_{\{q''/\exists a' \in A, a' < a, \delta(q,a')=q''\}} Card(q'').$$

3.2   Example

We want to sort the names of day into alphabetic order (Figure 1-1). Since the state 15 (Figure 2) has no out-transition and it is a final state, its cardinal is 1. With the exception of states 0, 1 and 16, all states have just one out-transition and they are not final states. Thus:

- $Card(15) = 1.$

- $Card(7) = Card(8) = Card(9) = Card(10) = Card(11) = Card(12) = Card(13) = Card(14) = Card(15) = 1.$

- $Card(2) = Card(3) = Card(4) = Card(12) = 1.$

- $Card(5) = Card(6) = Card(12) = 1.$

- $Card(17) = Card(18) = Card(11) = 1.$

- $Card(19) = Card(20) = Card(12) = 1.$

- $Card(1) = Card(2) + Card(6) = 1 + 1 = 2.$

- $Card(16) = Card(10) + Card(17) = 1 + 1 = 2.$

- $Card(0) = Card(1) + Card(5) + Card(7) + Card(16) + Card(19) = 2 + 1 + 1 + 1 + 1 = 6.$

With the exception of the out-transitions of states 0, 1 and 16, the cardinal of the transitions of this automaton is 0. We compute below the seven other transitions (Figure 2):

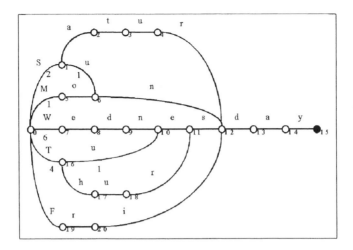

Figure 2: Transition cardinals.

- $Card((0, F, 19)) = 0.$

- $Card((0, M, 5)) = Card(19) = 1.$

- $Card((0, S, 1)) = Card(19) + Card(5) = 1 + 1 = 2.$

- $Card((0, T, 16)) = Card(19) + Card(5) + Card(1) = 1 + 1 + 2 = 4.$

- $Card((0, W, 7)) = Card(19) + Card(5) + Card(1) + Card(16) = 1 + 1 + 2 + 1 = 5.$

- $Card((1, a, 2)) = 0.$

- $Card((1, u, 6)) = Card(2) = 1.$

- $Card((16, h, 17)) = 0.$

- $Card((16, u, 10)) = Card(17) = 1.$

We consider this automaton to be a transducer with the cardinals of transition as output. When we read a word, we add the corresponding cardinals of transition to compute its place in alphabetic order. For instance, the word *Tuesday*:

$$\text{T (4) u (+1) e (+0) s (+0) d (+0) a (+0) y (+0)} \leftarrow 5.$$

First we present the recursive procedure *ComputeCardinalState* (Algorithm 1) and then the computation of the transition cardinals (Algorithm 2).

Algorithm 1 State cardinal computation.

FUNCTION $ComputeCardinalState(q)$

if $q \in F$ **then**

  $Card(q) \leftarrow 1$

else

  $Card(q) \leftarrow 0$

end if

while $\exists (a, q') \in A \times Q, \delta(q, a, q') \neq \epsilon$ **do**

  if $Card(q') = \epsilon$ **then**

    $Card(q') \leftarrow ComputeCardinalState(q')$

  end if

  $Card(q) \leftarrow Card(q) + Card(q')$

end while

RETURN$(Card(q))$

Algorithm 2 Transition cardinal computation.

$ComputeCardinalState(q_0)$

for $q \in Q$ **do**

  if $q \in F$ **then**

    $Cardinal = 1$

  else

    $Cardinal = 0$

  end if

  while $\exists (a, q') \in A \times Q, \delta(q, a, q') \neq \epsilon$ **do**

    (*We assume transitions in alphabetic order*)

    $Card((q, a, q')) \leftarrow Cardinal$

    $Cardinal \leftarrow Cardinal + Card(q')$

  end while

end for

---

4 To Build an MFA [6]

4.1 Definition

Stoyan Mihov built an MFA word by word, using an intermediate automaton, an DFA minimal except for a word:

Definition 3 *Let (A, Q, q_0, F, δ) be an acyclic DFA and L its recognized language. Let $\omega = \omega_1 \omega_2 ... \omega_n \in L$ be a word and $T = \{t_1 = \delta(q_0, \omega_1), t_2 = \delta(t_1, \omega_2)..., t_n = \delta(t_{n-1}, \omega_n)\}$. This automaton is minimal except for the word ω iff:*

- *There are no different equivalent states in the set $Q \setminus T$.*

- *The states of T have one and only one in-transition.*

4.2 Example

We assume that we are building the MFA that recognizes the days of the week and that our list is sorted from the first day to the last (Figure 3-1):

- For the first word, *Sunday*, we create seven states $q_0, t_1, ..., t_6$.

- There is no common prefix between *Sunday* and *Monday*; so, state t_6 is numbered 1, state t_5 is numbered 2, and so on: we also create six new states, $t_1, t_2, ..., t_6$.

- Neither is there a common prefix between *Tuesday* and the first words of the list. Thus, we merge equivalent states: t_6 and 1, t_5 and 2, and so on; then we number state t_1 7.

- We do the same for the word *Wednesday*.

- The common prefix of the word *Tuesday* and the first words of the list is T (state 10 in $S \setminus T$. Thus, we merge equivalent states (Figure 3-2): t_8 and 1, t_7 and 2, and so on. We number t_1 the state 10, 10 the state t_3, 14 the state t_2 and 15 the state t_1.

- The common prefix of the word *Saturday* and the first words of the list is S. See Figure 3-3.

- Finally, we merge the last equivalent states of T and we number the others to obtain the MFA (Figure 3-4).

Algorithm 3 first uses a function, *ReadPrefix*, to read the common prefix of the word and the automaton. If a state of the common prefix has more than one in-transition, this function duplicates it[1]. Then

---

[1] In [6], the common prefixes of the words are sorted, but the duplication of these states allows us to forget this condition.

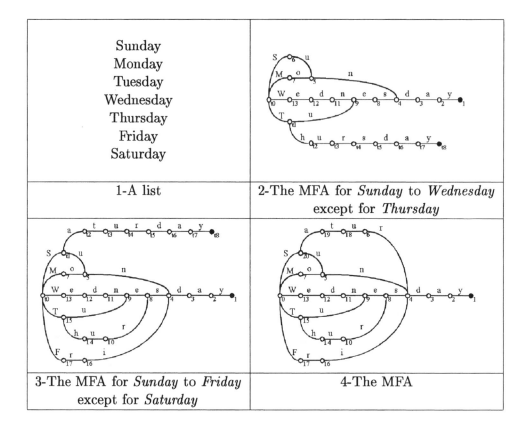

1-A list	2-The MFA for *Sunday* to *Wednesday* except for *Thursday*
3-The MFA for *Sunday* to *Friday* except for *Saturday*	4-The MFA

Figure 3: MFAs except for a word.

it uses two procedures: the first merges the equivalent states of S and $T \setminus \{..., t_i\}$ (*MergeEquivalentStates*) and numbers the other states; the second creates the states and the transitions to recognize the suffix of the word, *CreateOthersStatesT*.

5   A Compact Representation of an MFA [16, 5, 12]

First, the letters of A are coded. The best code is obtained in descending order of token number in the automaton. The letter u appears twice in our example (Figure 4) and the other letters appear once:

ϵ	u	a	d	e	h	r	s	T	y
0	1	2	3	4	5	6	7	8	9

Figure 4: Letter code.

Algorithm 3 To build the MFA.

---

    for $\omega \in L$ **do**

        $i \leftarrow ReadPrefix(q_0, T, \omega)$

        $MergeEquivalentStates(i + 1)$

        $i \leftarrow ReadPrefix(t_i, S, \omega)$

        $CreateOthersStatesT(i, \omega)$

    end for

    $MergeEquivalentStates(1)$

Then the states are numbered again, trying to keep the links to each state to a minimum (Figure 5):

- State 0 is numbered 0; the out-put transition is put on the cell $0 + T = 0 + 8 = 8$ where the letter T and the state (2) reached are noted.

- State 1 is numbered 1.

- A new number for state 2 is sought; we need the cells located at distance h (5) and u (1) to be free. The number 2 can be kept.

- We do the same for states 3, 4, 5 and 6.

- But, if a (2) is added to 7, we obtain cell 9 which is not free. Similarly, cell $8 + a = 8 + 2 = 10$ is not free. State 7 must be numbered 9, because $9 + a = 9 + 2 = 11$ and cell 11 is free.

- State 8 is numbered 7, because $7 + y = 7 + 9 = 16$ and cell 16 is free.

- $8 + e = 8 + 4 = 12$ and cell 12 is not free, but $10^2 + e = 10 + 4 = 14$ and cell 14 is free. So state 9 is numbered 10.

Finally, an array of seventeen cells is obtained. If we want to read *Tuesday*, we compute $0 + T = 0 + 8 = 8$ and we check there is an T on cell 8; then, we compute $2 + u = 2 + 1 = 3$, $10 + e = 10 + 4 = 14$, and so on. But, if we are reading *Turkey*, we compute $10 + r = 10 + 6 = 16$. We note that cell 16 does not contain the letter r and we stop.

Algorithm 4 builds the automaton compact representation. It uses a function, $FirstNewState(q)$, that searches the first free new state number, i, as, for each out-transition (q, a, q'), the cell $i + a$ is free and, if q is final, cell i is also free.

---

[2]Number 9 is already allocated.

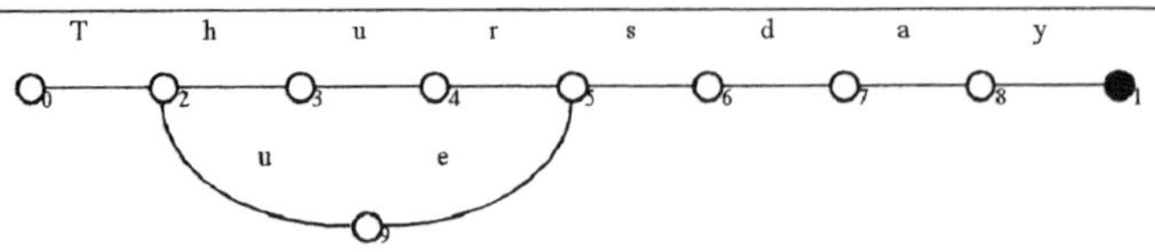

T   h   u   r   s   d   a   y
0 — 2 — 3 — 4 — 5 — 6 — 7 — 8 — 1
u ... e ... 9

	0	1	2	3	4	5	6	7	8	9	10	11	12	13	14	15	16
0 → 0									T 2								
1 → 1 2 → 2		ε		u 9					h 3	T 2							
3 → 3 4 → 4 5 → 5 6 → 6		ε		u 9	u 4				h 3	T 2	d 7	r 5		s 6			
7 → 9 8 → 7 9 → 10		ε		u 10	u 4				h 3	T 2	d 9	r 5	a 7	s 6	e 5		y 1

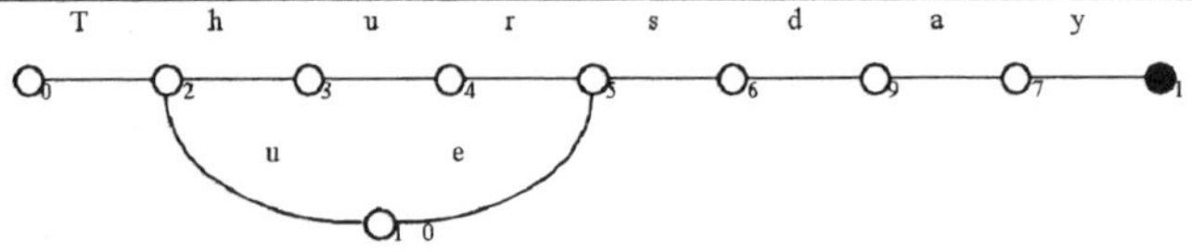

T   h   u   r   s   d   a   y
0 — 2 — 3 — 4 — 5 — 6 — 9 — 7 — 1
u ... e ... 10 — 0

Figure 5: The new state numbers.

Algorithm 4 To build a compact representation.

```
for i >= 0 do
    NewStateAllocated[i] ← FALSE
    CellAllocated[i] ← FALSE
end for
for q ∈ Q do
    i ← FirstNewState(q)
    NewStateAllocated[i] ← TRUE
    while ∃(a, q') ∈ A × Q, δ(q, a, q') ≠ ε do
        CellAllocated[i + code[a]] ← TRUE
        CompactRepresentation[i + code[a]][letter] ← a
        CompactRepresentation[i + code[a]][state] ← q'
    end while
    if q ∈ F then
        CellAllocated[i] ← TRUE
        CompactRepresentation[i][letter] ← ε
    end if
end for
```

References

[1] B. Courtois, Dictionnaire Électronique des Mots Simples du Français DELAS V07-E1, Research Report LADL 33, Université Paris 7, 1992.

[2] J. Daciuk, B.W. Watson and R.E. Watson, Incremental construction of minimal acyclic finite state automata and transducers. In *Proceedings of Finite State Methods in Natural Language Processing, FSMNLP'98*, 1998.

[3] M. Gross and D. Perrin, *Electronic Dictionaries and Automata in Computational Linguistics*, Lecture Notes in Computer Science, 377. Springer, Berlin, 1989.

[4] G.A. Kiraz, Compressed storage of sparse finite-state transducers. In *Proceedings of the Workshop on Implementing Automata, WIA'99*. Springer, Berlin, to appear.

[5] F.M Liang, Word Hyphenation by Computer, PhD dissertation, Computer Science Department, Stanford University, Research Report STAN-CS-83-977, 1983.

[6] S. Mihov, Direct construction of minimal acyclic finite states automata, *Annuaire de l'Université de Sofia St. Kl. Ohridski. Faculté de Mathématique et Informatique*, 91, livre 1, 1998.

[7] S. Mihov, Direct construction of minimal acyclic finite states automata, *Annuaire de l'Université de Sofia St. Kl. Ohridski. Faculté de Mathématique et Informatique*, 92, livre 2, 1999.

[8] M. Mohri, Minimization of sequential transducers. In Lecture Notes in Computer Science, 807. Springer, Berlin, 1994.

[9] M. Mohri, On some applications of finite-state automata theory to natural language processing, *Natural Language Engineering*, 2 (1996), 1–20.

[10] E.F. Moore, Gedanken experiments on sequential machines. In *Automata Studies*. Princeton University Press, Princeton, N.J., 1956, 129–156.

[11] J. Myhill, Finite automata and representation of events. In *Fundamental Concepts in the Theory of Systems*, Wright Air Development Center, 1957.

[12] D. Revuz, Dictionnaires et Lexiques: Méthodes et Algorithmes, PhD dissertation, Université Paris 7, 1991.

[13] D. Revuz, Minimization of acyclic deterministic automata in linear time, *Theoretical Computer Science*, 92 (1992), 181–189.

[14] E. Roche and Y. Schabes (eds.), *Finite State Language Processing*. MIT Press, Cambridge, Mass., 1997.

[15] M. Silberztein, INTEX 4.1 for Windows: a walkthrough. In *Proceedings of the Workshop on Implementing Automata, WIA'98*. Springer, Berlin, 1999, 230–243.

[16] R.E. Tarjan and A.C. Yao (1975), Storing a sparce table, *Journal of the ACM*, 22 (1975), 606–611.

[17] B.W. Watson, A fast new semi-incremental algorithm for the construction of minimal acyclic DFAs. In *Proceedings of the Workshop on Implementing Automata, WIA '98*. Springer, Berlin, 1999, 121–132.

A New Recursive Incremental Algorithm for Building Minimal Acyclic Deterministic Finite Automata

Bruce W. Watson

Department of Computer Science
University of Pretoria
South Africa
bwatson@cs.up.ac.za

Abstract. This chapter presents a new algorithm for incrementally building minimal acyclic deterministic finite automata. Such minimal automata are a compact representation of a finite set of words (e.g. in a spell checker). The incremental aspect of such algorithms (where the intermediate automaton is minimal) facilitates the construction of very large automata in limited computer memory where other (nonincremental) algorithms would fail with intermediate data structures too large to fit in memory.

1   Introduction

This chapter presents a new algorithm for incrementally building minimal acyclic deterministic finite automata. Such minimal automata are a compact representation of a finite set of words (e.g. in a spell checker). The incremental aspect of such algorithms (where the intermediate automaton is minimal) facilitates the construction of very large automata in limited computer memory where other (nonincremental) algorithms would fail with intermediate data structures too large to fit in memory. Although several similar algorithms have recently been published [2, 1, 5, 6, 8], the one presented here is unique in several aspects:

1. It is recursive and short (both aspects enhance understanding and ease implementing the algorithm).

2. It is derived and presented in a number of easily-followed steps, using formal methods to ensure its correctness.

3. Preliminary benchmarking (not presented here) indicates that the algorithm is competitive with the existing ones.

2 Mathematical Preliminaries

Throughout this chapter, we assume the audience has a working knowledge of automata and formal languages. For such a background, see [4, 7].

We make the following definitions: FA is the set of all *finite automata*; DFA is the set of all *deterministic* FAs; MDFA is the set of all *minimal* DFAs; ADFA is the set of all *acyclic* DFAs; MADFA is the set of all *minimal* ADFAs. Throughout this chapter, we assume a DFA $(Q, \Sigma, \delta, q_0, F)$, where Q is the set of states, Σ the alphabet, δ the transition function (which yields value $\bot$ when it is undefined), q_0 as the start state, and F as the set of final states. A *confluence* state is one which has more than one in-transition. *Cloning* a state q makes a new state p with all of the same out-transitions (to the same destination states) as q but no in-transitions.

In this chapter, we are primarily interested in algorithms which build MADFAs. The algorithms are readily extended to work with acyclic deterministic *transducers*, though such an extension is not considered in this chapter.

For any $M \in$ FA, $|M|$ is the number of states in M and $\mathcal{L}(M)$ is the language (set of words) accepted by M. The primary definition of minimality of an $M \in$ DFA is:

$$(\forall\, M' \in \text{DFA} : \mathcal{L}(M') = \mathcal{L}(M) : |M| \leq |M'|).$$

In that case, we can write $Min(M)$. A useful DFA property is: $\mathcal{L}(M)$ is finite $\wedge\, Min(M) \equiv M \in$ MADFA.

For any state p in an FA, define the right language of p, written $\overrightarrow{\mathcal{L}}(p)$ to be the set of words on all paths from p to a final state. There is also an inductive definition of right language, in terms of the right languages of other states (we define $\overrightarrow{\mathcal{L}}(\bot) = \varnothing$):

$$\overrightarrow{\mathcal{L}}(p) = (\cup\, a : a \in \Sigma : \overrightarrow{\mathcal{L}}(\delta(p, a))) \cup \begin{cases} \{\varepsilon\} & \text{if } p \in F, \\ \varnothing & \text{otherwise.} \end{cases}$$

Note that if p is a final state then $\varepsilon \in \overrightarrow{\mathcal{L}}(p)$. With this notion of a right language, we can present predicate Min in a more usable form:

$$Min(Q, \Sigma, \delta, q_0, F) \equiv (\forall\, p, q : p \in Q \wedge q \in Q \wedge p \neq q : \overrightarrow{\mathcal{L}}(p) \neq \overrightarrow{\mathcal{L}}(q))$$

If two distinct states had the same right language, one could be eliminated in favour of the other.

We define *head* to be a function which takes a string w over Σ and returns the first symbol of w; the function is undefined if there is no such symbol (if $w = \varepsilon$). Similarly, *tail* takes a string and returns the string without the first symbol; it is also undefined if there is no first symbol. If $w \neq \varepsilon$ then $w = head(w)\,tail(w)$.

All of the algorithms presented here are in the guarded command language, a type of pseudo-code — see [3].

3   A First Algorithm

We begin with a procedure which manipulates a global MADFA $(Q, \Sigma, \delta, q_0, F)$ which is initially $(\{q_0\}, \Sigma, \emptyset, q_0, \emptyset)$. (Strictly speaking, this is not the minimal MADFA accepting the empty language — though this is the only time it is not minimal.) It takes a word w and a state q and modifies the MADFA to add w to the right language of q; it returns q. (To denote this modification, the algorithm below contains a *shadow* variable L which serves no purpose other than annotation.) The algorithm's precondition is that q is not a confluence state — if it were, the algorithm may inadvertently add more words than just w since there would be more than one path from q_0 to q. The algorithm considers the two basic cases for w:

1. $w = \varepsilon$, in which case q is simply made into a final state.

2. $w \neq \varepsilon$, in which case w is of the form $head(w)\,tail(w)$. A transition of q on $head(w)$ may already exist:

    - If such a transition does not exist, a new transition (on $head(w)$) and state are created and $tail(w)$ is added to the new state's right language.

    - If the transition does exist, the destination state q' is found and $tail(w)$ is added to the q' right language. Care is taken that q' is cloned if it is a confluence state.

The algorithm assumes two external procedures: *create* makes a new state with no transitions, while *clone* takes a state and creates a new state with the same finality (whether or not it is final) and out-transitions as the argument state. The algorithm is:

Algorithm 3.1:

---

proc $add(q : Q, w : \Sigma^*) : Q$
    $\{\ q$ is not a confluence state $\}$

$$L := \overrightarrow{\mathcal{L}}(q);$$
$$\textbf{if } w = \varepsilon \rightarrow F := F \cup \{q\}$$
$$\| \quad w \neq \varepsilon \rightarrow$$
$$\quad q' := \delta(q, head(w));$$
$$\quad \textbf{if } q' = \bot \rightarrow \delta(q, head(w)) :=$$
$$\quad\quad add(create(), tail(w))$$
$$\quad \| \quad q' \neq \bot \rightarrow$$
$$\quad\quad \textbf{if } q' \text{ is a confluence} \rightarrow \delta(q, head(w))$$
$$\quad\quad\quad := add(clone(q'), tail(w))$$
$$\quad\quad \| \quad q' \text{ is not a confluence} \rightarrow \delta(q,$$
$$\quad\quad\quad head(w)) := add(q', tail(w))$$
$$\quad\quad \textbf{fi}$$
$$\quad \textbf{fi}$$
$$\textbf{fi};$$
$$\{ \overrightarrow{\mathcal{L}}(q) = L \cup \{w\} \}$$
$$\textbf{return } q$$
$$\textbf{corp}$$

□

To add an entirely new word w to the ADFA, the initial invocation is $add(q_0, w)$, where the start state q_0 is assumed to pre-exist. Some observations about the algorithm's behavior:

1. Once a *create* has been done, more will have to be done (until w is exhausted) while processing the rest of the string w, since the newly created states will have no out-transitions.

2. Once a confluence state has been encountered, all subsequent states (along that path, while they exist) will also be confluences since the cloning process ensures that transitions from q and $clone(q)$ will share destination states.

3. The two above situations are disjoint, since a newly created state cannot be a confluence state.

The final modification to this initial algorithm is to make it incrementally minimizing. We assume that the ADFA is initially minimal (before *add* is invoked). The only states whose right languages may have changed are those visited (and, of course, those which are newly created or cloned) — in other words, those states which are returned during invocations of *add*. It follows that, on return, we need only ensure that each state visited (while adding a word) is still unique (it does not have the same right language as some other state); if any such state is not unique, it can be eliminated. We can change the above algorithm into an incremental minimizing one by changing the return value from q to $possibly\_merge(q)$. Procedure $possibly\_merge$ takes a state and:

1. Determines whether there is a pre-existing equivalent state (with the same right language).

2. If there is, it destroys the argument state and returns the pre-existing equivalent one.

3. If there is not, it returns the argument state.

4   Two Specialized Algorithms

We can use the two observations in the previous section to specialize the algorithms in the following sections.

4.1   Specializing for Calls to *create*

In this section, we create a specialized version of *add* (which is called *add_create*) to deal with the first of the observations in the previous section. Additionally, to avoid redundantly passing a newly created state into the specialization, we push the *create* call into the beginning of the specialized procedure. Naturally, specializing also necessitates modifying *add*, which we now call *add'*:

Algorithm 4.1:

---

proc $add'(q : Q, w : \Sigma^*) : Q$
    { q is not a confluence state }
    $L := \overrightarrow{\mathcal{L}}(q)$;
    if $w = \varepsilon \rightarrow F := F \cup \{q\}$
    ❙ $w \neq \varepsilon \rightarrow$
        $q' := \delta(q, head(w))$;
        if $q' = \bot \rightarrow \delta(q, head(w)) :=$
        $add\_create(tail(w))$
        ❙ $q' \neq \bot \rightarrow$
            if q' is a confluence $\rightarrow \delta(q, head(w))$
                $:= add'(clone(q'), tail(w))$
            ❙ q' is not a confluence $\rightarrow \delta(q,$
                $head(w)) := add'(q', tail(w))$
        fi
    fi
  fi;

$$\{ \overrightarrow{\mathcal{L}}(q) = L \cup \{w\} \}$$
return *possibly_merge*(q)
corp
proc *add_create*(w : Σ*) : Q
    q : = *create*();
    { q is not a confluence state by observation 1 }
    $$\{ \overrightarrow{\mathcal{L}}(q) = \varnothing \}$$
    if w = ε → F : = F ∪ {q}
    ▌ w ≠ ε → δ(q, head(w)) : =
        add_create(tail(w))
    fi;
    $$\{ \overrightarrow{\mathcal{L}}(q) = \{w\} \}$$
    return *possibly_merge*(q)
corp

□

In *add_create*, the two inner **if-fi** statements have been eliminated because clearly δ(q, head(w)) = ⊥ (since q was just created).

4.2   Specializing for Confluence States

We can apply our second observation to create a specialized version of *add'* (which is called *add_clone*), which specifically deals with cases where q' (the destination state δ(q, head(w))) is a confluence state. In the following, we rename our main procedure *add''* and we do not reproduce *add_create* since it has not changed.

Algorithm 4.2:

proc *add''*(q : Q, w : Σ*) : Q
    { q is not a confluence state }
    L : = $\overrightarrow{\mathcal{L}}(q)$;
    if w = ε → F : = F ∪ {q}
    ▌ w ≠ ε →
        q' : = δ(q, head(w));
        if q' = ⊥ → δ(q, head(w)) : =
            add_create(tail(w))
        ▌ q' ≠ ⊥ →
            if q' is a confluence →
                δ(q, head(w)) : = *add_clone*(clone(q'),
                tail(w))

$\llbracket$ q' is not a confluence $\rightarrow \delta(q,$
$head(w)) := add''(q', tail(w))$
  fi
 fi
fi;
$\{ \overrightarrow{\mathcal{L}}(q) = L \cup \{w\} \}$
return $possibly\_merge(q)$
corp
proc $add\_clone(q : Q, w : \Sigma^*) : Q$
  $\{ q$ is not a confluence state $\}$
  $L := \overrightarrow{\mathcal{L}}(q);$
  if $w = \varepsilon \rightarrow F := F \cup \{q\}$
  $\llbracket$ $w \neq \varepsilon \rightarrow$
   $q' := \delta(q, head(w));$
   if $q' = \bot \rightarrow \delta(q, head(w)) :=$
   $add\_create(tail(w))$
   $\llbracket$ $q' \neq \bot \rightarrow$
    $\{ q'$ is a confluence by observation 2 $\}$
    $\delta(q, head(w)) := add\_clone(clone(q'),$
    $tail(w))$
   fi
  fi;
  $\{ \overrightarrow{\mathcal{L}}(q) = L \cup \{w\} \}$
  return $possibly\_merge(q)$
corp

□

5   Implementing *possibly_merge*

This procedure takes state q and finds *the* state p (if there is one) which is equivalent to q. If p exists and it is distinct from q, q is eliminated and p is returned. Since the entire algorithm is to be incremental and minimizing, we will assume the invariant that the ADFA is minimal, except for the states currently being processed or created by *add* (or one of the specializations). Procedure *possibly_merge* must compute the set:

$$\{ p \mid \overrightarrow{\mathcal{L}}(p) = \overrightarrow{\mathcal{L}}(q) \}.$$

(In the set comprehension predicate, we could also stipulate that $p \neq q$; for simplicity, we leave that out in this chapter.) We can rework this set into a more (computationally) manageable form as follows:

$$\{\, p \mid \overrightarrow{\mathcal{L}}(p) = \overrightarrow{\mathcal{L}}(q)\,\}$$

$= \quad$ { using the inductive definition of $\overrightarrow{\mathcal{L}}$ }

$$\{\, p \mid (p \in F \equiv q \in F) \land (\forall\, a : a \in \Sigma : \overrightarrow{\mathcal{L}}(\delta(p,a)) = \overrightarrow{\mathcal{L}}(\delta(q,a)))\,\}$$

$= \quad$ { invariant: the rest of the ADFA is minimal }

$$\{\, p \mid (p \in F \equiv q \in F) \land (\forall\, a : a \in \Sigma : \delta(p,a) = \delta(q,a))\,\}$$

$= \quad$ { splitting set comprehension predicate }

$$\{\, p \mid (p \in F \equiv q \in F)\,\} \cap \{\, p \mid (\forall\, a : a \in \Sigma : \delta(p,a) = \delta(q,a))\,\}$$

$= \quad$ { expanding universal quantifier in set comprehension predicate }

$$\{\, p \mid (p \in F \equiv q \in F)\,\} \cap (\cap\, a : a \in \Sigma : \{\, p \mid \delta(p,a) = \delta(q,a)\,\})$$

$= \quad$ { dropping dummy p and noting $\{\, p \mid \delta(p,a) = \delta(q,a)\,\} = \delta^{-1}(\delta(q,a),a)$ }

$$\{\, p \mid (p \in F \equiv q \in F)\,\} \cap (\cap\, a : a \in \Sigma : \delta^{-1}(\delta(q,a),a))$$

We can now define our procedure:

Algorithm 5.1:

```
proc possibly_merge(q : Q) : Q
    A, C := Σ, {p | (p ∈ F ≡ q ∈ F)};
    do A ≠ ∅ ∧ C ≠ ∅ →
        let a : a ∈ A;
        A := A \ {a};
        C := C ∩ δ⁻¹(δ(q,a),a)
    od;
    if C = ∅ → return q
    ▯ C = {p} → return p
    fi
corp
```

$\square$

5.1 A Further Optimization of *add_create'*

Given our algorithm, an important observation is that if q has *no* out-transitions, then q is a final state (otherwise it would be a *dead* state). Since all such *leaf* states are equivalent, only one is needed and q could then be eliminated in favour of the existing leaf state, if there is one.

We can use this observation to rewrite *add_create*. In the new version, we can eliminate a redundant call to *create* when q is a leaf state. Similarly, we can move the invocation of *possibly_merge* up into the **if-fi** statement.

Algorithm 5.2:

---

proc $add\_create'(w : \Sigma^*) : Q$
    if $w = \varepsilon \wedge$ there is already a leaf state $\rightarrow$
        { no need to call $create$ }
        return *the* final leaf state
    $[\!]$ $w \neq \varepsilon \vee$ there is no leaf state $\rightarrow$
        $q := create();$
        { q is not a confluence state by observation 1 }
        { $\overrightarrow{\mathcal{L}}(q) = \emptyset$ }
        if $w = \varepsilon \rightarrow$
            { there is no leaf state }
            $F := F \cup \{q\};$
            { $\overrightarrow{\mathcal{L}}(q) = \{\varepsilon\}$ }
            { there is no need to return $possibly\_merge(q)$ — it is unique }
            return q
        $[\!]$ $w \neq \varepsilon \rightarrow$
            { $w = head(w)tail(w)$ }
            $\delta(q, head(w)) := add\_create'(tail(w))$
            ;
            { $\overrightarrow{\mathcal{L}}(q) = \{w\}$ }
            return $possibly\_merge(q)$
        fi
    fi
corp

$\square$

6   The Final Algorithm

The collected algorithm is:

Algorithm 6.1:

---

proc $add''(q : Q, w : \Sigma^*) : Q$
    { q is not a confluence state }
    $L := \overrightarrow{\mathcal{L}}(q);$
    if $w = \varepsilon \rightarrow F := F \cup \{q\}$
    $[\!]$ $w \neq \varepsilon \rightarrow$
        $q' := \delta(q, head(w));$

> **if** $q' = \bot \rightarrow \delta(q, head(w)) :=$
> $add\_create'(tail(w))$
> **⫾** $q' \neq \bot \rightarrow$
>   **if** q' is a confluence $\rightarrow$
> $\delta(q, head(w)) := add\_clone(clone(q'),$
> $tail(w))$
>   **⫾** q' is not a confluence $\rightarrow \delta(q,$
> $head(w)) := add''(q', tail(w))$
>   **fi**
>   **fi**
> **fi**;
> $\{ \overrightarrow{\mathcal{L}}(q) = L \cup \{w\} \}$
> **return** $possibly\_merge(q)$

corp

proc $add\_create'(w : \Sigma^*) : Q$
> **if** $w = \varepsilon \wedge$ there is already a leaf state $\rightarrow$
> $\{$ no need to call $create$ $\}$
>   **return** the final leaf state
> **⫾** $w \neq \varepsilon \vee$ there is no leaf state $\rightarrow$
> $q := create();$
> $\{ q$ is not a confluence state by observation 1 $\}$
> $\{ \overrightarrow{\mathcal{L}}(q) = \emptyset \}$
>   **if** $w = \varepsilon \rightarrow$
> $\{$ there is no leaf state $\}$
> $F := F \cup \{q\};$
> $\{ \overrightarrow{\mathcal{L}}(q) = \{\varepsilon\} \}$
> $\{$ there is no need to return $possibly\_merge(q)$ — it is unique $\}$
>   **return** q
>   **⫾** $w \neq \varepsilon \rightarrow$
> $\{ w = head(w)tail(w) \}$
> $\delta(q, head(w)) := add\_create'(tail(w))$
>   ;
> $\{ \overrightarrow{\mathcal{L}}(q) = \{w\} \}$
>   **return** $possibly\_merge(q)$
>   **fi**
> **fi**

corp

proc $add\_clone(q : Q, w : \Sigma^*) : Q$
> $\{ q$ is not a confluence state $\}$
> $L := \overrightarrow{\mathcal{L}}(q);$
> **if** $w = \varepsilon \rightarrow F := F \cup \{q\}$
> **⫾** $w \neq \varepsilon \rightarrow$
> $q' := \delta(q, head(w));$

```
          if q' = ⊥ → δ(q, head(w)) : =
             add_create'(tail(w))
          ▯ q' ≠ ⊥ →
             { q' is a confluence by observation 2 }
             δ(q, head(w)) : = add_clone(clone(q'),
             tail(w))
          fi
       fi;
       { L⃗(q) = L ∪ {w} }
       return possibly_merge(q)
corp
proc possibly_merge(q : Q) : Q
       A, C : = Σ, { p | (p ∈ F ≡ q ∈ F) };
       do A ≠ ∅ ∧ C ≠ ∅ →
          let a : a ∈ A;
          A : = A \ {a};
          C : = C ∩ δ⁻¹(δ(q, a), a)
       od;
       if C = ∅ → return q
       ▯ C = {p} → return p
       fi
corp
```

□

7   Conclusions

This chapter presented a new recursive algorithm (consisting of several short procedures) for incrementally constructing acyclic deterministic finite automata, while maintaining the minimality of the resulting automaton. The presentation is unique in several respects:

1. The formal derivation aids in understanding the algorithm and provides a measure of confidence in its correctness.

2. The algorithm is easily understood and implemented thanks to its surprising conciseness.

In acknowledgement, I would like to thank Nanette Y. Saes for proofreading this paper.

References

[1] J.D. Daciuk, S. Mihov, B.W. Watson and R.E. Watson, Incremental construction of minimal acyclic finite state automata, *Computational Linguistics*, 2000, to appear.

[2] J.D. Daciuk, B.W. Watson and R.E. Watson, An incremental algorithm for constructing acyclic deterministic transducers. In *Proceedings of the International Workshop on Finite State Methods in Natural Language Processing*, Ankara, 1998.

[3] E.W. Dijkstra, *A Discipline of Programming*. Prentice-Hall, Englewood Cliffs, N.J., 1976.

[4] J.E. Hopcroft and J.D. Ullman, *Introduction to Automata Theory, Languages, and Computation*. Addison-Wesley, Reading, Mass., 1979.

[5] S. Mihov, Direct building of minimal automaton for given list. Available at stoyan@lml.acad.bg

[6] D. Revuz, Minimisation of acyclic deterministic automata in linear time, *Theoretical Computer Science*, 92 (1992), 181–189.

[7] B.W. Watson, Taxonomies and Toolkits of Regular Language Algorithms, PhD dissertation, Eindhoven University of Technology, 1995. www.OpenFIRE.org

[8] B.W. Watson, A fast new semi-incremental algorithm for the construction of minimal acyclic DFAs. In *Proceedings of the Third Workshop on Implementing Automata*, Rouen, 1998.

III

LOGICS, LANGUAGES
AND COMBINATORICS

Syntactic Calculus and Pregroups

Wojciech Buszkowski

Faculty of Mathematics and Computer Science
Adam Mickiewicz University
Poznań, Poland
buszko@amu.edu.pl

Abstract. Pregroups were introduced in Lambek [12] as a modification of residuated monoids. They are algebraic models of a calculus of syntactic types (syntactic calculus) from Lambek [11]. The formal calculus of pregroups is more easily treatable than the original syntactic calculus and seems to have a greater linguistic expressibility. In this paper, I characterize its models (i.e. pregroups) and study interrelations between both calculi.

1   Introduction

The syntactic calculus in [11], called the Lambek calculus in literature, was designed as a type change system for categorial grammars. *Types* are formed out of *atomic* types by means of binary operation symbols $\cdot$, $\backslash$ and $/$. The calculus operates with sequents of the form $\Gamma \rightarrow A$ such that Γ is a finite sequence of types, and A is a type. The axioms and rules of the calculus are:

(Ax) $A \rightarrow A$,
($\backslash$l) from $\Gamma, B, \Delta \rightarrow C$ and $\Phi \rightarrow A$ infer $\Gamma, \Phi, A\backslash B, \Delta \rightarrow C$,

(\r) from $A, \Gamma \to B$ infer $\Gamma \to A\backslash B$,

(/l) from $\Gamma, B, \Delta \to C$ and $\Phi \to A$ infer $\Gamma, B/A, \Phi, \Delta \to C$,

(/r) from $\Gamma, A \to B$ infer $\Gamma \to B/A$,

(·l) from $\Gamma, A, B, \Delta \to C$ infer $\Gamma, A \cdot B, \Delta \to C$,

(·r) from $\Gamma \to A$ and $\Delta \to B$ infer $\Gamma, \Delta \to A \cdot B$,

(CUT) from $\Gamma, A, \Delta \to B$ and $\Phi \to A$ infer $\Gamma, \Phi, \Delta \to B$.

The above formulation of the Lambek calculus is a Gentzen style logical system. The rule (CUT) can be eliminated in the pure calculus [11], but it is necessary in its axiomatic extensions. Actually, the original syntactic calculus from [11] is more restrictive, since it does not admit empty antecedents of sequents (thus, Γ must be nonempty in rules (\r) and (/r)), and the above system is a straightforward strengthening of the former and corresponds to languages which can contain the null string [5].

Nowadays, the Lambek calculus and its variants have been extensively investigated from different points of view: (1) in mathematical linguistics as basic systems for flexible categorial grammars, (2) in logic as noncommutative substructural logics which anticipate Girard's linear logics [9]. Good references are [8], [3], [7] and [13].

In categorial grammars, types are interpreted as languages, i.e. sets of strings over an alphabet (a lexicon). $A \cdot B$ is the type of all strings ab such that a is of type A and b is of type B. $A\backslash B$ is the type of all strings c such that ac is of type B, for every a of type A (c is a left-directed functor from type A to type B). B/A is the type of all strings c such that ca is of type B, for all a of type A (c is a right-directed functor from type A to type B). For instance, if S is the type of sentences and PN is the type of proper nouns, then $VP = PN\backslash S$ corresponds to intransitive verb phrases, $NP = S/VP$ to full noun phrases in the subject position, $NP_o = (S/PN)\backslash S$ to full noun phrases in the object position, $TV = VP/PN$ to transitive verb phrases, S/S to sentence modifiers, $(S\backslash S)/S$ to sentential connectives, and so on.

Classical categorial grammars, stemming from Ajdukiewicz [1] and Bar-Hillel et al. [2], are based on simple reductions $A, A\backslash B \to B$ and $B/A, A \to B$ (which can be axiomatized as above by (Ax), (\l) and (/r) only). Thus, *John invites Jane* is represented by $PN, (PN\backslash S)/PN, PN$, which reduces to S. Classical reductions, however, cannot parse *some boy invites every girl*, since the string NP, TV, NP_o does not reduce to S. On the other hand, the sequent $NP, TV, NP_o \to S$ is provable in Lambek calculus, which yields the parsing. Accordingly, Lambek categorial grammars are more flexible than the classical ones.

The most general algebraic models of Lambek calculus are *residuated monoids*, i.e. structures $(G, \leq, \cdot, \backslash, /, 1)$ such that $(G, \cdot, 1)$ is a monoid (i.e. a semigroup with a unit), $\leq$ is a partial ordering on G, and $\backslash, /$ are binary operations on G, satisfying the equivalences:

(RES) $a \cdot b \leq c$ iff $b \leq a\backslash c$ iff $a \leq c/b$, for all $a, b, c \in G$. It can easily be shown that $a \leq b$ entails $ca \leq cb$, $ac \leq bc$; hence, $(G, \leq, \cdot, 1)$ is a partially ordered monoid. It is known that the sequents that are provable in the Lambek calculus are precisely the inequalities $A_1, \ldots, A_n \leq A$. These are true in all residuated monoids, if types are interpreted as members of the monoid in an obvious way (the sequent $\rightarrow A$ with the empty antecedent is supposed to be true, if $1 \leq A$).

For linguistic applications, it is reasonable to consider more special models, such as those consisting of all subsets of a monoid $(G, \cdot, 1)$, with $\subseteq$ as the ordering and the operations defined as above (see the explanation about the meaning of types in categorial grammars). Again, the Lambek calculus yields all sequents that are generally true in these models [5, 7]. The same holds for models consisting of languages, i.e. subsets of a free monoid [15]. Axiomatic extensions of the Lambek calculus remain complete with respect to monoid models (fulfilling the extra axioms), but not with respect to free monoid models. Notice that monoid models, i.e. powerset models over monoids, are residuated monoids. Actually, every residuated monoid is embeddable into a powerset model over a monoid.

Now, we define the notion of a pregroup (after [12]). A *pregroup* is a structure $(G, \leq, \cdot, l, r, 1)$ such that $(G, \leq, \cdot, 1)$ is a partially ordered monoid, and l and r are unary operations on G, satisfying the inequalities:

(PRE) $a^l a \leq 1 \leq a a^l$ and $a a^r \leq 1 \leq a^r a$,

for all $a \in G$. a^l (resp. a^r) is called the *left* (resp. *right*) *adjoint* of a. Trivial examples of pregroups are partially ordered groups in which $a^l = a^r = a^{-1}$. Notice that, if $\cdot$ is commutative, then the latter equalities must hold and, consequently, there are no nontrivial commutative pregroups. Therefore, the notion of a pregroup is a generalization of that of a p.o. group which can be really innovatory for the noncommutative case only.

The only example of a nontrivial pregroup, given in [12], is the following. G consists of all (downward and upward) unbounded monotone mappings of the set of integers into itself, $\cdot$ is the superposition, and 1 is the identity mapping. One defines: $f \leq g$ if $f(x) \leq g(x)$, for all integers x. Then, the mapping $f(x) = 2x$ yields $f^l(x) = [(x+1)/2]$ and $f^r(x) = [x/2]$, where $[x]$ stands for the greatest integer $\leq x$. Consequently, $f^l \neq f^r$. The fact that this structure is a pregroup follows from our previous considerations.

As shown in [12], in every pregroup G, the operations $a \backslash b = a^r b$ and $b/a = ba^l$ can be defined, and the resulting structure $(G, \leq, \cdot, \backslash, /, 1)$ is a residuated monoid. It is enough to show (RES). Assume $ab \leq c$. Then, $a \leq abb^l \leq cb^l$, by (PRE) and monotonicity. Assume $a \leq cb^l$. Then, $ab \leq cb^l b \leq c$, by monotonicity and (PRE). For $a \backslash b$ we proceed in a dual way.

As a consequence, all type change laws provable in the Lambek calculus are also generally true inequalities in pregroups. Lambek [12] provides many interesting illustrations of parsing based on the formalism of pregroups, which cannot be reported here in detail. We only sketch a general outline. Personal pronouns are typed as follows: I - type π_1, *you, we, they* - type π_2, *he, she, it* - type π_3. Sentences in the present tense are typed s_1, and those in the past tense s_2. Three non-finite forms of intransitive verbs are typed: the infinitive (*go*) - type i, the present participle (*going*) - type p_1, the past participle (*gone*) - type p_2. The following types of conjugated forms can be derived: in the present tense *go* - types $\pi_1^r s_1$ and $\pi_2^r s_1$, *goes* - type $\pi_3 s_1$, and in the past tense *went* - types $\pi_k^r s_2$, for $k = 1, 2, 3$. Thus, the sentence *he goes* is parsed:

$$\pi_3 \pi_3^r s_1 \leq s_1,$$

and assigning $\pi_3^r s_1 i^l$ to *does* enables one to parse *he does go*:

$$\pi_3 \pi_3^r s_1 i^l i \leq s_1 i^l i \leq s_1.$$

Further, Lambek considers parsing with adverbs, transitive verbs, auxiliary verbs, bitransitive verbs, negative sentences, interrogatives, wh-questions, discontinuous dependencies, and other syntactic constructions.

In this paper, we are mainly interested in the mathematical aspects of the apparatus of pregroups. In section 2, we characterize pregroups of functions over a poset (all pregroups are isomorphic to pregroups of that kind), we provide more examples and we prove the nonexistence of finite, noncyclic pregroups. In section 3, we compare the formalism of pregroups with the Lambek syntactic calculus: although the former are a nonconservative extension of the latter, they are equally strong for first-order types. Hence, categorial grammars based on pregroups generate precisely context-free languages.

2 Pregroups

Let $(X, \leq)$ be a poset (i.e. a partially ordered set). We consider functions $f : X \mapsto X$ with the ordering: $f \leq g$ iff, for all $x \in X$, $f(x) \leq g(x)$. Let $\mathcal{F}$ be a set of functions from X into X, closed under composition ($f \circ g)(x) = f(g(x))$. We consider the structure $(\mathcal{F}, \leq, \circ, I)$, where I is the identity mapping $I(x) = x$. Clearly, this structure is a partially ordered monoid, if all functions f in $\mathcal{F}$ are order preserving: $x \leq y$ entails $f(x) \leq f(y)$. So, we define *a p.o. monoid of functions* as a structure of the above form with $\mathcal{F}$ satisfying the latter assumption. By *a pregroup of functions*, we mean a structure $(\mathcal{F}, \leq, \circ, l, r, I)$ such that $(\mathcal{F}, \leq, \circ, I)$ is a p.o. monoid of functions, and l, r are unary operations on $\mathcal{F}$, satisfying:

$$f^l(f(x)) \leq x \leq f(f^l(x)) \text{ and } f(f^r(x)) \leq x \leq f^r(f(x)), \tag{1}$$

for all $x \in X$. The latter inequalities amount to: $f^l \circ f \leq I \leq f \circ f^l$ and $f \circ f^r \leq I \leq f^r \circ f$ and, consequently, every pregroup of functions is a pregroup.

It is easy to show that every pregroup is isomorphic to a pregroup of functions. Let $(G, \leq, \cdot, l, r, 1)$ be a pregroup. Let $(X, \leq)$ be equal to $(G, \leq)$. To element $a \in G$, assign function f_a from G into G, defined by: $f_a(x) = ax$. Let $\mathcal{F}$ consist of all functions f_a, for $a \in G$. The operations l and r are given by: $f_a^l = f_{a^l}$ and $f_a^r = f_{a^r}$. Then, $(\mathcal{F}, \leq, \circ, l, r, I)$ is a pregroup of functions, and the mapping $F(a) = f_a$ is the required isomorphism.

Our first proposition gives a useful characterization of pregroups of functions.

Proposition 1 *Let $(\mathcal{F}, \leq, \circ, I)$ be a p.o. monoid of functions (on X), and let l and r be unary operations on $\mathcal{F}$. Then, $(\mathcal{F}, \leq, \circ, l, r, I)$ is a pregroup of functions if, and only if, the following equalities hold: $f^l(x) = \min\{z : x \leq f(z)\}$, $f^r(x) = \max\{z : f(z) \leq x\}$, for all $x \in X$.*

Proof. Assume $(\mathcal{F}, \leq, \circ, l, r, I)$ to be a pregroup of functions. By (1), $x \leq f(f^l(x))$, hence $f^l(x) \in \{z : x \leq f(z)\}$. Take an arbitrary $z \in X$ such that $x \leq f(z)$. Then, $f^l(x) \leq f^l(f(z)) \leq z$, by (1), and consequently $f^l(x) = \min\{z : x \leq f(z)\}$. For f^r, the reasoning is dual.

Now, assume that the equalities from proposition 1 hold. Since $f^l(x) \in \{z : x \leq f(z)\}$, then $x \leq f(f^l(x))$. Also $f^l(f(x)) = \min\{z : f(x) \leq f(z)\}$ and $x \in \{z : f(x) \leq f(z)\}$, which yields $f^l(f(x)) \leq x$. For f^r, the reasoning is dual. □

A function $f : X \mapsto X$ is called *a permutation* if f is one-to-one and onto.

Proposition 2 *Let $(\mathcal{F}, \leq, \circ, l, r, I)$ be a pregroup of functions, and let $f \in \mathcal{F}$ be a permutation. Then, $f^l = f^r = f^{-1}$.*

Proof. Since $x = f(f^{-1}(x))$, then $f^{-1}(x) \in \{z : x \leq f(z)\}$. By proposition 1, $f^l(x) \leq f^{-1}(x)$. Suppose $f^l(x) < f^{-1}(x)$. Then, $f(f^l(x)) < f(f^{-1}(x)) = x$, which contradicts (1). Consequently, $f^l(x) = f^{-1}(x)$. In a dual way, one shows $f^r(x) = f^{-1}(x)$. □

A pregroup $(G, \leq, \cdot, l, r, 1)$ is said to be *cyclic*, if $a^l = a^r$, for all $a \in G$. Clearly, every cyclic pregroup is a partially ordered group with $a^l = a^r = a^{-1}$. In any finite partially ordered group the ordering relation must be the identity relation. If $a < b$, then $1 = a^{-1}a < a^{-1}b$, hence $1 < c$, for some c; then, $c^n < c^{n+1}$, for all $n \geq 0$, which yields an infinite sequence of elements.

Theorem 1 *Every finite pregroup is cyclic.*

Proof. It suffices to show that every pregroup of functions on a finite poset $(X, \leq)$ is cyclic. Let $(\mathcal{F}, \leq, \circ, l, r, I)$ be a pregroup of functions on the finite poset $(X, \leq)$. By proposition 2, it suffices to show that every $f \in \mathcal{F}$ is a permutation. We proceed by induction on the number of elements in X. Assume our claim be true, for all sets Y the cardinality of which is less than that of X. Define X_1 as the set of $\leq$-minimal elements in X, and $X_2 = X - X_1$. Fix $f \in \mathcal{F}$. We show that f restricted to X_1 is a permutation on X_1. Let $x \in X_1$. By (1), $f(f^r(x)) \leq x$, hence $f(y) \leq x$, for some $y \in X_1$, and this yields $f(y) = x$, for some $y \in X_1$. Then, f restricted to X_1 (as a domain) admits all elements of X_1 as values, which means that f restricted to X_1 is a permutation on X_1, and this is true for every $g \in \mathcal{F}$. We show that $f(x) \in X_2$, for every $x \in X_2$. Suppose $x \in X_2$ and $f(x) \in X_1$. Then $f^r(f(x)) \geq x$, by (1). Hence, for $g = f^r$, we have $g(y) \in X_2$, for some $y \in X_1$, which contradicts the fact that g restricted to X_1 is a permutation on X_1. Accordingly, every function $f \in \mathcal{F}$ can be represented as the join $f_1 \cup f_2$, where f_i is f restricted to X_i $(i = 1, 2)$, f_1 is a permutation on X_1, and f_2 maps X_2 into X_2. It follows that, for $x \in X_2$, $f^l(x) \in X_2$ and $f^r(x) \in X_2$. Consequently, the set of all functions f_2, for $f \in \mathcal{F}$, is a pregroup of functions on the poset $(X_2, \leq)$. By the induction hypothesis, every function f_2 is a permutation on X_2. Therefore, $f = f_1 \cup f_2$ is a permutation on X. $\square$

According to theorem 1, noncyclic pregroups must be infinite structures. The situation is different for the case of residuated monoids, witness powerset models over finite monoids. A simple example of a noncyclic pregroup is Lambek's structure consisting of all unbounded order-preserving mappings from the set $\mathbf{Z}$ of integers into itself. The unboundedness conditions are the following:

$$\forall x \; \exists y \; f(y) \leq x \quad \text{and} \quad \forall x \; \exists y \; x \leq f(y). \qquad (2)$$

These conditions are equivalent to the nonemptiness of sets $\{z : f(z) \leq x\}$ and $\{z : x \leq f(z)\}$, appearing in the equations from proposition 1. Hence, they are necessary for the existence of left and right adjoints. Consequently, the universe $\mathcal{F}$ of any pregroup of functions must consist of order-preserving functions, fulfilling (2). Further, the first set is upward bounded, and the second set is downward bounded, which also follows from (2). The properties of the ordering of integers mean that the first set contains the greatest element, and the second set the smallest element, which enables left and right adjoints to be defined by the equations from proposition 1; evidently, they are order-preserving and unbounded. Since unbounded order-preserving functions are closed under composition, and the identity function belongs to this class, then Lambek's structure is a pregroup of functions (apply proposition 1). It is noncyclic, since $f^l \neq f^r$, for $f(x) = 2x$. For $f^l(x) = \min\{z : x \leq 2z\} = [x + 1/2]$ and $f^r(x) = \max\{z : 2z \leq x\} = [x/2]$. The reader can easily show that $f^l \neq f^r$, for, say, $f(x) = 3x + 5$.

Other examples of noncyclic pregroups are not difficult to find. Consider the poset $(Z_1 \cup Z_2, \leq)$ such that Z_1, Z_2 are two disjoint copies of the set of integers, and the ordering is the join of natural orderings on Z_1 and Z_2. Again, the set of all order-preserving functions on this poset which fulfil (2) gives rise to a nocyclic pregroup. Observe that if f is order-preserving and fulfils (2), then it maps either Z_1 into Z_1 and Z_2 into Z_2, or Z_1 into Z_2 and Z_2 into Z_1. In a similar way, one can build noncyclic pregroups based on joins of more copies of the set of integers. Another construction leads to non-Archimedean models. Take $(Z_1 \cup Z_2, \leq)$ with Z_1, Z_2 as above, but $\leq$ being the linear ordering which extends natural orderings on Z_1 and Z_2 and puts all elements of Z_2 after those of Z_1. Now, order-preserving functions which map Z_1 into Z_1 and Z_2 into Z_2, and fulfil (2) both in Z_1 and Z_2, give rise to a noncyclic pregroup. All the examples considered above are uncountable. Countable noncyclic pregroups can be defined as certain substructures of the above structures; for instance, one can consider only those unbounded order-preserving functions which are definable in some first-order language.

All the pregroups mentioned above contain Lambek's structure or a countable substructure of Lambek's structure. This seems to be unavoidable, but a detailed analysis of this matter must be deferred to another paper. A linguist may notice (with disgust) that they all are pretty far from normal linguistic algebras. Actually, there is no chance of defining an interesting pregroup whose elements are languages. Let λ denote the empty string. The inequalities $L^l L \subseteq \{\lambda\} \subseteq LL^l$ imply $L = \{\lambda\}$ as the only solution. Although the algebra of languages cannot be supplied with adjoint operations, it is still possible to define pregroups on larger universes, containing languages and some extra-elements. Lambek [12] discusses a special construction of this kind, leading to a so-called free pregroup, which will be presented in the section below.

3 Free Pregroups and Lambek Calculus

In p.o. monoids, for any element a, there is at most one element b such that $ba \leq 1 \leq ab$, and there is at most one element c such that $ac \leq 1 \leq ca$. To show this, assume $ba \leq 1 \leq ab$ and $b"a \leq 1 \leq ab'$. Then, $b \leq bab' \leq b'$ and $b' \leq b'ab \leq b$, which yields $b = b'$, and the second part can be shown in a similar way. Consequently, for any element a of a pregroup, a^l and a^r are unique elements which fulfil (PRE) with a. Using this observation, it can be shown that the following conditions must hold in every pregroup:

$$1^l = 1^r = 1, \ a^{lr} = a^{rl} = a, \ (ab)^l = b^l a^l, \ (ab)^r = b^r a^r, \tag{3}$$

$$\text{if } a \leq b \text{ then } b^l \leq a^l \text{ and } b^r \leq a^r. \tag{4}$$

Given a poset $(P, \leq)$, a *free pregroup* over this poset is defined as follows. For an integer n and $a \in P$, the power a^n is defined by recursion: $a^0 = a$,

$a^{n-1} = (a^n)^l$, for $n \leq 0$, $a^{n+1} = (a^n)^r$, for $n \geq 0$. The elements of the free pregroup are formal sequences $a_1^{n_1} \dots a_k^{n_k}$ such that $k \geq 0$, $a_1, \dots, a_k \in P$ and $n_1, \dots, n_k \in \mathbf{Z}$ (for $k = 0$, the empty sequence identified with 1 is obtained). The operation $\cdot$ is the concatenation of sequences; the left adjoint of the above sequence is $a_k^{n_k-1} \dots a_1^{n_1-1}$, and the right adjoint is $a_k^{n_k+1} \dots a_1^{n_1+1}$ (see (3)). The ordering relation in the free pregroup is the reflexive and transitive closure of the following relation: $x \leq y$ iff either $x = za^n z'$, $y = zb^n z'$ and $a \leq b$ (resp. $b \leq a$) in P, if n is even (resp. odd), or $x = za^n a^{n+1} z'$, $y = zz'$, or else $x = zz'$, $y = za^n a^{n-1} z'$, for some sequences z, z' and some integer n. The transition from x to y is called *an induced step*, for the first case, *a contraction*, for the second case, and *an expansion*, for the third case. This structure is denoted by $F(P)$. It can easily be shown that $F(P)$ is a (noncyclic) pregroup.

From the examples in section 1, the reader can see that the parsing procedure based on pregroups consists of formal derivations of inequalities in a free pregroup over a poset whose elements are atomic types. *A derivation* of $x \leq y$ is a sequence $x_0, \dots, x_n$ $(n \geq 0)$ such that $x_0 = x$, $x_n = y$, and, for $i = 1, \dots, n$, x_i arises from x_{i-1} by an induced step, a contraction or an expansion. If P is finite, and the ordering on P is explicitly given, then this procedure can be treated as a formal calculus. Induced steps can be joined with contractions and expansions to form a single step. *A generalized contraction* is the transition from $x = za^n b^{n+1} z'$ to $y = zz'$ with $a \leq b$ (resp. $b \leq a$), if n is even (resp. odd). *A generalized expansion* is the transition from $x = zz'$ to $y = za^n b^{n-1} z'$ with $a \leq b$ (resp. $b \leq a$), if n is odd (resp. even). Lambek [12] proves the following lemma.

Lemma 1 *If $x \leq y$ in a free pregroup, then there exist sequences x', y' such that $x \leq x'$ can be derived by generalized contractions only, $x' \leq y'$ can be derived by induced steps only, and $y' \leq y$ can be derived by generalized expansions only.*

Proof. The proof is based on the possibility of changing the type of transitions. Let a generalized expansion of $x = zz'$ to $y = za^n b^{n-1} z'$ be followed by a generalized contraction of $y = uc^m d^{m+1} u'$ to $w = uu'$. We consider several cases: (1) if $c^m d^{m+1}$ occurs in z or z', then the two transitions can be performed in the reverse order; (2) if $d^{m+1} = a^n$, then $a = d$, $n = m + 1$, $z = uc^m$ and $u' = b^{n-1} z'$. We consider two subcases: (2.1) n is odd, so $a \leq b$ and $c \leq d$ in P. Hence $c \leq b$. The transition from $x = uc^m z'$ to $w = ub^{n-1} z'$ is a single induced step. (2.2) n is even, so $b \leq a$ and $d \leq c$ in P. Hence $b \leq c$, and again the latter transition is a single induced step. (3) If $c^m = b^{n-1}$, then we proceed in a similar way. Let an induced step from $x = za^n z'$ to $y = zb^n z'$ be followed by a generalized contraction of $y = uc^m d^{m+1} u'$ to $w = uu'$. Again, several cases are to be considered. (1) is analogous to (1) above. (2) If $d^{m+1} = b^n$, then $b = d$, $m + 1 = n$, $z = uc^m$ and $u' = z'$. (2.1)

n is even. Then, $a \leq b$ and $d \leq c$ in P. Hence $a \leq c$. The transition from $x = uc^m a^n z'$ to $w = uz'$ is a single generalized contraction. (2.2) n is odd. Then, $b \leq a$ and $c \leq d$ in P. Hence, $c \leq a$, and again the latter transition is a single generalized contraction. (3) If $c^m = b^n$, then, analogously, the transition from x to w can be treated as a generalized contraction. Finally, the situation of a generalized expansion followed by an induced step can be reduced to a single generalized expansion or to an induced step followed by a generalized expansion. Accordingly, each formal derivation of $x \leq y$ can be transformed into a derivation in the required form.   □

Pentus [14] showed that Lambek categorial grammars generate precisely the (λ-free) context-free languages. We prove an analogous theorem for grammars based on free pregroups.

A *pregroup grammar* is a quintuple $G = (V, P, \leq, I, s)$ such that V is a finite lexicon, $(P, \leq)$ is a finite poset, I is a mapping which to any $v \in V$ assigns a finite set of elements of $F(P)$, and $s \in S$. I is called *the lexical assignment* and s *the principal type* of G. We say that G *assigns* an element $x \in F(P)$ to a string $v_1 \ldots v_n$ with all v_i in V, if there are $x_i \in I(v_i)$, $i = 1, \ldots, n$, such that $x_1 \ldots x_n \leq x$ holds in $F(P)$. The set of all strings assigned s by G is called *the language of G* and denoted $L(G)$.

Theorem 2 *The languages of pregroup grammars are precisely the context-free languages.*

Proof. The fact that every pregroup grammar generates a context-free language easily follows from lemma 1. Let $E(G)$ denote the set of all elements a^n appearing in I. In derivations of inequalities $x_1 \ldots x_k \leq s$ in $F(P)$, for $x_i \in E(G)$, the only induced steps and generalized contractions that can be used are those whose exponents n have an absolute value no greater than the maximal absolute value of exponents appearing in $E(G)$. Such derivations can be simulated by derivations in a context-free grammar whose nonterminals are 1, whose elements are a^n with $a \in P$, whose n is restricted as above, and whose production rules are: $b^n \to a^n$ with $a \leq b$ (resp. $b \leq a$) if n is even (resp. odd), $1 \to a^n b^{n+1}$ with the same constraints, $1 \to 1a^n$ and $1 \to a^n 1$. Then, $L(G)$ is context-free.

We show that every context-free language can be generated by a pregroup grammar G with the ordering $\leq$ on P being the identity relation. We use the Gaifman theorem [2]: every context-free language can be generated by a classical categorial grammar whose lexical assignment involves, at most, types of the form p, p/q, $(p/q)/r$, with $p, q, r \in P$. *The order* of a type is a nonnegative integer defined by the following recursion: the order of $p \in P$ is 0, the order of x/y and $y\backslash x$ is maximum of the order of x and "the order of $y{+}1$". Thus, types in the Gaifman form are of order ≤ 1.

Buszkowski [7] showed that, for sequents $x_1 \ldots x_n \to s$, with types $x_1, \ldots, x_n$ of order ≤ 1, the Lambek calculus has the same power as the

Ajdukiewicz calculus (i.e. the calculus of classical categorial grammars). Accordingly, in the Gaifman theorem, "classical categorial grammar" can be replaced by "Lambek categorial grammar".

The translation of types into elements of $F(P)$ is given by: $t(p) = p$, for $p \in P$, $t(x/y) = t(x)t(y)^l$ (it is enough to consider types formed out of atomic types by means of $/$ only).

Lemma 2 *For types $x_1 \ldots x_n$ of order ≤ 1, the sequent $x_1 \ldots x_n \to s$ is derivable in the Ajdukiewicz calculus if, and only if, the inequality $t(x_1) \ldots t(x_n) \leq s$ holds in $F(P)$ with the identity ordering on P.*

Proof. The "only if" part follows from the fact that Ajdukiewicz reductions $(x/y)y \to x$ are translated into inequalities $t(x)t(y)^l t(y) \leq t(x)$, which hold in all pregroups. For the "if" part, if x is of order 1, then $t(x)$ must be of the form $pq_1^{-1} \ldots q_k^{-1}$ with $p, q_i \in P$. Since the ordering on P is identity, the only possible transitions in derivations of $t(x_1) \ldots t(x_n) \leq s$ in $F(P)$ are contractions, and these contractions reduce pairs $q^{-1}q$ to λ. The first contraction must involve elements $t(x_i)t(x_{i+1})$ such that $t(x_i) = zq^{-1}$, $t(x_{i+1}) = qq_1^{-1} \ldots q_k^{-1}$ and change them into $zq_1^{-1} \ldots q_k^{-1}$. Clearly, z must be nonempty, and $z = t(y)$, for some type y that is less complex than x_i. If $k = 0$, then this contraction is a translation of Ajdukiewicz reduction $(y/q)q \to y$. If $k > 0$, then it is a translation of the sequent:

$$(y/q)(\ldots ((q/q_1)/q_2)/ \ldots /q_k) \to (\ldots ((y/q_1)/q_2)/ \ldots /q_k),$$

which is derivable in Lambek calculus. Accordingly, by a straightforward induction on the complexity of $x_1 \ldots x_n \to s$, we prove that, if $t(x_1) \ldots t(x_n) \leq s$ is derivable in $F(P)$, then $x_1 \ldots x_n \to s$ is derivable in the Lambek calculus and, consequently, the latter sequent is derivable in the Ajdukiewicz calculus, due to the order restriction upon $x_1, \ldots, x_n$. This finishes the proof of lemma 2.   □

Lemma 2 shows that the pregroup grammar is the translation of the Graifman grammar and generates the same language. Consequently, every context-free language can be generated by a pregroup grammar.   □

For types of order greater than 1, Lambek calculus is essentially weaker than the calculus of free pregroups. Lambek [12] gives the example $(p \cdot q)/r \leq p \cdot (q/r)$; this holds in pregroups, since both sides are equal to pqr^l, but $(p \cdot q)/r \to p \cdot (q/r)$ is not derivable in Lambek calculus. For product-free types, we find the following example. In pregroups, $1 \leq pp^l = p \cdot (1/p) \leq p \cdot ((p/p)/p)$, and consequently, $p/(p \cdot ((p/p)/p)) = (p/((p/p)/p))/p \leq p/1 = p$, but the sequent $(p/((p/p)/p))/p \to p$ is not derivable in Lambek calculus.

Therefore, one cannot directly infer theorems about Lambek categorial grammars from analogous theorems about pregroup grammars. In particular, the Pentus theorem is not a consequence of theorem 2. On the other hand, methods for handling pregroup grammars seem to be simpler than those for Lambek categorial grammars. In particular, our proof of theorem 2 is simpler than that of the Pentus theorem, and similarly, the proof of lemma 1 is simpler than the proof of an analogous normalization lemma for nonassociative Lambek calculus, given in [6, 10]. Then, pregroup grammars seem to be an attractive alternative to categorial grammars for axiomatizing deductive parsing of type-theoretic style.

References

[1] K. Ajdukiewicz, Die syntaktische Konnexität, *Studia Philosophica*, 1 (1935), 1–27.

[2] Y. Bar-Hillel, C. Gaifman and E. Shamir, On categorial and phrase structure grammars, *Bulletin of the Research Council of Israel*, F9 (1960), 155–166.

[3] J. van Benthem, *Language in Action: Categories, Lambdas and Dynamic Logic*. North-Holland, Amsterdam, 1991.

[4] J. van Benthem and A. ter Meulen (eds.), *Handbook of Logic and Language*. Elsevier/MIT Press, Amsterdam, 1997.

[5] W. Buszkowski, Completeness results for Lambek syntactic calculus, *Zeitschrift für Mathematische Logik und Grundlagen der Mathematik*, 32 (1986), 13–28.

[6] W. Buszkowski, Generative capacity of nonassociative Lambek calculus, *Bulletin of the Polish Academy of Sciences. Mathematics*, 34 (1986), 507–516.

[7] W. Buszkowski, Mathematical linguistics and proof theory. In J. van Benthem and A. ter Meulen (eds.), *Handbook of Logic and Language*. Elsevier/MIT Press, Amsterdam, 1997, 683–736.

[8] W. Buszkowski, W. Marciszewski and J. van Benthem (eds.), *Categorial Grammar*. John Benjamins, Amsterdam, 1988.

[9] J.Y. Girard, Linear logic, *Theoretical Computer Science*, 50 (1987), 1–102.

[10] M. Kandulski, The equivalence of nonassociative Lambek categorial grammars and context-free grammars, *Zeitschrift für Mathematische Logik und Grundlagen der Mathematik*, 34 (1988), 41–52.

[11] J. Lambek, The mathematics of sentence structure, *American Mathematical Monthly*, 65 (1958), 154–170.

[12] J. Lambek, Type grammar revisited. In A. Lecomte, F. Lamarche and G. Perrier (eds.), *Logical Aspects of Computational Linguistics*. Springer, Berlin, 1999, 1–27.

[13] M. Moortgat, Categorial type logics. In J. van Benthem and A. ter Meulen (eds.), *Handbook of Logic and Language*. Elsevier/MIT Press, Amsterdam, 1997, 93–177.

[14] M. Pentus, Lambek Grammars are Context-Free, Prepublication Series Mathematical Logic and Theoretical Computer Science 8, Steklov Mathematical Institute, Moscow, 1992.

[15] M. Pentus, Lambek Calculus is L-Complete, ILLC Report, University of Amsterdam, 1993.

Homomorphic Characterizations of Linear and Algebraic Languages

Virgil E. Căzănescu

Faculty of Mathematics
University of Bucharest
Romania
vec@funinf.cs.unibuc.ro

Manfred Kudlek

Computer Science Department
University of Hamburg
Germany
kudlek@informatik.uni-hamburg.de

Abstract. In this paper we show that algebraic and linear languages defined by their corresponding systems of equations over an ω-complete semiring with a special property can be characterized by homomorphic images of intersections of a regular and a (linear) Dyck language (over the free monoid with catenation), respectively.

1   Systems of Equations

In this section we introduce the definitions of rational, linear and algebraic languages as the least fixed points of corresponding systems of equations.

Let $\mathcal{M}$ be a monoid with binary operation $\circ$ and unit element $\mathbf{1}$, or with a binary operation $\circ : \mathcal{M} \times \mathcal{M} \to \mathcal{P}(\mathcal{M})$ with unit element $\mathbf{1}$, i.e. $\mathbf{1} \circ \alpha = \alpha \circ \mathbf{1} = \{\alpha\}$.

Extend $\circ$ to an associative binary operation $\circ : \mathcal{P}(\mathcal{M}) \times \mathcal{P}(\mathcal{M}) \to \mathcal{P}(\mathcal{M})$, which is distributive with union $\cup$ ($A \circ (B \cup C) = (A \circ B) \cup (A \circ C)$ and $(A \cup B) \circ C = (A \circ B) \cup (B \circ C)$), with the unit element $\{\mathbf{1}\}$ ($\{\mathbf{1}\} \circ A = A \circ \{\mathbf{1}\} = A$), and the zero element $\emptyset$ ($\emptyset \circ A = A \circ \emptyset = \emptyset$).

Then $\mathcal{S} = (\mathcal{P}(\mathcal{M}), \cup, \circ, \emptyset, \{\mathbf{1}\})$ is an ω-complete semiring, i.e. if $A_i \subseteq A_{i+1}$ for $0 \le i$ then $B \circ \bigcup_{i \ge 0} A_i = \bigcup_{i \ge 0} (B \circ A_i)$ and $(\bigcup_{i \ge 0} A_i) \circ B = \bigcup_{i \ge 0} (A_i \circ B)$.

Define also $A^{(0)} = \{\mathbf{1}\}$, $A^{(1)} = A, A^{(k+1)} = A \circ A^{(k)}$, $A^\circ = \bigcup_{k \ge 0} A^{(k)}$.

Let $\mathcal{X} = \{X_1, \ldots, X_n\}$ be a set of variables such that $\mathcal{X} \cap \mathcal{M} = \emptyset$. A *monomial* over $\mathcal{S}$ with variables in $\mathcal{X}$ is a finite string of the form : $A_1 \circ A_2 \circ \ldots \circ A_k$, where $A_i \in \mathcal{X}$ or $A_i \subseteq \mathcal{M}, |A_i| < \infty, i = 1, \ldots, k$. Without loss of generality, $A_i = \{\alpha_i\}$ with $\alpha_i \in \mathcal{M}$ suffices. The α_i (or $\{\alpha_i\}$) will be called *constants*. A *polynomial* $p(\underline{X})$ over $\mathcal{S}$ is a finite union of monomials, where $\underline{X} = (X_1, \cdots, X_n)$.

In the following, the symbol $\prod$ will be used to denote finite products with operation $\circ$:

$$\prod_{i=1}^{m} A_i = A_1 \circ \cdots \circ A_m$$

and the symbol $\sum$ to denote finite unions:

$$\sum_{i=1}^{n} A_i = \bigcup_{i=1}^{n} A_i = A_1 \cup \cdots \cup A_n.$$

A *system of equations* over $\mathcal{S}$ is a finite set of equations $\mathcal{E} := \{X_i = p_i(\underline{X}) \mid i = 1, \ldots, n\}$, where $p_i(\underline{X})$ are polynomials. This will also be denoted by $\underline{X} = \underline{p}(\underline{X})$.

The *solution* of $\mathcal{E}$ is a n-tuple $\underline{L} = (L_1, \ldots, L_n)$ of sets over $\mathcal{M}$, with $L_i = p_i(L_1, \ldots, L_n)$ and the n-tuple is the least one with this property, i.e. if $\underline{L}' = (L_1', \ldots, L_n')$ is another n-tuple satisfying $\mathcal{E}$, then $\underline{L} \le \underline{L}'$ (where the order is defined componentwise with respect to inclusion, i.e. $\underline{A} = (A_1, \cdots, A_n) \le (B_1, \cdots, B_n) = \underline{B} \Leftrightarrow \forall_{i=1}^{n} : A_i \subseteq B_i$).

From the theory of semirings follows that any system of equations over $\mathcal{S}$ has a unique solution, and this is the least fixed point starting with:

$$\underline{X}^{(0)} = (X_1^{(0)}, \cdots, X_n^{(0)}) = (\emptyset, \cdots, \emptyset) = \underline{\emptyset}, \text{ and } \underline{X}^{t+1} = \underline{p}(\underline{X}^{(t)}).$$

Then the following fact holds: $\underline{X}^{(t)} \le \underline{X}^{(t+1)}$ for $0 \le t$.

This is seen by induction and the property of the polynomial with respect to inclusion, as $\underline{\emptyset} \le \underline{X}^{(1)}$ and $\underline{X}^{(t+1)} = \underline{p}(\underline{X}^{(t)}) \le \underline{p}(\underline{X}^{(t+1)}) = \underline{X}^{(t+2)}$. For the theory of semirings, see [5, 8].

A general system of equations is called *algebraic, linear* if all monomials are of the form $A \circ X \circ B$ or A, and *rational* if they are of the form $X \circ A$ or A, with $A \subseteq \mathcal{M}$ and $B \subseteq \mathcal{M}$. Corresponding families of languages (solutions of such systems of equations) are denoted by $\underline{ALG}(\circ)$, $\underline{LIN}(\circ)$, and $\underline{RAT}(\circ)$.

If $\circ$ is commutative, then all families are identical: $\underline{ALG(\circ)} = \underline{LIN(\circ)} = \underline{RAT(\circ)}$.

Note that the algebraic case corresponds to context-free languages if $\circ$ is normal catenation.

1.1   Grammars

Interpreting an equation $X_i = p_i(\underline{X})$ as a set of rewriting productions $X_i \to m_{ij}$, with $m_{ij} \in M(X_i)$, where $M(X_i)$ denotes the set of monomials of $p_i(\underline{X})$, *regular*, *linear*, and *context-free* grammars $G_i = (\mathcal{X}, \mathcal{C}, X_i, P)$ using the operation $\circ$, can be defined. Here $\mathcal{C}$ stands for the set of all constants in the system of equations, and P for all productions defined as above. As the productions are *context-free (terminal)*, derivation trees can also be defined. Note that the interior nodes are labelled by variables, and the leafs by constants from $\mathcal{C}$.

2   Normal Forms

In the following lemma, forests of terminal trees are constructed representing approximations of the least fixed point, and it is shown that the sets of terminal derivation trees with respect to $\circ$ are equivalent.

Lemma 1 (Approximation of the least fixed point)

Terminal trees for the approximation of the least fixed point and terminal derivation trees are equivalent.

Proof. $\underline{X}^{(0)} - \underline{\emptyset}$, $\underline{X}^{(t+1)} = \underline{p}(\underline{X}^{(t)})$. Thus:

$$X_i^{(t+1)} = \sum_j \prod_k X_{ijk}^{(t)} + \sum_j \{\alpha_{ij}\},$$

especially:

$$X_i^{(0)} = \emptyset , \ X_i^{(1)} = \sum_j \{\alpha_{ij}\}.$$

Construct forests $\mathcal{T}$ of terminal trees as follows:

- $\mathcal{T}^{(1)}$ consists of all trees with roots $X_i^{(1)}$ and children (only leaves) $\{\alpha_{ij}\}$ $(1 \leq i \leq n)$.

- $\mathcal{T}^{(t+1)}$ is constructed, from trees in $\mathcal{T}^{(1)}$, as the set of trees with roots $X_i^{(t+1)}$ and their children which are either $X_{ijk}^{(t)}$, roots of trees from $\mathcal{T}^{(t)}$, or $\{\alpha_{ij}\}$.

Thus, the set of frontiers of leaves of all trees in $\mathcal{T}^{(t)}$ with root $X_i^{(t)}$ is just the approximation $X_i^{(t)}$.

On the other hand, any terminal derivation tree for X_i is contained in $\mathcal{T}$. For this, interpret a deepest non-terminal vertex (i.e. with greatest distance from the root) as $X_j^{(1)}$ for some j, and the root as $X_i^{(t+1)}$ for some i. Then all non-terminal vertices get some step number s with $1 \leq s \leq t+1$.   □

Lemma 2 *Any linear system of equations can be transformed, with additional variables, into another where all monomials are of the form $X \circ \alpha$, $\alpha \circ X$, or α, and the new system has identical minimal solutions in the old variables.*

Proof. Consider any monomial $\alpha \circ X \circ \beta$. Replace it by $\alpha \circ Y$, and add a new equation $Y = X \circ \beta$. Then it is obvious that the new system has identical solutions in the old variables.   □

Below it will be shown that any algebraic system of equations can be transformed, with additional variables, into a system of equations where all monomials have the form $X \circ Y$ or $\{\alpha\}$, and the new system has identical minimal solutions for the old variables. To prove this, some lemmata have to be shown first. For that, the ω-complete semiring has to have the following:

Property

Let $\mathcal{S} = (\mathcal{P}(\mathcal{M}), \cup, \circ, \emptyset, \{1\})$ be an ω-complete semiring, where $\mathcal{M}$ is a monoid. $\mathcal{S}$ has property $(\otimes)$ if:

$(\otimes)$ $1 \in A \circ B \Leftrightarrow (1 \in A \wedge 1 \in B)$.

This property is some kind of *nondivisibility* of the unit.

The following example does not have this property.

Example 1 *Consider the operation $\circ$ defined on $\mathcal{P}(\Sigma^*)$ by:*
$\{\lambda\} \circ \{u\} = \{u\} \circ \{\lambda\} = \{u\}$ *for* $u \in \Sigma^*$,
$\{u\} \circ \{v\} = \{\lambda, u, v, uv\}$ *for* $u, v \in \Sigma^+$.
As usual:

$$A \circ B = \bigcup_{u \in A, v \in B} \{u\} \circ \{v\},$$

where $\circ$ is an associative operation.

$(\{\lambda\} \circ \{u\}) \circ \{v\} = \{u\} \circ \{v\} = \{\lambda, u, v, uv\}$
$\{\lambda\} \circ (\{u\} \circ \{v\}) = \{\lambda\} \circ \{\lambda, u, v, uv\} = \{\lambda, u, v, uv\}$
$(\{u\} \circ \{\lambda\}) \circ \{v\} = \{u\} \circ \{v\} = \{\lambda, u, v, uv\}$
$\{u\} \circ (\{\lambda\} \circ \{v\}) = \{u\} \circ \{v\} = \{\lambda, u, v, uv\}$
$(\{u\} \circ \{v\}) \circ \{\lambda\} = \{\lambda, u, v, uv\} \circ \{\lambda\} = \{\lambda, u, v, uv\}$
$\{u\} \circ (\{v\} \circ \{\lambda\}) = \{u\} \circ \{v\} = \{\lambda, u, v, uv\}$
$(\{u\} \circ \{v\}) \circ \{w\} = \{\lambda, u, v, uv\} \circ \{w\} = \{w, \lambda, u, uw, v, vw, uv, uvw\}$

$$\{u\} \circ (\{v\} \circ \{w\}) = \{u\} \circ \{\lambda, v, w, vw\} = \{u, \lambda, v, uv, w, uw, vw, uvw\}$$

Thus, $\mathcal{S} = (\mathcal{P}(\Sigma^*), \emptyset, \{\lambda\}, \cup, \circ)$ is an ω-complete semiring.

But the example does not have the property ($\otimes$) since $\lambda \in \{a\} \circ \{b\} = \{\lambda, a, b, ab\}$.

Lemma 3 *If ($\otimes$) holds then:*

$$1 \in \prod_{i=1}^{k} A_i \Leftrightarrow \forall_{i=1}^{k} : 1 \in A_i.$$

Proof. $\Leftarrow$ is trivial.

$\Rightarrow$: $\forall_{i=1}^{k} : 1 \in A_i$ implies $1 \in A_1 \wedge \forall_{i=2}^{k} : 1 \in A_i$ by property ($\otimes$), and then induction. $\qquad \square$

Let $\mathcal{X} = \{X_1, \cdots, X_n\}$ be a set of variables. To each variable $X \in \mathcal{X}$ in an algebraic system of equations there exists a set of monomials $M(X)$ such that $X = \sum_{m \in M(X)} m$.

Lemma 4 *(Separation of variables and constants)*

For any algebraic system of equations there exists another, possibly with additional variables, with the same (partial) solution in the original variables, and for which the following property holds:

If $X_i = \prod_{j=1}^{r(i)} m_{ij}$ then each monomial is either of the form $\prod_{k=1}^{s(ij)} X_{ijk}$ or $\{\alpha_{ij}\}$ (a constant).

Proof. If m_{ij} is not of that form and not a constant, then $m_{ij} = \prod_{k=1}^{s(ij)} A_{ijk}$ with A_{ijk} either a variable or a constant β_{ijk}. Replace each constant β_{ijk} in it by a new variable Y_{ijk}, and add a new equation $Y_{ijk} = \{\beta_{ijk}\}$.

Trivially, the new system of equations has the same solution in the original variables. $\qquad \square$

Lemma 5 *(Removal of $\{1\}$)*

To each algebraic system of equations there exists another with the same set of variables such that no monomial has the form $\{1\}$ and the solutions are $L_i - \{1\}$ if L_i are the solutions of the old system.

Proof. Let $\mathcal{Y}$ be a set of variables and $\mathcal{F}(\mathcal{Y})$ the set of all (formal) terms on $\mathcal{Y}$ with operation $\circ$.

Define inductively:

$$\mathcal{Y}_1 = \{X \in \mathcal{X} \mid \{1\} \in M(X)\}, \quad \mathcal{Y}_{i+1} = \mathcal{Y}_i \cup \{X \in \mathcal{X} \mid \exists m \in \mathcal{F}(\mathcal{Y}_i) : m \in M(X)\}.$$

Note that all monomials m consist only of variables.

Trivially $\mathcal{Y}_i \subseteq \mathcal{Y}_{i+1}$, and therefore there exists a k with $\mathcal{Y}_k = \mathcal{Y}_{k+j} = \mathcal{Y}$ for all $0 \leq j$ since $\mathcal{X}$ is finite.

Let L_k be the solution for X_k. Then the following fact holds:
$\mathbf{1} \in L_k \Leftrightarrow X_k \in \mathcal{Y}$.

$\Leftarrow$) If $X_k \in \mathcal{Y}$, then $\mathbf{1} \in L_k$ is seen by induction. Trivially, if $X_k \in \mathcal{Y}_1$, then $\mathbf{1} \in M(X)$ and therefore $\mathbf{1} \in L_k$. Assume $\mathbf{1} \in L_k$ for all $X_k \in \mathcal{Y}_j$ for $1 \leq j$. If $X_k \in \mathcal{Y}_{j+1}$ then by definition there exists a monomial $m \in \mathcal{F}(\mathcal{Y}_j)$ such that $m \in M(X)$. Therefore, $\mathbf{1} \in L_k$.

$\Rightarrow$) If $\mathbf{1} \in L_k$, then $\{\mathbf{1}\} \in X_k^{(t)}$ for some $t \geq 1$. We prove by induction on t that $\mathbf{1} \in X_i^{(t)}$ implies $X_i \in \mathcal{Y}$.

If $t = 1$, then $\mathbf{1} \in M(X_i)$ and therefore $X_i \in \mathcal{Y}_1 \subseteq \mathcal{Y}$.

Let $t > 1$. If $\mathbf{1} \in M(X_i)$ then again $X_i \in \mathcal{Y}_1 \subseteq \mathcal{Y}$.

If $\mathbf{1} \notin M(X_i)$, then $\mathbf{1} \in Y_1^{(t-1)} \circ \cdots \circ Y_r^{(t-1)} = m_1 \in M(X_i)$. The property $(\otimes)$ implies $\mathbf{1} \in Y_j^{(t-1)}$ for $1 \leq j \leq r$. By induction $Y_j \in \mathcal{Y}$ for all $1 \leq j \leq r$ Therefore, there is a u such that $m_1 \in \mathcal{F}(\mathcal{Y}_u)$. We deduce $X_i \in \mathcal{Y}_{u+1}$. Hence, $X_i \in \mathcal{Y}$.

Now construct a new system of equations $\mathcal{E}'$ in which in all monomials m_{ij} 0 or more variables $Y_j \in \mathcal{Y}$ are deleted such that the new monomials $m'_{ij} \neq \{\mathbf{1}\}$.

Then the system $\mathcal{E}'$ has the solution $L_i - \{\mathbf{1}\}$. $\square$

Lemma 6 *For each algebraic system of equations there exists another with additional variables X'_i for each old X_i such that the monomials in $p'_i(X, X')$ are either of the form $\{\mathbf{1}\}$ or do not contain X'_j. The solutions of the new system for the new variables X'_i are $L'_i = L_i$.*

Proof. By Lemma 5 let $\mathcal{E}'$ be a system of equations with $L'_i = L_i - \{\mathbf{1}\}$.

Construct a new system $\mathcal{E}''$ in which for each variable X_i a new one X'_i is defined. Let $p_i(X, X') = p_i(\underline{X})$ for X_i and define $p'_i(X, X') = \{\mathbf{1}\} + p_i(\underline{X})$ if $\mathbf{1} \in L_i$, and for $\mathbf{1} \notin L_i$ $p'_i(X, X') = p_i(\underline{X})$. Then the solutions for the new variables are $L'_i = L_i$. $\square$

Lemma 7 *(Removal of monomials of the form Y)*

For each algebraic system of equations there exists another one with the same variables such that no monomial is of the form Y and the solutions is identical to the old one.

Proof. Assume that the system already has the form of lemmata 4, 5, and 6.

Construct inductively sets of variables for $X \in \mathcal{X}$:

$\mathcal{Y}_1(X) = \{X\}$,
$\mathcal{Y}_{j+1}(X) = \mathcal{Y}_j(X) \cup \{Y \in \mathcal{X} \mid \exists Z \in \mathcal{Y}_j(X) : Y \in M(Z)\}$.

Since $\mathcal{X}$ is finite, there exists a k with $\mathcal{Y}_k(X) = \mathcal{Y}_{k+j}(X) = \mathcal{Y}(X)$ for $j \geq 0$.

If $L(X)$, $L(Y)$ are the solutions for X, Y, respectively, then obviously the following fact holds: $Y \subseteq L(X) \Leftrightarrow Y \in \mathcal{Y}(X)$.

Now construct the new system by taking all monomials which are constants and consider all monomials $m = Y_1 \circ \cdots \circ Y_k \in M(X)$ with $k \geq 2$. Construct new monomials $m' = Z_i \circ \cdots \circ Z_k \in M(Y)$ with $X \in \mathcal{Y}(Y)$ and $Z_j \in \mathcal{Y}(Y_j)$.

Then $L'_i = L_i$. □

Lemma 8 (Normal form)

For each algebraic system there exists another one with additional variables such that all monomials have only the forms $1 \in M(X)$ (then no other monomial contains X), or $Y \circ Z$, or $\{\alpha\}$ with $\alpha \neq 1$ and the solutions for the old variables are identical.

Proof. Assume that the system of equations has the form of the previous lemmata.

Consider an arbitrary monomial $m = Y_1 \circ \cdots \circ Y_k \in M(X)$ with $k \geq 2$. Replace it by $Y_1 \circ Z_1 \in M(X)$ and the new equations $Z_1 = Y_2 \circ Z_2, \cdots, Z_{k-2} = Y_{k-1} \circ Y_k$.

Then the new system of equations obviously has the same solutions in the old variables. □

3   Results

Definition 1 *Consider two ω-complete semirings $S_i = (\mathcal{P}(\mathcal{M}_i), \cup, \circ_i, \emptyset, \{1_i\})$, $i - 1, 2$, where $\mathcal{M}_i$ are monoids with operations $\circ_i$ and unit elements 1_i. A homomorphism $h : S_1 \rightarrow S_2$ is a mapping with $h(A \circ_1 B) = h(A) \circ_2 h(B)$, $h(A \cup B) = h(A) \cup h(B)$, $h(\emptyset) = \emptyset$ and $h(\{1_1\}) = \{1_2\}$.*

Definition 2 *Let Σ be an alphabet and $\Gamma = \Sigma \cup \Sigma'$. Then the linear Dyck language over Γ is defined by $D_{\mathcal{L}} = \{w(w')^{rev} \mid w \in \Sigma^*\}$, where w^{rev} is the mirror image of w.*

Theorem 1 *Each linear set L (over a semiring $S = (\mathcal{P}(\mathcal{M}), \cup, \circ, \emptyset, \{1\})$) can be characterized by $L = h(D_{\mathcal{L}} \cap R)$ where $D_{\mathcal{L}}$ is a linear Dyck language and R a regular set, both with respect to catenation, i.e. the semiring $S_0 = (\mathcal{P}(\Sigma^*), \cup, \cdot, \emptyset, \{\lambda\})$, and a homomorphism $h : S_0 \rightarrow S$.*

Proof. a) The class of linear languages is closed under intersection with regular sets. Construct a linear grammar in normal form for $L_0 = D_{\mathcal{L}} \cap R$. From this a linear system $\mathcal{E}_0$ of equations in normal form can be constructed easily, one of the components of the solution $\underline{X}_0$ being L_0.

From this, a corresponding linear system $\mathcal{E}$ in normal form can be constructed, replacing catenation · by ∘ and every constant a by $\{h(a)\}$. Then the solution of $\mathcal{E}$ is just $\underline{X} = h(\underline{X_0})$.

Thus, $h(D_{\mathcal{L}} \cap R)$ is a linear set.

b) Let the linear set L be a component of the solution of a linear system $\mathcal{E}$ of equations. It can be assumed that this system is in normal form, i.e. the monomials have the forms $\{\alpha\} \circ Y$, $Y \circ \{\alpha\}$, $\{\alpha\}$, or $\{\mathbf{1}\}$.

Let $(\{\alpha_1\}, \cdots, \{\alpha_k\})$ enumerate all occurrences of constants in the linear system of equations. Note that some α_i, α_j may be identical.

Now define the alphabet $\Sigma = \{a_1, \cdots, a_k\}$, and the alphabet $\Gamma = \Sigma \cup \Sigma' \cup \bar{\Sigma} \cup \bar{\Sigma}' \cup \{d, d'\}$, where $\Sigma' = \{a' \mid a \in \Sigma\}$, $\bar{\Sigma} = \{\bar{a} \mid a \in \Sigma\}$, $\bar{\Sigma}' = \{\bar{a}' \mid a \in \Sigma\}$, and $d \notin \Sigma$.

The linear grammar $G_\Gamma = (\{\sigma\}, \Gamma, \sigma, \{\sigma \to \lambda\} \cup \{\sigma \to x\sigma x' \mid x \in \Sigma \cup \bar{\Sigma} \cup \{d\}\}$ generates the linear Dyck language $D_{\mathcal{L}}(\Gamma) \subseteq \Gamma^*$.

Now define a regular grammar $G_R = (V_N \cup \{X\}, \Gamma, S, P_R)$, where $X \notin V$, and:

$$P_R = \{A \to a_i B \mid \{\alpha_i\} \circ B \in M(A)\} \cup \{A \to \bar{a}B \mid B \circ \{\alpha_i\} \in M(A)\}$$
$$\cup \{A \to a_i dd' a_i' X \mid \{\alpha_i\} \in M(A)\} \cup \{A \to dd' X \mid \{\mathbf{1}\} \in M(A)\}$$
$$\cup \{X \to xX \mid x \in \Sigma' \cup \bar{\Sigma}'\} \cup \{X \to \lambda\}.$$

Define the homomorphism $h : \Gamma^* \to \mathcal{P}(\mathcal{M})$ by $h(a_i) = h(\bar{a}_i') = \{\alpha_i\}$ for $1 \le i \le k$, and $h(x) = \{\mathbf{1}\}$ for $x \in \bar{\Sigma} \cup \Sigma' \cup \{d, d'\}$.

If $R = L(G_R)$, then $L = h(D_{\mathcal{L}}(\Gamma) \cap R)$. $\square$

The following theorem was proved, using another method, before 1983. Unfortunately, it has only been published in Romanian [4].

Theorem 2 *Each algebraic set L (over a semiring $\mathcal{S} = (\mathcal{P}(\mathcal{M}), \cup, \circ, \emptyset, \{\mathbf{1}\})$) can be characterized by $L = h(D \cap R)$, where D is a Dyck language and R a regular set, both with respect to catenation, i.e. the semiring $\mathcal{S}_0 = (\mathcal{P}(\Sigma^*), \cup, \cdot, \emptyset, \{\lambda\})$, and a homomorphism $h : \mathcal{S}_0 \to \mathcal{S}$.*

Proof. The proof is a modification of the proof for context-free languages in [9].

Let L be a component of an algebraic system of equations in normal form. If $\mathbf{1} \in L$ and if $L - \{\mathbf{1}\} = h(D \cap R)$, then $L = h(D \cap (R \cup \{\lambda\}))$ since $h(\{\lambda\}) = \{\mathbf{1}\}$.

Enumerate all monomials that are constants $\{\alpha_j\}$, i.e. $\{\alpha_j\} \subseteq X_i$ for some i, by $1 \le j \le r$. Note that some of them may be identical. Similarly, enumerate all monomials $Y_j \circ Z_j$ occurring in equations, i.e. $Y_k \circ Z_k \subseteq X_i$ for some i, by $1 \le k \le s$. Again, some of them may be identical.

Let D_{r+s} be the Dyck language over the alphabet:

$$\Sigma_{r+s} = \{u_1, \cdots, a_{r+s}, a_1', \cdots, a_{r+s}'\}$$

Define the homomorphism $h : \Sigma_{r+s}^* \to \mathcal{M}$ by:

$h(a_j) = \alpha_j, 1 \leq j \leq r,$
$h(a_j) = \mathbf{1}, r+1 \leq j \leq r+s,$
$h(a'_j) = \mathbf{1}, 1 \leq j \leq r+s,$

and extend it to $\mathcal{S}$. Consider the grammars $G_X = (V, V_T, X, P)$ with $V = \mathcal{X} \cup \Sigma_{r+s}$ and $V_T = \Sigma_{r+s}$ and:

$$P = \{X \to a_j a'_j \mid \{\alpha_j\} \in M(X)\}$$
$$\cup \{X \to a_j a'_j a'_{r+k} Z_k \mid \{\alpha_j\} \in M(X), 1 \leq k \leq s\}$$
$$\cup \{X \to a_{r+k} Y_k \mid Y_k \circ Z_k \in M(X)\}.$$

By Lemma 1, *context-free* derivations can be used.

$\Rightarrow$) Let $X = W_0 \to W_1 \to \cdots \to W_t = W$ with $X \in \mathcal{X}$ and $W_i \in \mathcal{A}(\mathcal{M})$ be a derivation, where $\mathcal{A}(\mathcal{M})$ denotes all algebraic expressions with variables from $\mathcal{X}$ and constants from $\mathcal{C}$. Then there exists $w \in D_{r+s}$ such that $X \overset{*}{\to} w$ and $W = h(w)$. This is proved by induction.

$t = 1$. Then $W_1 = \{\alpha_i\} \in \mathcal{C}$, thus $X \to \{\alpha_i\}$, and therefore $X \to a_i a'_i nP$. Choose $w = a_i a'_i$ which yields $h(w) = \{\alpha_i\} \circ \{\mathbf{1}\} = \{\alpha_i\}$.

Assume the assumption for all $p \leq t$ $(t \geq 1)$. Consider $X = W_1 \to W_2 \to \cdots W_t \to W_{t+1}$. Since the productions are in normal form, there exists an i $(1 \leq i \leq s)$ with $X_i \to Y_i \circ Z_i$. Since the system is algebraic (*context-free*), there exist W, W' with $Y_i \overset{*}{\to} W$ and $Z_i \overset{*}{\to} W'$, derivations of lengths $\leq t$, and $W_{t+1} = W \circ W'$. By the induction hypothesis there exist $w, w' \in D_{r+s}$ with $W = h(w), W' = h(w')$, and $Y_i \overset{*}{\to} w, Z_i \overset{*}{\to} w'$.

Since $X_i \to Y_i \circ Z_i$ is a production, it follows that $X_i \to a_{r+i} Y_i \in P$. Hence:

$$X_i \to a_{r+i} Y_i \overset{*}{\to} a_{r+i} w a'_{r+i} Z_i \to a_{r+i} w a'_{r+i} w'$$

(where in the last step a production $X \to a_j a'_j a'_{r+i} Z_i$ instead of $X \to a_j a'_j$ is used).

Choose $w'' = a_{r+i} w a'_{r+i} w'$ yielding:
$$h(w'') = h(a_{r+i}) \circ h(w) \circ h(a'_{r+i}) \circ h(w') = \{\mathbf{1}\} \circ h(w) \circ \{\mathbf{1}\} \circ h(w')$$
$$= h(w) \circ h(w') = W \circ W' = W_{t+1}.$$

$\Leftarrow$)

Let $X = w_1 \to w_2 \to \cdots w_t = w$ with $X \in \mathcal{X}$, $w \in D_{r+s}$ be a derivation, and $X \overset{*}{\to} h(w)$.

$t = 1$. Then $X \to w_1 = a_j a'_j$. Hence $X \to \{\alpha_j\}$ and $\{\alpha_j\} = h(a_j a'_j)$.

Assume the assumption for $p \leq t$ and consider $X \to w_1 \to w_2 \to \cdots \to w_{t+1} = w$ with $w \in D_{s+t}$. Since $t + 1 \geq 2$, the first production $X \to w_1$ is of the form $X \to a_j a'_j$ and not of the form $X \to a_j a'_j a'_{r+i} Z_i$ (since otherwise $w = a_j a'_j a' r + i w'' \in D_{r+s}$, a contradiction). Therefore, $X = X_i$, $w_1 = a_{r+i} Y_i$ and $X \to Y_i \circ Z_i$. This implies $Y_i \to u_2 \to \cdots u_{t+1}$ giving $X \overset{*}{\to} a_{r+i} u_{t+1}$.

Since $a_{r+i} u_{t+1} \in D_{r+s}$, there exist $v, v' \in D_{r+s}$ with $w = a_{r+i} v a'_{r+i} v'$. Therefore, there exists some $k \geq 2$ with $w_k = v a'_{r+i} Z_i$, and the production used for $w_{k-1} \to w_k$ is $Y \to a_j a'_j a'_{r+i} Z_i$. Thus, $Y \to \{\alpha_j\}$ and therefore $Y \to a_j a'_j \in P$. Replacing in step $k-1$ the production $Y \xrightarrow{*} a_j a'_j a'_{r+i} Z_i$ with $Y \to a_j a'_j$ gives $Y_i \xrightarrow{*} v$ of length $\leq t$. Obviously, also $Z_i \xrightarrow{*} v'$ (starting at $w_k = v a'_{r+i} Z_i$) is of length $\leq t$.

By the induction hypothesis, $Y_i \xrightarrow{*} h(v)$, $Z_i \xrightarrow{*} h(v')$.

Hence, $X \to Y_i \circ Z_i \xrightarrow{*} h(v) \circ h(v') = h(a_{r+i} v a'_{r+i} v')$.   □

References

[1] A. Baranga, Z-continuous posets, *Discrete Mathematics*, 152 (1996), 33–45.

[2] A. Baranga, The contraction principle as a particular case of Kleene's fixed point theorem, *Discrete Mathematics*, 98 (1991), 75–79.

[3] V.E. Căzănescu, Parties algébriques d'un monoïde, *Revue Roumaine des Mathématiques Pures et Appliquées*, XXIII (1978).

[4] V.E. Căzănescu, *Introducere în Teoria Limbajelor Formale*. Editura Academiei RSR, Bucureşti, 1983.

[5] J.S. Golan, *The Theory of Semirings with Application in Mathematics and Theoretical Computer Science*. Longman Scientific and Technical, London, 1992.

[6] M. Kudlek, Generalized iteration lemmata, *PU.M.A.*, 6.2 (1995), 211–216.

[7] M. Kudlek, Iteration lemmata for certain classes of word, trace and graph languages, *Fundamenta Informaticae*, 34 (1999), 249–264.

[8] W. Kuich and A. Salomaa, *Semirings, Automata, Languages*. Springer, Berlin, 1986.

[9] A. Salomaa, *Formal Languages*. Academic Press, New York, 1973.

Using Alternating Words to Describe Symbolic Pictures

Gennaro Costagliola

Vincenzo Deufemia

Filomena Ferrucci

Carmine Gravino

Marianna Salurso

Department of Mathematics and Computer Science
University of Salerno
Baronissi, Italy
{gencos,filfer}@unisa.it

Abstract. In this chapter we present the concepts of *drawn symbolic picture* and *symbolic picture*. We provide an elegant string description for such models consisting of *alternating words*, i.e. strings whose letters are in alternation from an alphabet of symbols and an alphabet of directions. We characterize the grammars for alternating words and define the generative model for (drawn) symbolic pictures.

1   Introduction

In recent years, there has been considerable interest in studying rewriting systems for the generation and description of *pictures* (e.g. [4, 5, 7, 12, 13]). Each of these studies considers a particular data structure for representing the pictures (a string, an array, a tree, a plex, a graph). Pioneering work in suggesting and applying string languages was done by Freeman [6], who introduced the notion of 'chain code'. One of the most elegant and successful chain code methods for working with line-drawing was introduced in [13], where a drawn picture is defined as a connected set of unit lines from the Cartesian plane. A picture description is a word over $\Pi = \{r, l, u, d\}$ describing the picture by the graphical representation of the movement sequence *right*, *left*, *up*, and *down*. Languages of chain-encoded pictures have been intensively studied also from a computational point of view (see for example [10, 9, 11, 13, 14, 15]).

In this chapter, we present two types of symbolic pictures: *drawn symbolic pictures* and *symbolic pictures*. Informally, a symbolic picture is given by a set of symbols disposed on a two-dimensional plane, while a drawn symbolic picture is a symbolic picture where pairs of symbols on the plane are connected by lines to form a connected graph. A (drawn) symbolic picture may be obtained by evaluating a string of alternating symbols from an alphabet Σ, with symbols from an alphabet indicating the moves *right*, *left*, *down* and *up*. This definition of (drawn) symbolic pictures extends both the concept of the traditional string and that of picture as defined in [13]. In order to define a (drawn) symbolic picture language we define a (drawn) symbolic picture grammar as a pair that consists of a grammar for alternating words and an evaluation function. When the evaluation function is applied to an alternating word, it generates the two-dimensional layout corresponding to a (drawn) symbolic picture. A canonical form for grammars for alternating words is also provided.

2 Preliminaries: Drawn Pictures

We assume that the reader is familiar with the basic definitions of string languages as can be found for example in [8]. Here we simply recall the basic notions about picture languages from [13]. Let A be a finite set of symbols called an alphabet. A^* denotes the monoid generated by A with the operation of concatenation (the neutral element is the empty string ϵ). As usual, **Z** denotes the set of integers. The *universal point set*, denoted by M_0, is the Cartesian product of **Z** with itself. For each point $v = (m, n) \in M_0$, the *up-neighbor* of v, denoted by $u(v)$, is the point $(m, n + 1)$, the *down-neighbor* of v, denoted by $d(v)$, is the point $(m, n - 1)$, the *left-neighbor* of v, denoted by $l(v)$, is the point $(m - 1, n)$, and the *right-neighbor* of v, denoted by $r(v)$, is the point $(m + 1, n)$. The *neighborhood* of v is defined as $N(v) = \{u(v), d(v), l(v), r(v)\}$. The *universal line set* M_1 is defined as the set of lines of length 1 and ends in M_0. Formally, $M_1 = \{ \{v, v'\} \mid v, v' \in M_0$ and $v' \in N(v)\}$.

A *drawn picture* q is a triple $q = \langle b, s, e \rangle$, where b is a connected finite subset of M_1, and the points $s = (0,0)$ and e are called *start* and *end* point of q, respectively. If b is nonempty, then s and e are points in $W(q) = \{v \in M_0 \mid$

$\{v, v'\}$ is in p, for some $v \in M_0\}$. If b is empty, then $s = e = (0,0)$, and $W(q) = \{(0,0)\}$. Thus, the empty drawn picture is denoted by $<\emptyset, (0,0), (0,0)>$. As an example, let us consider the drawn picture $q = <\{\{(0,0), (1,0)\}, \{(1,0), (1,1)\}, \{(1,1), (2,1)\}, \{(2,1), (2,0)\}, \{(2,0), (3,0)\}\}, (0,0), (1,1)>$ which is depicted in Fig. 1, where the start and end points are marked with a circle and a square, respectively.

Since any point v in the plane has four neighbors (namely, $u(v)$, $d(v)$, $l(v)$, and $r(v)$), a natural way to describe a drawn picture $q = <b, s, e>$ is to describe a walk through the picture. Such a walk starts from the start point s, touches at least once each line in b, and ends at the end point e. Each move in the walk from a point v to its neighbor v' is represented in a string by a single symbol: 'u' if $v' = u(v)$, 'd' if $v' = d(v)$, 'l' if $v' = l(v)$, 'r' if $v' = r(v)$. Thus, a walk is described by a string on the alphabet $\Pi = \{u, d, l, r\}$. For example, the string $w = \textbf{rurdrlul}$ describes the drawn picture of Fig. 1. The empty drawn picture $<\emptyset, (0,0), (0,0)>$ is described by the empty string ϵ.

The drawn picture described by a string w over Π, denoted by $dpic(w)$, is defined inductively as follows:

- if $w = \epsilon$, then $dpic(w) = <\emptyset, (0,0), (0,0)>$;

- if $w = z\pi$ for some $z \in \Pi^*$ and $\pi \in \Pi$, with $dpic(z) = <b, s, e>$, then $dpic(w) = <b \cup \{e, \pi(e)\}, s, \pi(e)>$.

Every string over Π is called a *picture description*, and every language over Π is called a *picture description language*.

3 Drawn Symbolic Pictures and Symbolic Pictures

In this section, we formalize the notions of *drawn symbolic pictures* and *symbolic pictures* as an extension of the drawn pictures. The extension is based on the following observation. A drawn picture is a set of points which are connected by unit lines. If we associate a symbol from an alphabet Σ to each point of the picture, then we have a drawn symbolic picture. In turn, a symbolic picture can be considered as a drawn symbolic picture where the unit lines which represent the spatial relations 'up', 'down', 'left' and 'right' are not visualized. So a symbolic picture is defined as a set of symbols associated to positions on the Cartesian plane.

Figure 1: A drawn picture.

In order to define a drawn symbolic picture we need the notion of invisible symbol ϕ as a special symbol that has no visual representation when used in a picture. In the sequel, we indicate with Σ an arbitrary alphabet which does not contain ϕ, with Σ_ϕ an alphabet which contains ϕ, and with δ_ϕ a function which may contain ϕ in its range. Thus, the definition of drawn symbolic picture can be formalized as follows:

Definition 1 *A* drawn symbolic picture *is a triple* dsp $= <$dp$, \Sigma_\phi, \delta_\phi >$, *where* dp *is a drawn picture,* Σ_ϕ *is an alphabet of symbols, and* δ_ϕ *is a function* δ_ϕ: $W(dp) \to \Sigma_\phi$. *The start and end points of* dsp *are the start and end points of* dp, *respectively. An empty drawn symbolic picture is denoted by* $<< \emptyset, (0,0), (0,0) >, \Sigma_\phi, \eta >$, *where* η *is an undefined function on (0,0). A* drawn symbolic picture language *is a set of drawn symbolic pictures.*

Example 1 *Let* dsp $= <$dp$, \Sigma_\phi, \delta_\phi >$ *be a drawn symbolic picture, where* dp $= < \{\{(0,0), (1,0)\}, \{(1,0), (1,1)\}, \{(1,1), (2,1)\}, \{(2,1), (2,0)\}, \{(2,0), (3,0)\}\}, (0,0), (3,0)>, \Sigma_\phi = \{a, b, c, \phi\}$, *and* δ_ϕ *is the function such that* $\delta_\phi(0,0) = a$, $\delta_\phi(1,0) = \phi$, $\delta_\phi(1,1) = b$, $\delta_\phi(2,1) = b$, $\delta_\phi(2,0) = \phi$, $\delta_\phi(3,0) = a$. *The visual representation of this picture is depicted in Fig. 2:*

Figure 2: A drawn symbolic picture.

Informally, a symbolic picture consists of a set of symbols, each of which is associated with a point of the plane. This notion is formalized by a set P of points, a function δ labeling the positions with symbols from an alphabet Σ, and the definition of a position as start point. More formally:

Definition 2 *A* symbolic picture *is a quadruple* sp $= <$P, s, Σ, $\delta >$, *where* $P \in M_0$ *is a set of points,* Σ *is an alphabet of symbols,* δ *is a function* δ: $P \to \Sigma$, *and* s $= (0,0)$ *is the* start point. *The start point belongs to* P *if* sp *is nonempty. An empty symbolic picture is denoted by* $< \emptyset, (0,0), \Sigma, \eta >$, *where* η *is undefined on (0,0). A* symbolic picture language *is a set of symbolic pictures.*

Example 2 *The arithmetic expression in Fig. 3 can be described by the symbolic picture* sp $= < \{(0,0), (1,0), (2,1), (2,0), (2,-1), (3,0), (4,1), (4,0), (4,-1)\}, (0,0), \Sigma, \delta >$, *where* $\Sigma = \{a, b, c, e, f\}$, *and* δ *is the function such that* $\delta(0,0) = a$, $\delta(1,0) = +$, $\delta(2,1) = b$, $\delta(2,0) = -$, $\delta(2,-1) = c$, $\delta(3,0) = +$, $\delta(4,1) = e$, $\delta(4,0) = -$, $\delta(4,-1) = f$.

It can easily be shown that symbolic picture languages strictly include the class of string languages. Moreover, they allow us to describe two-dimensional languages such as arithmetic expressions and colored bitmaps.

$$a + \frac{b}{c} + \frac{e}{f}$$

Figure 3: An arithmetic expression.

4 String-Based Representation

To define the string representations for (drawn) symbolic pictures we need the following notations. Given two alphabets Γ and Λ such that $\Gamma \cap \Lambda = \emptyset$, with $\Gamma \approx \Lambda$ (read as *tau alternated lambda*), we denote the set of *alternating words*, i.e. strings whose letters are taken from Γ and Λ in an alternate way. Thus, $\Gamma \approx \Lambda = \{\epsilon\} \cup \Gamma(\Lambda\Gamma)^*$. A string in the set $\Gamma \approx \Lambda$ is called a $\Gamma \approx \Lambda$-word. For example, the strings 'a' and '$arbdclarbdaub$' are words in the set $\Gamma \approx \Lambda = \{a, b, c\} \approx \{\mathbf{u}, \mathbf{d}, \mathbf{l}, \mathbf{r}\}$. A set of $\Gamma \approx \Lambda$-words is called a $\Gamma \approx \Lambda$-language.

4.1 Drawn Symbolic Picture Description

A drawn symbolic picture can be described by taking a walk through the picture. Such a walk starts from the start point, touches at least once each line and symbol in the drawn symbolic picture, and ends at the end point. Thus, such a walk is described by a $\Sigma_\phi \approx \Pi$-word w, where the symbols from Σ_ϕ are alternated with the moves from Π. As an example, the $\Sigma_\phi \approx \Pi$-word '$ar\phi ubr bd\phi r a$' describes the drawn symbolic picture depicted in Fig. 2. The drawn symbolic picture described by a $\Sigma_\phi \approx \Pi$-word w is denoted by *dspic(w)* and it is defined inductively in agreement with the following definition.

Definition 3 *Let* w *be a* $\Sigma_\phi \approx \Pi$-word. *The drawn symbolic picture described by* w, *denoted by* dspic(w), *is defined inductively as follows:*

- *if* w $= \epsilon$, *then* dspic(w) *is the empty drawn symbolic picture;*

- *if* w $= \sigma \in \Sigma_\phi$, *then* dspic(w) $= << \emptyset, (0,0), (0,0)>, \Sigma_\phi, \delta_\phi >$, *where* δ_ϕ *is only defined in (0,0) and* $\delta_\phi((0,0)) = \sigma$;

- *if* w $=$ w'τ *for some* w'$\in (\Sigma_\phi \approx \Pi$ -$\{\epsilon\})$, *with* dspic(w') $= <sc, \Sigma_\phi, \delta_\phi >$, sc $= <b, s, e>$, *and* $\tau = \pi\sigma$ *with* $\pi \in \Pi$ *and* $\sigma \in \Sigma_\phi$, *then* dspic(w) $= <<b \cup\{e, \pi(e)\}, s, \pi(e)>, \Sigma_\phi, \underline{\delta}_\phi >$, *where* $\underline{\delta}_\phi : W(sc) \cup \{\pi(e)\} \to \Sigma_\phi$ *is the labeling function such that:*

$$\underline{\delta}_\phi(v) = \begin{cases} \sigma & \text{if } (v = \pi(e) \text{ and } (v \notin W(sc) \text{ or } \delta_\phi(v) = \phi)), \\ \delta_\phi(v) & \text{otherwise.} \end{cases}$$

Informally, *dspic(w)* constructs a drawn symbolic picture by scanning the string w from left to right. The initial position in the plane is set to

(0,0) and a pointer p points to the first symbol in w. Whenever p points to a symbol $\sigma \in \Sigma_\phi$, and no visible symbol has been written on the current position on the plane, then σ is written on that position and p is moved to the next symbol in w. Whenever a visible symbol has already been written on the current position then p is moved to the next symbol in w. If p points to a symbol $\pi \in \Pi$, then the current position on the plane is updated in agreement with π and p points to the next symbol in w.

Observe that a conflict arises whenever p points to a symbol $\sigma \in \Sigma_\phi$, and a symbol $\rho \in \Sigma_\phi$ with $\rho \neq \sigma$ has already been written on the current position on the plane. In agreement with the above definition, $dspic()$ solves this type of conflict by overwriting ρ with σ in the case $\rho = \phi$ and by ignoring σ otherwise. A drawn symbolic picture description without conflicts is named a *consistent* description. As an example, the $\Sigma_\phi \approx \Pi$-word w = '$ar\phi ubd\phi u cr bd\phi r a$' describing the picture in Fig. 2 is not consistent. The same picture can be described by the consistent $\Sigma_\phi \approx \Pi$-word w = '$ar\phi ubd\phi u br bd\phi r a$'.

Given a drawn symbolic picture, there may exist many different $\Sigma_\phi \approx \Pi$-words describing it. The set of (consistent) words that describe a drawn symbolic picture dsp is the (*consistent*) *description language* of dsp. In [2] it has been proved that the (*consistent*) *description language* of dsp is a regular set.

4.2   Symbolic Picture Description

In order to give the string description of symbolic pictures, we recall the notion of invisible lines which was introduced in [14]. These lines correspond to moves which are executed by raising the pen. The set of invisible moves is denoted by $\Pi_b = \{r_b, l_b, u_b, d_b\}$. According to this definition each move m in Π_b is performed with the pen up so that no line is actually drawn on the plane. A symbolic picture can be described by taking a walk (with the pen up) through the picture. Such a walk starts from the start point and touches at least once each point in the symbolic picture. The walk between two non-contiguous points is described by exploiting invisible symbols. Thus, a symbolic picture can be described by a $\Sigma_\phi \approx \Pi_b$-word w, where the symbols from Σ_ϕ are alternated with the moves from Π_b. As an example, the $\Sigma_\phi \approx \Pi_b$-word '$ar_b + r_b - u_b bd_b\phi d_b cu_b\phi r_b + r_b - u_b ed_b\phi d_b f$' describes the picture depicted in Fig. 3. The symbolic picture described by a $\Sigma_\phi \approx \Pi_b$-word w is denoted by $spic(w)$ and it is defined inductively in agreement with Definition 4. This definition makes use of the technical notation $\theta(w)$ which indicates the position of the last symbol in w. $\theta(w)$ is defined as $\theta(w) = (\#r_b(w) - \#l_b(w), \#u_b(w) - \#d_b(w))$, where $\#m(w)$ denotes the number of symbols m in w, for $m \in \Pi_b$.

Definition 4 *Let w be a $\Sigma_\phi \approx \Pi_b$-word. The symbolic picture described by w, denoted by* spic(w), *is defined inductively as follows:*

- *if w $= \epsilon$, or w $= \phi$, then* spic(w) *is the empty symbolic picture $< \emptyset$, (0,0), Σ, $\nu >$;*

- *if w $= \sigma \in \Sigma$, then* spic(w) $= < \{(0,0)\}$, (0,0), Σ, $\delta >$, *where δ is only defined in (0,0) and $\delta((0,0)) = \sigma$;*

- *if w $=$ w'τ for some w'$\in (\Sigma_\phi \approx \Pi_b - \{\epsilon\})$, with* spic(w') $= <P$, (0,0), Σ, $\delta >$, *and $\tau = \pi\sigma$ with $\pi \in \Pi_b$, then:*

  if $\sigma = \phi$ then spic(w) $=$ spic(w'), *else*

  if $\sigma \in \Sigma$, then spic(w') $= <P\cup\{\pi(\theta(w))\}$, (0,0), Σ, $\underline{\delta} >$, *where $\underline{\delta}$: $P\cup\{\pi(\theta(w))\} \rightarrow \Sigma$ is the labeling function such that:*

$$\underline{\delta}(v) = \begin{cases} \delta(v) & \text{if } v \in P, \\ \sigma & \text{if } v \notin P. \end{cases}$$

Thus, a symbolic picture is depicted by scanning the string w from left to right in the same way as a drawn symbolic picture is constructed. Informally, symbolic picture languages can be considered as drawn symbolic picture languages where the spatial relations are not visualized. Even though this aspect may appear trivial, it introduces another degree of freedom in picture string descriptions which determines different complexity properties. As a matter of fact, it can be proved that the description language and the consistent description language of a symbolic picture are context-sensitive languages.

5   (Drawn) Symbolic Picture Grammars

A picture grammar generating (drawn) symbolic pictures must be based on a string grammar that can generate alternating words. Informally, a *grammar for alternating words* on two sets Γ and Λ ($\Gamma \approx \Lambda$-grammar for short) is a string grammar, with terminals in $\Gamma \cup \Lambda$, whose corresponding language is a subset of $\Gamma \approx \Lambda$ [3].

Definition 5 *Let Γ and Λ be two disjoint sets of symbols. An alternating grammar on the sets Γ and Λ, denoted by $\Gamma \approx \Lambda$-grammar, is specified by the 5-tuple $< \Gamma, \Lambda, N, P, S>$, and is defined as a $\Gamma \cup \Lambda$-grammar G$=< \Gamma \cup \Lambda$, N, P, S> for which L(G) $\subseteq \Gamma \approx \Lambda$.*

In particular, we will focus our attention on *context-free* $\Gamma \approx \Lambda$-grammars which are characterized by context-free $\Gamma\cup\Lambda$-grammars. An important form for context-free $\Gamma \approx \Lambda$-grammar is the *canonical form*, which is defined in agreement with the following definition:

Definition 6 *Let* G=< Γ, Λ, N, P, S > *be a* $\Gamma \approx \Lambda$-*grammar. We say that* G *is in canonical form if each production in* P *is of type* $A \to \alpha$, *where* $A \in N$ *and* $\alpha \in ((\Gamma \cup N) \approx \Lambda) - \{\epsilon\}$.

It can be proved that for each $\Gamma \approx \Lambda$-grammar G there exists a canonical $\Gamma \approx \Lambda$-grammar G' such that L(G')=L(G) – $\{\epsilon\}$.

In the sequel, a $\Sigma_\phi \approx \Pi$-grammar, denoted by $G_{\Sigma_\phi \approx \Pi}$, will be used to generate drawn symbolic picture descriptions. Thus, the drawn symbolic picture grammars and languages are defined as follows.

Definition 7 *A* drawn symbolic picture grammar *(DSP grammar, for short)* **G** *is a pair* < $G_{\Sigma_\phi \approx \Pi}$, dspic()>, *where* $G_{\Sigma_\phi \approx \Pi}$ *is a* $\Sigma_\phi \approx \Pi$-*grammar and* dspic() *is the function that translates a* $\Sigma_\phi \approx \Pi$-*word into a drawn symbolic picture. Given a DSP grammar* **G** = <$G_{\Sigma_\phi \approx \Pi}$, dspic()>, *the DSP language* **L** *generated by* **G**, *denoted by* **L(G)**, *is:*

$$\mathbf{L(G)} = \text{dspic}(L(G_{\Sigma_\phi \approx \Pi})) = \{dspic(x) | x \in L(G_{\Sigma_\phi \approx \Pi})\}.$$

Example 3 *Let us consider the drawn symbolic picture depicted in Fig. 4, which describes a section of a natural rubber molecule [4]. The symbols 'C' and 'H' represent the carbon atom and the hydrogen atom, resp., while the symbol '2' is used to denote a double link between two carbon atoms:*

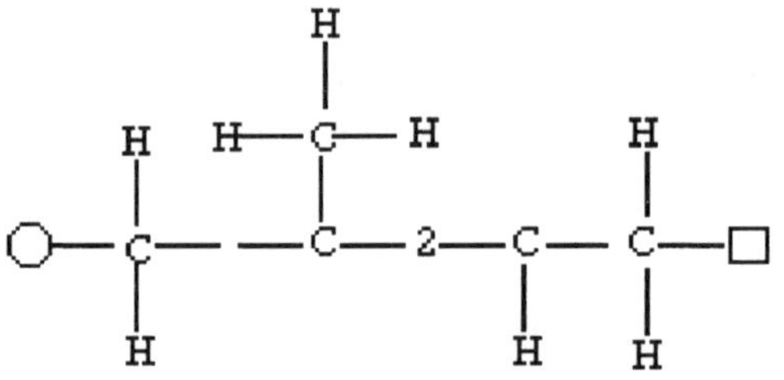

Figure 4: A drawn symbolic picture describing a section of a natural rubber molecule.

The repetitive chemical structure of a natural rubber molecule is described by the drawn symbolic picture language generated by the DSP grammar **RM** = <$RM_{\Sigma_\phi \approx \Pi}$, dspic()>, *with* $RM_{\Sigma_\phi \approx \Pi}$=< {C, H, 2, ϕ}, Π, {S, Section, $C_2\_H_3$, C_H_2, C_H}, P, S>, *where* S *is the initial symbol and* P *contains the productions:*

S $\to$ ϕ **r** Section **r** ϕ,

S $\to$ ϕ **r** Section **r** S,

Section $\to$ C_H_2 **r** ϕ **r** $C_2\_H_3$ **r** *2* **r** C_H **r** C_H_2,

C_H_2 $\to$ C **u** H **d** C_H,

$C_2\_H_3$ $\to$ C **u** C **r** H **l** C **u** H **d** C **l** H **r** C **d** C,

C_H $\to$ C **d** H **u** C.

The concept of consistency can be extended to DSP grammars in the natural way. A DSP grammar $\mathbf{G} = <G_{\Sigma_\phi \approx \Pi},\ dspic()>$, is *consistent* if any $\Sigma_\phi \approx \Pi$-word $w \in L(G_{\Sigma_\phi \approx \Pi})$ is consistent. The DSP grammar **RM** of Example 3 is a consistent DSP grammar. Indeed, any word $w \in L(RM_{\Sigma_\phi \approx \Pi})$ is consistent as can be easily verified by observing that any string on the right-hand side of any production is consistent and the subpictures generated by any nonterminal can never overlap.

It can be easily verified that when $G_{\Sigma_\phi \approx \Pi}$ is in canonical form, any sentential form *sf* is a sentence in $(\Sigma_\phi \cup N) \approx \Pi$, where N is the set of non-terminals. As a consequence, by applying the function *dspic()* to *sf*, we obtain a drawn symbolic picture called *pictorial sentential form*.

Now, we provide the definition of SP grammars and SP languages.

Definition 8 *A symbolic picture grammar (SP grammar, for short)* **G** *is a pair* $< G_{\Sigma_\phi \approx \Pi},\ \mathrm{spic}()>$, *where* $G_{\Sigma_\phi \approx \Pi_b}$ *is a* $\Sigma_\phi \approx \Pi_b$*-grammar and* $\mathrm{spic}()$ *is the function that translates a* $\Sigma_\phi \approx \Pi_b$*-word into a symbolic picture. Given an SP grammar* $\mathbf{G} = <G_{\Sigma_\phi \approx \Pi_b},\ \mathrm{spic}()>$, *the SP language* **L** *generated by* **G**, *denoted by* **L(G)**, *is:*
$$\mathbf{L(G)} = \mathrm{spic}(L(G_{\Sigma_\phi \approx \Pi_b})) = \{spic(x) | x \in L(G_{\Sigma_\phi \approx \Pi_b})\}.$$

In the next example we provide an SP grammar in canonical form which describe some simple arithmetic expressions, and show some pictorial sentential forms.

Example 4 *Let* $\mathbf{G} = << \Sigma_\phi = \{\mathrm{n},\ +,\ ^*,\ -,\ \phi\},\ \Pi_b,\ \mathrm{N} = \{E, T, F\},\ \mathrm{P},\ \mathrm{S}>,$ $\mathrm{spic}()>$, *be the symbolic picture grammar in canonical form with* $\mathrm{P} = \{E \rightarrow E\ \mathbf{r}_b + \mathbf{r}_b\ T,\quad E \rightarrow T,\quad F \rightarrow \mathrm{n},\quad T \rightarrow T\ \mathbf{r}_b\ ^* \mathbf{r}_b\ F,\quad T \rightarrow\ -\ \mathbf{u}_b\ \mathrm{n}\ \mathbf{d}_b\phi$ $\mathbf{d}_b\mathrm{n}\ \mathbf{u}_b\phi,\quad T \rightarrow F\}$. *A rightmost pictorial derivation of* **G** *is:*

$$E \rightarrow E + T \rightarrow E + F \rightarrow E + \mathrm{n} \rightarrow T + \mathrm{n} \rightarrow \frac{n}{n} + \mathrm{n}.$$

In [2] the canonical normal form of DSP and SP grammars has been exploited to prove their equivalence with some subclasses of Positional Grammars. Such formalism has been proposed to specify visual languages and has successfully been used for the automatic generation of visual programming environments [1]. The most appealing feature of the model is its capability of inheriting and extending to the visual field concepts and techniques of a traditional string languages, because it represents a natural extension of context-free grammars. We believe that the study of symbolic picture languages can be usefully exploited to provide insight into the features of the Positional grammar model.

6   Conclusions

In this chapter, we have presented models of drawn symbolic pictures and symbolic pictures. Several interesting questions may be studied in this context. In particular, the descriptions of symbolic pictures involve possible conflicts in the assignment of symbols to positions. The proposed definition solves the conflict by giving priority to the first visit of a position. So, it is interesting to investigate whether or not the lack of conflicts is decidable for a given (drawn) symbolic picture grammar. The problem has been partially addressed in [2], where it was shown that it is always possible to decide whether or not a $\Sigma_\phi \approx \Pi$-grammar generates only consistent descriptions for a drawn symbolic picture. The hypothesis of consistency plays an important role in the analysis of some decidability and complexity properties of (drawn) symbolic picture languages. As a matter of fact, in [2] it was shown that the membership problem for consistent regular drawn symbolic pictures within a stripe is decidable in linear time deterministically. Another interesting issue concerns the analysis of a variant of the proposed models where the conflicts can be resolved by taking into account priority values which could be assigned to the alphabet symbols.

References

[1] G. Costagliola, A. De Lucia, S. Orefice and G. Tortora, Automatic generation of visual programming environments, *IEEE Computer*, 28 (1995), 56–66.

[2] G. Costagliola and F. Ferrucci, Symbolic picture languages and their decidability and complexity properties, *Journal of Visual Languages and Computing*, Special Issue on the Theory of Visual Languages, 10 (1999), 381–419.

[3] G. Costagliola and M. Salurso, Drawn symbolic picture languages. In *Proceedings of the International Workshop on Theory of Visual Languages*, Capri, 1997, 1–14.

[4] J. Feder, Plex languages, *Information Sciences*, 3 (1971), 225–241.

[5] F. Ferrucci et al., Symbol-relation grammars: a formalism for graphical languages, *Information and Computation*, 131 (1996), 1–46.

[6] H. Freeman, Computer processing of line-drawing images, *Computer Surveys*, 6 (1974), 57–97.

[7] D. Giammarresi and A. Restivo, Two-dimensional languages. In G. Rozenberg, A. Salomaa (eds.), *Handbook of Formal Languages*. Springer, Berlin, 1997, vol. 3, 215–267.

[8] J.E. Hopcroft and J.D. Ullman, *Introduction to Automata Theory, Languages, and Computation*. Addison-Wesley, Reading, Mass. 1979.

[9] C. Kim, Complexity and decidability for restricted class of picture languages, *Theoretical Computer Science*, 73 (1990), 295–311.

[10] C. Kim, Picture iteration and picture ambiguity, *Journal of Computer and System Sciences*, 40 (1990), 289–306.

[11] C. Kim and I.H. Sudborough, The membership and equivalence problems for picture languages, *Theoretical Computer Science*, 52 (1987), 177–191.

[12] K. Marriott, Constraint multiset grammars. In *Proceedings of the IEEE Symposium on Visual Languages*, St. Louis, Mi., 1994, 118–125.

[13] H.A. Maurer, G. Rozenberg and E. Welzl, Using string languages to describe picture languages, *Information and Control*, 54 (1982), 155–185.

[14] K. Slowinski, Picture words with invisible lines, *Theoretical Computer Science*, 108 (1993), 357–363.

[15] I.H. Sudborough and E. Welzl, Complexity and decidability for chain code picture languages, *Theoretical Computer Science*, 36 (1985), 173–202.

What Is the Abelian Analogue of Dejean's Conjecture?[1]

James D. Currie

Department of Mathematics and Statistics
University of Winnipeg
Manitoba, Canada
currie@uwinnipeg.ca

Abstract. We motivate the study of Abelian repetitive thresholds and give interesting open problems and some (weak) bounds.

We say that a word w *encounters* yy if we can write $w = XY_1Y_2Z$, where $Y_1 = Y_2 \neq \epsilon$. For example, *banana* = *b an an a* encounters yy, with $X = b$, $Y_1 = Y_2 = na$, $Z = a$. We say that w *avoids* yy if w does not encounter yy. Thue [14] proved that the language $L = \{w \in \{a, b, c\}^* : w \text{ avoids } yy\}$ is infinite.

We define an equivalence relation $\sim$ on words by saying that $Y_1 \sim Y_2$ if Y_2 can be obtained from Y_1 by reordering the letters. For example *stops* $\sim$ *posts*. We say that a word w *encounters* yy *in the Abelian sense* if we can write $w = XY_1Y_2Z$, where $Y_1 \sim Y_2 \neq \epsilon$. For example, *disproportionate* = *dis pro por tionate* encounters yy in the Abelian sense, with $X = dis$, $Y_1 = pro$, $Y_2 = por$, $Z = tionate$. We say that w *avoids* yy *in the Abelian sense* if w does not encounter yy. In 1961, Erdös [9] asked whether the language $L = \{w \in \Sigma^* : w \text{ avoids } yy \text{ in the Abelian sense}\}$ was infinite for some finite alphabet Σ. Evdokimov [10] answered this question in the affirmative in

[1]Research supported by an NSERC Operating Grant.

1968 for $|\Sigma| = 25$. Erdös had mentioned the possibility that L might be infinite even with $|\Sigma|$ as small as 4, and the problem of showing that 4 letters sufficed to avoid yy in the Abelian sense became well-known. Finally, in 1992, Keränen [11] showed that 4 letters sufficed.

These problems of avoiding yy generalize into interesting decision problems. Let $p = y_1 y_2 \cdots y_n$ be a word, where the y_i are letters. We say that word w *encounters* p if we can write $w = X Y_1 Y_2 \cdots Y_n Z$, where $Y_i = Y_j \neq \epsilon$ if $y_i = y_j$. Thus (departing from English words, alas):

$$12345672347568 = 1\ 234\ 56\ 7\ 234\ 7\ 56\ 8$$

encounters $p = abcacb$ with $X = 1$, $Y_1 = 234 = Y_4$, $Y_2 = 56 = Y_6$, $Y_3 = 7 = Y_5$, $Z = 8$. We say that p is *k-avoidable* if there is some alphabet Σ, $|\Sigma| = k$, such that the language $L = \{w \in \Sigma^* : w \text{ avoids } p\}$ is infinite. Evidently this depends only on k. If p is k-avoidable for some finite k, then we say that p is *avoidable*.

In 1979, Bean, Ehrenfeucht & McNulty [2], and independently Zimin [15] gave a decision procedure for determining whether p is avoidable. The problem of deciding, given k and p, whether p is k-avoidable has remained open, but a solution to it has been pursued vigorously. For example, Cassaigne's PhD thesis [4] includes an inventory of words over $\{a, b, c\}$, which are classified according to whether they are known to be 2-avoidable, 3-avoidable or 4-avoidable. Here is one intriguing open problem:

Problem 1 *Is there an avoidable pattern which is not 4-avoidable?* [1]

As in the case of yy, we have natural Abelian analogues: We say that word w *encounters* p *in the Abelian sense* if we can write $w = X Y_1 Y_2 \cdots Y_n Z$, where $Y_i \sim Y_j \neq \epsilon$ if $y_i = y_j$. Thus, for example:

$$12345672437658 = 1\ 234\ 56\ 7\ 243\ 7\ 65\ 8$$

encounters $p = abcacb$ in the Abelian sense with $X = 1$, $Y_1 = 234 \sim 243 = Y_4$, $Y_2 = 56 \sim 65 = Y_6$, $Y_3 = 7 = Y_5$, $Z = 8$. Proceeding in the obvious way, define *k-avoidability (avoidability) in the Abelian sense*. In 1993 [6], I asked which patterns are avoidable or k-avoidable in the Abelian sense. From Keränen's work we know that yy is 4-avoidable in the Abelian sense, but what more can be said?

At this point we should point out that Abelian k-avoidability can have useful implications for 'ordinary' k-avoidability:

Lemma 1 *The pattern $p = abcacb$ is 4-avoidable.*

Proof. Any word avoiding yy in the Abelian sense must avoid $abcacb$ in the ordinary sense. Since yy is 4-avoidable in the Abelian sense, $abcacb$ is 4-avoidable. □

Remark. In fact, $abcacb$ is 3-avoidable. Whether $abcacb$ is 2-avoidable is an open problem [4].

Ordinary and Abelian avoidability are nevertheless not identical, as shown by the following lemma from [7]:

Lemma 2 *The pattern $p = abcabdabcba$ is avoidable, but not in the Abelian sense.*

The problem of ordinary k-avoidability has been studied since 1979. Another tool which can be brought to bear on this problem is the study of *repetitive thresholds*, introduced by Dejean [8].

Fix r, $1 < r \leq 2$. We say that w *encounters* y^r if we can write a subword of w as uvu, where $|uvu|/|uv| = r$. We say that y^s *is avoidable on an n letter alphabet* if there are infinitely many words on n letters not encountering y^r for any $r \geq s$. Define the *repetitive threshold function* by:

$$RT(n) = \inf\{s : y^s \text{ is avoidable on } n \text{ letters}\}.$$

Dejean conjectured that:

$$RT(n) = \begin{cases} 2, & n = 2 \\ 7/4, & n = 3 \\ 7/5, & n = 4 \\ n/(n-1), & n > 4 \end{cases}$$

Dejean's conjecture has been verified up to $n = 11$ [13, 12], and is startling in that she made the conjecture having only the values $RT(2) = 2$ (found by Thue) and $RT(3) = 7/4$ (found by Dejean)! She did, however, show that $RT(n) \geq n/(n-1)$. Dejean's RT function gives another tool for studying k-avoidability:

Lemma 3 *The pattern $p = abacbab$ is 4-avoidable.*

Since $RT(4) = 7/5$, it will suffice to show that any word encountering $abacbab$ encounters y^s for some $s > 7/5$. Suppose then that w contains some subword $ABACBAB$. Our proof can be broken into cases depending on which of $|A|$ and $|B|$ is greater:

1. $|A| > |B|$: In this case, w contains the word uvu, where $u = A$, $v = B$. Then, $|uvu|/|uv| = |ABA|/|AB| = 1 + |A|/|AB| \geq 1 + 1/2 > 7/5$.

2. $|B| > |A|$: In this case, w contains the word uvu, where $u = B$, $v = A$. Again, $|uvu|/|uv| = |BAB|/|BA| = 1 + |B|/|BA| \geq 1 + 1/2 > 7/5$.

In either case, w encounters y^r, and $r > 7/5$.□

The problem of which words p are k-avoidable, or avoidable in the Abelian sense, is obviously of intrinsic interest. We see also that Abelian avoidability gives insight into the open problem of 'ordinary' k-avoidability. Since repetitive thresholds also give insight into k-avoidability, it is natural to seek the Abelian version of Dejean's conjecture.

Fix r, $1 < r \leq 2$. We say that w *encounters* y^r *in the Abelian sense* if we can write a subword of w as uvu_1, where $|uvu|/|uv| = r$ and $u \sim u_1$. We say that y^s *is avoidable in the Abelian sense on an* n *letter* if there are infinitely many words on a fixed n-letter alphabet not encountering y^r for any $r \geq s$. Define the *Abelian repetitive threshold function* by:

$$ART(n) = \inf\{s : y^s \text{ is avoidable on } n \text{ letters in the Abelian sense}\}.$$

The function $ART(n)$ was introduced in [5], where it is shown that $\lim_{n \to \infty} ART(n) = 1$. If Dejean's conjecture is correct, we will have $ART(n) \geq n/(n-1)$ for $n > 4$. The behaviour of a sort of dual repetitive threshold function would also be useful: Define the *dual Abelian repetitive threshold function* on $(1, 2]$ by letting:

$$DART(r) = \min\{n \in \mathbb{N} : y^r \text{ is avoidable in the Abelian sense on } n \text{ letters.}\}$$

For example, by [11], $DART(2) = 4$. Only very weak information about $DART(r)$ is known for most r. In [5] it is shown that:

$$DART(r) \leq (3 + \frac{7}{2(k-1)})5^{((3+\frac{7}{2(k-1)})^2+1)^2} + 2(3 + \frac{7}{2(k-1)})^2 + 1. \quad (1)$$

Considering the number of years it took to show that $DART(2) = 4$, it seems unlikely that showing $ART(4) < 2$ will be easy. Nevertheless, here is the challenge:

Problem 2 *What are the values of* $ART(n)$ *and* $DART(r)$?

We will end this paper by showing that one can certainly do better than our bounds on ART and $DART$. First of all, we offer the following lemma without proof:

Lemma 4 *Let* $m \in \mathbb{N}$. *The word* $h^m(0)$ *avoids* $y^{1.9}$ *in the Abelian sense, where* h *is the morphism on* $\{0, 1, 2, \ldots, 8\}$ *given by:*

$$
\begin{aligned}
h(0) &= 0740103050260, \\
h(1) &= 1051214161371, \\
h(2) &= 2162325272402,
\end{aligned}
$$

$$
\begin{aligned}
h(3) &= 3273436303513, \\
h(4) &= 4304547414624, \\
h(5) &= 5415650525735, \\
h(6) &= 6526761636046, \\
h(7) &= 7637072747157.
\end{aligned}
$$

It follows that $DART(1.9) \leq 8$. This is (somewhat) better than the bound given by Equation (1), which is around 10^{1642}. Exhaustive searching shows that no word of length greater than 14 on $\{0,1,2,3,4,5,6,7\}$ avoids $y^{1.5}$ in the Abelian sense, so that $ART(8) \geq 1.5$. We thus have $1.5 \leq ART(8) \leq 1.9$. By way of comparison, $RT(8) = 8/7$.

References

[1] K.A. Baker, G.F. McNulty and W. Taylor, Growth problems for avoidable words, *Theoretical Computer Science*, 69.3 (1989), 319–345.

[2] D.R. Bean, A. Ehrenfeucht and G. McNulty, Avoidable patterns in strings of symbols, *Pacific Journal of Mathematics*, 85 (1979), 261–294.

[3] J. Berstel, Axel Thue's papers on repetitions in words: a translation.

[4] J. Cassaigne, *Motifs Évitables et Régularités dans les Mots*. PhD dissertation, L.I.T.P., Université Paris 6, 1994.

[5] J. Cassaigne and J.D. Currie, Words strongly avoiding fractional powers, *European Journal of Combinatorics*, to appear.

[6] J.D. Currie, Open problems in pattern avoidance, *American Mathematical Monthly*, 100 (1993), 790–793.

[7] J.D. Currie and V. Linek, Avoiding patterns in the Abelian sense. Submitted to *Canadian Journal of Mathematics*.

[8] F. Dejean, Sur un théorème de Thue, *Journal of Combinatorial Theory. Series A*, 13 (1972), 90–99.

[9] P. Erdös, Some unsolved problems, *Magyar Tud. Akad. Mat. Kutato. Int. Kozl.*, 6 (1961), 221–254.

[10] A.A. Evdokimov, Strongly asymmetric sequences generated by a finite number of symbols, *Dokl. Akad. Nauk. SSSR*, 179 (1968), 1268–1271; *Soviet Math. Dokl.*, 9 (1968), 536–539.

[11] V. Keränen, Abelian squares are avoidable on 4 letters. In *Automata, Languages and Programming*. Springer, Berlin, 1992, 41–52.

[12] J. Moulin Ollagnier, Proof of Dejean's conjecture for alphabets with 5, 6, 7, 8, 9, 10 and 11 letters, *Theoretical Computer Science*, 95 (1992), 187–205.

[13] J.J. Pansiot, A propos d'une conjecture de F. Dejean sur les répétitions dans les mots, *Discrete Applied Mathematics*, 7 (1984), 297–311.

[14] A. Thue, Über unendliche Zeichenreihen, *Norske Vid. Selsk. Skr. I. Mat.-Nat. Kl. Christiana*, 7 (1906).

[15] A. Zimin, Blocking sets of terms, *Mat. Sb. (N.S.)*, 119, 161 (1982); *Math. USSR Sbornik*, 47 (1984), 353–364.

Threshold Locally Testable Languages
in Strict Sense

Pedro García

José Ruiz

Department of Informatic Systems and Computation
Polytechnical University of Valencia
Spain
{pgarcia,jruiz}@dsic.upv.es

Abstract. The family of Locally Testable languages in Strict Sense ($LTSS$) was characterized by de Luca and Restivo and has been widely used in disciplines like Pattern Recognition. The aim of this paper is to define a new family of rational languages, the *Locally Threshold Testable languages in Strict Sense* ($LTTSS$) which includes the family of $LTSS$. As it happens in $LTSS$ languages, membership of a word to a $LTTSS$ language can be decided by means of local scanning, using a sliding window of a fixed length k, although, in this case, we have to take care that the number of occurrences of certain segments of length $\leq k$ in the words is not greater than a level of restriction, less than a threshold r. We generalize the results obtained by de Luca and Restivo for $LTSS$ languages and a syntactic characterization of $LTTSS$ languages is proposed.

1 Introduction

The family of Locally Testable Languages (LT) is one of the cornerstones in the literature about formal languages. The membership of a word to a

LT language is determined by the set of factors of a fixed length k, and by the prefixes and suffixes of length less than k of the word. The number of occurrences of the factors in the word or the order in which they appear are not relevant. Local Languages are known either for their ability to generate the family of regular languages by means of morphisms [9] or by Chomsky-Schützenberger's theorem for Context-Free Languages, and they are included in the family of LT. In fact they constitute a particular instance of the so-called Locally Testable Languages in Strict Sense ($LTSS$).

The words of a $LTSS$ language L, for a given value of $k \geq 1$, are defined by means of three finite sets, the set of their prefixes (suffixes) of length less than k and the set of their forbidden factors of length k. If we call those sets A, B and C, then we have $\Sigma^{k-1}\Sigma^* \cap L = A\Sigma^* \cap \Sigma^* B - \Sigma^* C\Sigma^*$. Membership of a word either to a LT or to a $LTSS$ language, for a given value of k (resp. k-T or k-TSS), can be decided by means of special types of automata called scanners [1].

The family of LT languages is a variety of languages and the corresponding variety of semigroups has been characterized by Brzozowski-Simon [2], Zalcstein [17] and McNaughton [8] as being the variety of locally idempotent and locally commutative semigroups. The family of $LTSS$ languages, which is not a variety, has been syntactically characterized by de Luca and Restivo [7] using the concept of the constant. Intuitively, a word w is a constant for a language L iff every prefix of the language that ends by w has the same set of suffixes in L. A language L is k-TSS iff every word of length $\geq k - 1$ is a constant for L. From the point of view of the canonical acceptor of L, that means that the state of the automaton reached by any prefix u of L depends only on the last $k - 1$ symbols of u. The concept of constant can be translated to the image of the language in its syntactic semigroup and so, a language L is $LTSS$ iff every idempotent of the syntactic semigroup of L is a constant for the syntactic image of L.

An extension of the family of LT languages is the family of Threshold Locally Testable Languages (LTT). If a word x belongs to a LTT language L for given values of k and r, any word y will also belong to L iff it meets the three following conditions:

1. Begins and ends with the same factors of length $k - 1$ as x.

2. The number of times each factor of length $\leq k$ occurs in y is the same as it occurs in x, if this number is less than r.

3. If the number of times that a factor z of length $\leq k$ occurs in x is $\geq r$, then so is the number of times that the factor z occurs in y.

LTT languages include the family of LT as a particular instance (the case $r = 1$) and have been syntactically characterized in [12], [14]. On the other hand, the class of LTT languages corresponds to the class of languages that can be defined by first order sentences in which numerical predicates are of the form $x = y$ and $y = x + 1$ [15]. These results are treated in a systematic manner in [13].

It seems natural, in this context, to define the family of threshold locally testable languages in strict sense ($LTTSS$), which, on the one hand, is a restriction of the family of LTT and, on the other, is an extension of $LTSS$ languages. We define $LTTSS$ languages by means of the sets of prefixes and suffixes of length $< k$ and by a set of restricted factors of length $\leq k$. Each factor in the set of restricted factors is associated with a level of restriction, which is less than a fixed threshold r. The factors for which this level is zero are forbidden. The language so defined contains the words that begin and end with elements of the indicated sets and such that none of the restricted factors appears in them a number of times beyond its level of restriction. For each value of k, if we set r to 1 we obtain the family of k-testable languages in strict sense (k-TSS). Using a generalization of the concept of constant, we propose a syntactic characterization of $LTTSS$ languages that generalizes what de Luca and Restivo obtained for $LTSS$ languages.

Besides its theoretical interest, another reason to study $LTTSS$ languages is the possibility of being used in Pattern Recognition (PR) [3], [5]. Stochastic $LTSS$ languages, also known as N-grams, are frequently used in PR, particularly in Speech Recognition, both in Acoustic-Phonetics Decoding as in Language Modeling [16]. Both k-TSS languages [4] and (k, r)-$TTSS$ languages are learnable from positive data [11].

2 Preliminaries and Notation

We assume the reader is familiar with the rudiments of formal languages and finite semigroup theory. Any concept not mentioned here can be found in [6] (formal languages) and [10] (finite semigroups).

Let Σ be a finite alphabet and Σ^* the free monoid generated by Σ with concatenation as the binary operation and λ as neutral element. A *language* L over Σ is a subset of Σ^*, its elements will be referred as *words*. The set of all words of length k will be represented as Σ^k, also $\Sigma_1^k = \bigcup_{i=1}^k \Sigma^i$. Given $x \in \Sigma^*$, if $x = uvw$ with $u, v, w \in \Sigma^*$, then u (w) is called *prefix* (*suffix*) of x, while v is said to be a *factor* of x. $\Pr(L)$ ($\mathrm{Suf}(L)$) denotes the set of prefixes (suffixes) of L. The *length* of a word x is represented by $|x|$. The number of times that a word w occurs as a factor of x is denoted as $|x|_w$. Given $u \in \Sigma^*$ and $L \subseteq \Sigma^*$, $u^{-1}L = \{v \in \Sigma^* : uv \in L\}$. Also, if $R \subseteq \Sigma^*$ is a finite set, $|R|$ denotes its cardinal.

A deterministic finite state automaton (DFA) is a 5-tuple $A = (Q, \Sigma, \delta, q_0, F)$ where Q is a finite set of states, Σ is a finite alphabet, $q_0 \in Q$ is the initial state, $F \subseteq Q$ is the set of final states and δ is a partial function mapping $Q \times \Sigma$ to Q that can be extended to words by defining $\delta(q, \lambda) = q$ and $\delta(q, xa) = \delta(\delta(q, x), a)$, $\forall q \in Q$, $\forall x \in \Sigma^*$, $\forall a \in \Sigma$. A word x is accepted by A if $\delta(q_0, x) \in F$. The language accepted by an automaton A is represented by $L(A)$.

Let $L \subseteq \Sigma^*$ and let $\equiv$ be an equivalence relationship defined in Σ^*. We say that $\equiv$ saturates L if L is the union of equivalence classes module $\equiv$. An equivalence relation is called a congruence if it is both sides compatible with the operation of the monoid. The congruence $\equiv_L$ defined as $x \equiv_L y \Leftrightarrow (\forall u, v \in \Sigma^*, uxv \in L \Leftrightarrow uyv \in L)$ is called the *syntactic congruence* of L and it is the coarsest congruence that saturates L. $\Sigma^* / \equiv_L$ is called the *syntactic monoid* of L and is represented by $S(L)$. The morphism $\varphi : \Sigma^* \to S(L)$, that maps each word to its equivalence class module $\equiv_L$ is called the *syntactic morphism* of L. The set of idempotents of $S(L)$ is denoted as $E(S(L))$. If S is a semigroup, we give S^1 to the monoid $S \cup \{1\}$ with the operation $\circ$ defined as $x \circ y = xy$ if $x, y \in S$, and $x \circ 1 = 1 \circ x = x$. Of course, $S^1 = S$ if $1 \in S$.

3   Threshold Locally Testable Languages in Strict Sense

Given two positive integers k and r the equivalence relation $\approx_{k,r}$ over Σ^* is defined as follows:

1. If $|x| < k$, then $x \approx_{k,r} y$ iff $x = y$.

2. Otherwise $x \approx_{k,r} y$ iff:

   (a) $\mathrm{Pr}(x) \cap \Sigma^{k-1} = \mathrm{Pr}(y) \cap \Sigma^{k-1}$.

   (b) $Suf(x) \cap \Sigma^{k-1} = Suf(y) \cap \Sigma^{k-1}$.

   (c) $\forall w \in \Sigma_1^k (|x|_w = |y|_w < r \vee (|x|_w > r \wedge |y|_w > r))$.

The relationship $\approx_{k,r}$ is a congruence of finite index. A language is (k, r)-*TT* iff it is saturated by $\approx_{k,r}$. A language L is *Threshold Locally Testable* (*LTT*) if there exist integers $k, r \geq 1$ such that L is (k, r)-*TT*. If $r = 1$ this family coincides with the family of Locally Testable Languages (*LT*).

A subclass of the *LT* languages is the class of *Locally Testable Languages in Strict Sense* (*LTSS*). A language L over Σ is *LTSS* iff there exists an integer $k \geq 1$ and sets $A, B \subseteq \Sigma^{k-1}$ and $C \subseteq \Sigma^k$ such that $\Sigma^{k-1}\Sigma^* \cap L = A\Sigma^* \cap \Sigma^* B - \Sigma^* C\Sigma^*$.

We are going to define a new family of languages, called *Threshold Locally Testable Languages in Strict Sense* (*LTTSS*) that is related to *LTT* languages in the same way as *LTSS* are related to *LT*. The case $r = 1$ constitutes the family of *LTSS* languages.

Given a language L over an alphabet Σ and given integers $k, r \geq 1$, we define the 4-tuple $(I_k, F_k, T_{k,r}, g)$ where:

- $I_k, F_k \subseteq \Sigma^{k-1}$ are respectively the sets of prefixes and suffixes of length $k-1$ of L,

- $T_{k,r} \subseteq \Sigma_1^k$ is the set of restricted factors of length k of the words of L, and

- $g : T_{k,r} \rightarrow \{0, 1, ...r-1\}$ is a function that defines the level of restriction of each of the restricted factors.

Definition 1 *We say that L is a k-Testable Language in Strict Sense with Threshold r (in the sequel this family will be referred as (k,r)-TTSS) iff:*

$$\Sigma^{k-1}\Sigma^* \cap L = I_k\Sigma^* \cap \Sigma^* F_k - \left(\bigcup_{w \in T_{k,r}} \{x \in \Sigma^* : |x|_w > g(w)\}\right).$$

For example, the language a^+ba^+b is $(2,3)$-TTSS, as it fulfils the definition with $I_2 = \{a\}$, $F_2 = \{b\}$, $g(ab) = 2$, $g(ba) = 1$, $g(bb) = 0$.

L is obviously regular. Language L, except for a finite number of words of length less than k, is the set of all the words that begin with a prefix of length $k-1$ which belongs to I_k, end with a suffix of length $k-1$ in F_k and, if they contain factors of length k belonging to $T_{k,r}$, the number of occurrences of those factors is less than or equal to the level of restriction given by the function g.

A language L is threshold locally testable in strict sense $(LTTSS)$ if there exist two positive integers k and r such that L is (k,r)-TTSS.

For every k, r greater than 0, the family of (k,r)-TTSS languages is included in the family of (k,r)-TT. It is obvious, from the definition of (k,r)-TTSS languages, that any language which belongs to that class is saturated by the relation $\approx_{k,r}$, so $LTTSS \subseteq LTT$. From the fact that (k,r)-TTSS languages are not closed under Boolean operations it follows that this inclusion is strict $(LTTSS \subset LTT)$.

4 Characterization of $LTTSS$ Languages

The concept of constant has been used by de Luca and Restivo [7] to characterize $LTSS$ languages. Given a semigroup S and a subset $B \subseteq S$, an element $c \in S$ is a constant for B if for any $s_1, s_2, s_3, s_4 \in S^1$, $s_1cs_2, s_3cs_4 \in B \Rightarrow s_1cs_4 \in B$.

If S is the free semigroup generated by Σ and L is a language over Σ, a word w is a constant for L iff $\forall uw, vw \in \Pr(L)$, it follows that $(uw)^{-1}L = (vw)^{-1}L$.

From the point of view of the minimum DFA accepting L, the fact that w is a constant means that all the prefixes of L ending with w reach the same state from the initial one.

We are going to generalize the concept of constant, making it dependent on k and r. We will obtain an analogous result to [7] which includes it as a particular case.

4.1   Semiconstants

Definition 2 *Given $k, r \geq 1$ and $L \subseteq \Sigma^*$, we define $T_{k,r}(L)$ as the set of factors of length between 1 and k such that the number of times each factor occurs in the words of L is bounded above by r, that is, $T_{k,r}(L) = \{w \in \Sigma_1^k : \forall x \in L, |x|_w < r\}$.*

Definition 3 *Given $x \in \Sigma^*$ and $L \subseteq \Sigma^*$, for every $w \in T_{k,r}(L)$ we define the components of $\mathbf{v}(x)$, the $|T_{k,r}(L)|$-dimensional vector associated to x, as $\mathbf{v}(x)_w = |x|_w$.*

Given $x, y \in \Sigma^*$, we say that $\mathbf{v}(x) \leq \mathbf{v}(y)$ iff $\forall w \in T_{k,r}(L), \mathbf{v}(x)_w \leq \mathbf{v}(y)_w$.

Definition 4 *Given $k, r \geq 1$ and $L \subseteq \Sigma^*$, we say that $v \in \Sigma^*$ is a (k,r)-constant for L iff $\forall x, y \in \Sigma^*(xv, yv \in \mathrm{Pr}(L) \wedge \mathbf{v}(xv) \leq \mathbf{v}(yv) \Rightarrow (yv)^{-1}L \subseteq (xv)^{-1}L)$.*

If φ is the syntactic morphism, the concept of (k,r)-constant can be extended to the syntactic semigroup $S(L)$ making $T_{k,r}(\varphi(L)) = \varphi(T_{k,r}(L))$ and defining, for $s \in S(L)$, $\mathbf{v}(s)$ as the $|\varphi(T_{k,r}(L))|$-dimensional vector in a similar way as above.

Definition 5 *We say that $w \in \Sigma^*$ is a semiconstant for L if there exist $k, r \geq 1$ such that w is a (k,r)-constant for L.*

4.2   Properties of (k,r)-Constants

In the sequel $C_{k,r}(L)$ and $C_{k,r}(\varphi(L))$ denote the sets of (k,r)-constants for L and $\varphi(L)$, respectively.

1. *$C_{k,r}(L)$ is a two-sided ideal of Σ^*.*

   Proof. Let $x \in \Sigma^* C_{k,r}(L)\Sigma^*$, then $x = x_1 v x_2$, being $x_1, x_2 \in \Sigma^*$ and $v \in C_{k,r}(L)$. Let $u_1, u_2 \in \Sigma^*$ such that $u_1 x, u_2 x \in \mathrm{Pr}(L)$ and $\mathbf{v}(u_1 x) \leq \mathbf{v}(u_2 x)$, then $\mathbf{v}(u_1 x_1 v x_2) \leq \mathbf{v}(u_2 x_1 v x_2)$ and so $\mathbf{v}(u_1 x_1 v) \leq \mathbf{v}(u_2 x_1 v)$. As $v \in C_{k,r}(L)$ it follows that $(u_2 x_1 v)^{-1}L \subseteq (u_1 x_1 v)^{-1}L$, and then $x_2^{-1}[(u_2 x_1 v)^{-1}L] \subseteq x_2^{-1}[(u_1 x_1 v)^{-1}L] \Rightarrow (u_2 x_1 v x_2)^{-1}L \subseteq (u_1 x_1 v x_2)^{-1}L \Rightarrow (u_2 x)^{-1}L \subseteq (u_1 x)^{-1}L \Rightarrow x \in C_{k,r}(L)$. $\square$

2. $\varphi(C_{k,r}(L)) \subseteq C_{k,r}(\varphi(L))$.

   Proof. Let $v \in C_{k,r}(L)$. We have to show that $\varphi(v) \in C_{k,r}(\varphi(L))$, that is, $\forall s_1, s_2, s_3, s_4 \in S(L)$ if $s_1\varphi(v)s_2, s_3\varphi(v)s_4$ and $\mathbf{v}(s_1\varphi(v)) \leq \mathbf{v}(s_3\varphi(v))$ it follows that $s_1\varphi(v)s_4 \in \varphi(L)$.

   Let $x_1, x_2, x_3, x_4 \in \Sigma^*$ such that $\varphi(x_i) = s_i$, $i = 1...4$ with $x_1vx_2, x_3vx_4 \in L$ and also $\mathbf{v}(x_1v) \leq \mathbf{v}(x_3v)$.

   As v is a *(k,r)*-constant for L, $x_1vx_4 \in L$ and thus $\varphi(x_1)\varphi(v)\varphi(x_4) = s_1\varphi(v)s_4 \in \varphi(L)$. □

3. $\varphi^{-1}(C_{k,r}(\varphi(L))) \subseteq C_{k,r}(L)$.

   Proof. Let $x_1, x_2, x_3, x_4 \in \Sigma^*$ such that $\varphi(x_i) = s_i$, $i = 1...4$, and let $u \in \varphi^{-1}(C_{k,r}(\varphi(L)))$, such that $x_1ux_2, x_3ux_4 \in L$ and $\mathbf{v}(x_1u) \leq \mathbf{v}(x_3u)$. Thus $s_1\varphi(u)s_2, s_3\varphi(u)s_4 \in \varphi(L)$, and $\mathbf{v}(s_1u) \leq \mathbf{v}(s_3u)$. As $\varphi(u) \in C_{k,r}(\varphi(L))$, it follows that $s_1\varphi(u)s_4 \in \varphi(L)$, and thus $x_1ux_4 \in L$. □

4. $\varphi^{-1}(\varphi(C_{k,r}(L))) = C_{k,r}(L)$.

   Proof. Let $u \in \varphi^{-1}(\varphi(C_{k,r}(L)))$, thus, by property 2, $\varphi(u) \in C_{k,r}(\varphi(L))$ and by 3, $u \in \varphi^{-1}(C_{k,r}(\varphi(L))) \subseteq C_{k,r}(L)$, then $\varphi^{-1}(\varphi(C_{k,r}(L))) \subseteq C_{k,r}(L)$. The inverse inclusion is obvious. □

5. $C_{k,r}(\varphi(L))$ *is a two-sided ideal of* $S(L)$.

   Proof. We have to show that $S(L)C_{k,r}(\varphi(L))S(L) \subseteq C_{k,r}(\varphi(L))$. Let $s = s_1vs_2$, with $s_1, s_2 \in S(L)$ and $v \in C_{k,r}(\varphi(L))$ and let $x_1, x_2, u \in \Sigma^*$ such that $\varphi(x_i) = s_i$, $i = 1, 2$ and $\varphi(u) = v$. Thus:

   $u \in \varphi^{-1}(v) \subseteq \varphi^{-1}(C_{k,r}(\varphi(L))) \subseteq C_{k,r}(L)$ and, by property 1, $x_1ux_2 \in C_{k,r}(L)$, thus $s_1vs_2 \in \varphi(C_{k,r}(L)) \subseteq C_{k,r}(\varphi(L))$. □

6. Let $w \in C_{k,r}(L)$, then if $k' \geq k$ or $r' \geq r$ then $w \in C_{k',r'}(L)$.

   Proof.

   (a) Let us show that if $w \in C_{k,r}(L)$ and $k' \geq k$ then $w \in C_{k',r}(L)$.

   As $T_{k,r}(L) \subseteq T_{k',r}(L)$, $\forall x, y \in \Sigma^*(v_{k'}(x) \leq v_{k'}(y) \Rightarrow v_k(x) \leq v_k(y))$ (v_k and $v_{k'}$ denote the vectors whose components are defined from $T_{k,r}(L)$ and $T_{k',r}(L)$, respectively).

   Let $x_1wx_2, x_3wx_4 \in L$ with $v_{k'}(x_1w) \leq v_{k'}(x_3w)$. As $w \in C_{k,r}(L)$, it follows that $x_1wx_4 \in L$, and thus $w \in C_{k',r}(L)$.

   (b) It can similarly be shown that if $w \in C_{k,r}(L)$ and $r' \geq r$ then $w \in C_{k,r'}(L)$.

   It is easy to prove a similar property for the (k,r)-constants of $S(L)$ for $\varphi(L)$. □

4.3   Characterization of $LTTSS$ Languages

Proposition 1 $L \subseteq \Sigma^*$ *is Threshold Locally Testable in Strict Sense if and only if* $\exists k, r \geq 1 : \forall w \in \Sigma^{k-1}\Sigma^*$, w *is (k,r)-constant for L.*

Proof. Let L be defined by the 4-tuple $(I, F, T_{k,r}, g)$, where I (F) is the set of prefixes (suffixes) of length $k-1$ of L, $T_{k,r}$ is the set of restricted factors of the words of L and g the function that defines the level of restriction of each factor.

Let $w \in \Sigma^{k-1}\Sigma^*$ and $x_1, x_2, x_3, x_4 \in \Sigma^*$ such that $x_1 w x_2, x_3 w x_4 \in L$ and $\mathbf{v}(x_1 w) \leq \mathbf{v}(x_3 w)$. As $|w| \geq k - 1$ it follows that $i_{k-1}(x_1 w x_2) = i_{k-1}(x_1 w)$, $f_{k-1}(x_3 w x_4) = f_{k-1}(w x_4)$ and $\mathbf{v}(x_1 w x_4) \leq \mathbf{v}(x_3 w x_4)$. Thus $x_1 w \in I\Sigma^*$, $w x_4 \in \Sigma^* F$, and $x_1 w x_4$ does not contain factors above their level of restriction, then $x_1 w x_4 \in L$ and w is a (k,r)-constant for L.

Conversely, let $I(L) = Pr(L) \cap \Sigma^{k-1}$, $F(L) = Suf(L) \cap \Sigma^{k-1}$ and $T_{k,r}(L) = \{w \in \Sigma_1^k : \forall x \in L, |x|_w < r\}$. We have to show that:

$$\Sigma^{k-1}\Sigma^* \cap L \;\; = \;\; I\Sigma^* \cap \Sigma^* F \;\; - \;\; \left(\bigcup - w \in T_k \{x \in \Sigma^* : |x|_w > g(w)\}\right),$$
$$\qquad\qquad\qquad (1) \qquad\qquad\qquad\qquad (2)$$

where $g(w) = \max_{x \in L} |x|_w$.

The left to right inclusion is obvious. To prove the opposite inclusion, if x belongs to the right side of the above equality, then $x = u_1 f = i u_2$, being $u_1, u_2 \in \Sigma^*$, $i \in I(L)$ and $f \in F(L)$ (by (1)), thus i and f are (k,r)-constants for L.

As x does not belong to (2), it is a completable word in L, that is, there exist words h_1 and h_2 such that $h_1 x h_2 \in L$, and thus $h_1 i u_2 h_2 \in L$. As $i \in I(L)$, there exists $h' \in \Sigma^*$ such that $i h' \in L$. Moreover $\mathbf{v}(i) \leq \mathbf{v}(h_1 i)$ thus, as i is a (k,r)-constant for L, $i u_2 h_2 = u_1 f h_2 \in L$.

As $f \in F(L)$, there exists $h'' \in \Sigma^*$ such that $h'' f \in L$ and $\mathbf{v}(h'' f) \geq \mathbf{v}(u_1 f)$. Since f *is* a (k,r)-constant for L it follows that $u_1 f = x \in L \cap \Sigma^{k-1}\Sigma^*$. $\qquad\square$

Corollary 1 *If L is (k, r)-TTSS then for* $k' \geq k$ *or* $r' \geq r$, *L is* (k',r')-*TTSS.*

Proof. By proposition 6, it follows that $\Sigma^{k-1}\Sigma^* \subseteq C_{k,r}(L)$ and, by property 6 of 4.2, with the current hypothesis we have $C_{k,r}(L) \subseteq C_{k',r'}(L)$. As $\Sigma^k \Sigma^* \subseteq \Sigma^{k-1}\Sigma^*$, applying proposition 6 again, the corollary holds. $\qquad\square$

Proposition 2 *Let* $L \subseteq \Sigma^*$ *be a regular language. L is Threshold Locally Testable in Strict Sense if and only if every idempotent of* $S(L)$ *is a semiconstant for* $\varphi(L)$.

Proof. If L is $LTTSS$ then there exist values of k and r such that every word of $\Sigma^{k-1}\Sigma^*$ is a (k,r)-constant for L. Let $e \in E(S(L))$ and let $x \in \Sigma^+$ such that $\varphi(x) = e$ and $|x| \geq k-1$, then x is a (k,r)-constant for L and, by property 2 of 4.2 $\varphi(x)$ is a (k,r)-constant for $\varphi(L)$.

Conversely, let us suppose that every idempotent of $S(L)$ is a semiconstant for $\varphi(L)$, then, there exist values for k and r such that every idempotent of $S(L)$ is a (k,r)-constant for $\varphi(L)$. By property 6 of 4.2, we can choose $k = |S(L)|$ and $r = \max_{w \in T_k}(\max_{x \in L} |x|_w) + 1$, where $T_k = \{w \in \Sigma_1^k : \exists n, \max_{x \in L} |x|_w \leq n\}$.

Let $x \in \Sigma^*$ with $|x| \geq k+1$, that is, $x = a_1 a_2 ... a_n$ being $n \geq k+1$. It is obvious that there exist p and q with $1 \leq p < q \leq k+1$ such that $a_1 ... a_p \equiv_L a_1 ... a_p (a_{p+1} ... a_q)^i \ \forall i \geq 1$. For a sufficiently large i $\varphi((a_{p+1} ... a_q)^i) \in E(S(L))$ and, so, it will be a (k,r)-constant for $\varphi(L)$. By property 3 of 4.2, $(a_{p+1} ... a_q)^i$ will be a (k,r)-constant for L and so will be $a_1 ... a_p(a_{p+1} ... a_q)^i a_{q+1} ... a_n$, since $C_{k,r}(L)$ is an ideal. As $x \equiv_L a_1 ... a_p(a_{p+1} ... a_q)^i a_{q+1} ... a_n$, we have that $x \in C_{k,r}(L)$. Hence by proposition 2 it follows that L is $LTTSS$. $\square$

Proposition 8 gives a decision procedure to determine if a language L is $LTTSS$. It is enough to check if $E(S(L)) \subseteq C_{k,r}(\varphi(L))$, k and r being the values used in the proof of proposition 2.

References

[1] D. Beauquier and J.E. Pin, Languages and scanners. *Theoretical Computer Science*, 84 (1991), 3–21.

[2] J.A. Brzozowski and I. Simon, Characterizations of locally testable events. *Discrete Mathematics*, 4 (1973), 243–271.

[3] K.S. Fu, *Syntactic Pattern Recognition and Applications*. Prentice-Hall, Englewood Cliffs, N.J., 1982.

[4] P. García and E. Vidal, Inference of k-testable languages in the strict sense and application to syntactic pattern recognition. *IEEE Transactions on Pattern Analysis and Machine Intelligence*, PAMI-12 (1990), 920–925.

[5] R.C. González and M.G. Thomason, *Syntactic Pattern Recognition: An Introduction*. Addison-Wesley, Reading, Mass., 1978.

[6] J. Hopcroft and J. Ullman, *Introduction to Automata Theory, Languages and Computation*. Addison-Wesley, Reading, Mass., 1979.

[7] A. de Luca and A. Restivo, A characterization of strictly locally testable languages and its application to subsemigroups of a free semigroup. *Information and Control*, 44 (1980), 300–319.

[8] R. McNaughton, Algebraic decision procedures for local testability. *Mathematics Systems Theory*, 8/1 (1971), 60–76.

[9] Y.T. Medvedev, On the class of events representable in a finite automaton. In E.F. Moore (ed.), *Sequential Machines: Selected Papers.* Addison-Wesley, Reading, Mass., 1964, 227–315.

[10] J.E. Pin, *Variétés de Langages Formels.* Masson, Paris, 1984.

[11] J. Ruiz, S. España and P. García, Threshold locally testable languages in strict sense. Application to the inference problem. In *Proceedings of the 4th International Conference on Grammatical Inference,* Springer, Berlin, 1998, 150–161.

[12] H. Straubing, Finite semigroup varieties of the form V*D. *Journal of Pure and Applied Algebra,* 36 (1985), 53–94.

[13] H. Straubing, *Finite Automata, Formal Logic and Circuit Complexity.* Birkhäuser, Boston, 1994.

[14] D. Thérien and A. Weiss, Graph congruences and wreath products. *Journal of Pure and Applied Algebra,* 35 (1985), 205–215.

[15] W. Thomas, Classifying regular events in symbolic logic. *Journal of Computer and System Sciences,* 25 (1982), 360–376.

[16] E. Vidal, F. Casacuberta and P. García, Grammatical inference and automatic speech recognition. In *Speech Recognition and Coding.* Springer, Berlin, 1995, 175–191.

[17] Y. Zalcstein, Locally testable languages. *Journal of Computer and System Sciences,* 6 (1972), 151–167.

Characterizations of Language Classes: Universal Grammars, Dyck Reductions, and Homomorphisms

Sadaki Hirose

Toyama University
Japan
hirose@ecs.toyama-u.ac.jp

Satoshi Okawa

The University of Aizu
Aizu-Wakamatsu, Japan
okawa@u-aizu.ac.jp

Abstract. This paper investigates characterizations of language classes with universal grammars, Dyck reductions, and homomorphisms, and discusses the relations between them.

1 Introduction

One of the main trends in formal language theory is to search for characterizations of language classes. There have been many characterizations from several viewpoints. Grammatical characterizations, characterizations by automata, homomorphic characterizations, characterizations with Dyck reductions, characterizations with equality set are just a few examples.

This paper discusses characterizations with universal grammars, Dyck reductions and homomorphisms and states that they have almost the same power of characterization. That is, the following three statements have the same meaning:

(1) *There exists a left universal context-free grammar for a language class $\mathcal{L}$ with respect to some language class $\mathcal{L}'$.*

(2) *A language class $\mathcal{L}$ can be characterized with Dyck reductions of some language class $\mathcal{L}'$.*

(3) *A language class $\mathcal{L}$ can be homomorphically characterized with a context-free language L_0 and some language class $\mathcal{L}'$.*

Throughout this paper, let $\mathcal{R}$, $\mathcal{ML}$, $\mathcal{LCF}$, $\mathcal{CF}$, $\mathcal{CS}$ and $\mathcal{RE}$ be the classes of languages that are regular, minimal linear (which is generated by a linear context-free grammar with only one nonterminal symbol (refer to Chomsky and Schützenberger [6])), linear context-free, context-free, context-sensitive and recursively enumerable, respectively.

Readers are expected to be familiar with the fundamental definitions and results of formal language theory. For background materials and additional details, refer to Salomaa [27] and Berstel [2], for example.

Sections 2, 3 and 4 give the results of characterization with universal grammars, Dyck reductions and homomorphisms, respectively. Section 5 discusses the relation between characterizations with universal grammars and homomorphisms, and Section 6 focusses on the relation between characterizations with Dyck reductions and homomorphisms.

2   Universal Grammars

Kasai [22] showed that, for each alphabet Σ, there exists a context-free grammar G such that for every context-free language L over Σ a regular control language C can be found such that G controlled by C generates L by leftmost derivations. Kasai called such a grammar *universal*.

Before giving the formal definitions of universal grammars, we define the terminologies and notations.

Let $G = (V, \Sigma, P, S)$ be a phrase structure grammar, where V is a finite set of nonterminals, Σ is a finite set of terminals, P is a finite set of production rules of the form $\pi : \alpha \to \beta$ with $\alpha \in (V \cup \Sigma)^* V (V \cup \Sigma)^*$ and $\beta \in (V \cup \Sigma)^*$, and $S \in V$ is a start symbol.

For any $\xi, \zeta \in (V \cup \Sigma)^*$, if $\xi = \omega_1 \alpha \omega_2, \zeta = \omega_1 \beta \omega_2$ and $\pi : \alpha \to \beta \in P$, then ζ can be derived from ξ with π and this derivation is denoted as $\xi \Rightarrow^\pi \zeta$. If $\omega_1 \in \Sigma^*$, then this derivation is called leftmost and denoted as $\xi \Rightarrow_l^\pi \zeta$.

For $\xi, \zeta \in (V \cup \Sigma)^*$ and $x = \pi_1 \pi_2 ... \pi_n$, with $\pi_i \in P$ for $i = 1, 2, ..., n$, we write $\xi \Rightarrow^x \zeta$ ($\xi \Rightarrow_l^x \zeta$) if there exist $\alpha_0, \alpha_1, ..., \alpha_n \in (V \cup \Sigma)^*$ such that $\alpha_0 = \xi$, $\alpha_n = \zeta$, and $\alpha_{i-1} \Rightarrow^{\pi_i} \alpha_i$ ($\alpha_{i-1} \Rightarrow_l^{\pi_i} \alpha_i$) for $i = 1, 2, ..., n$.

For a control language $C \subset P^+$ we define:

$$L^U(G) = \{w \in \Sigma^* \mid S \Rightarrow^x w \text{ } for \text{ } some \text{ } x \in U\},$$

$$L_l^C(G) = \{w \in \Sigma^* \mid S \Rightarrow_l^x w \text{ } for \text{ } some \text{ } x \in C\}.$$

Definition 1 *A grammar G is said to be universal (left universal) for a language class $\mathcal{L}$ with respect to some language class $\mathcal{L}'$ if for any language $L \in \mathcal{L}$ over Σ there exists a control language $C \in \mathcal{L}'$ such that $L^C(G) = L$ ($L_l^C(G) = L$).*

If $\mathcal{L}' = \mathcal{R}$, we simply call G universal or left universal for $\mathcal{L}$.

Theorem 1 (Kasai) *There exists a left universal context-free grammar for $\mathcal{CF}$.*

Rozenberg [26] gave a similar result for phrase structure grammars, recursively enumerable languages, and unrestricted derivations.

Theorem 2 (Rozenberg) *There exists a universal phrase structure grammar for $\mathcal{RE}$.*

An analogous result was independently proved by Hart [13].

Rozenberg, on the other hand, conjectured that no such universal context-sensitive grammar could exist, and Greibach [11] verified this conjecture.

Theorem 3 (Greibach) *There exists no universal context-sensitive grammar for $\mathcal{CS}$.*

Further studies have been made on universal grammars.

Theorem 4 (Greibach) *There exists a left universal context-free grammar for $\mathcal{RE}$ with respect to $\mathcal{LCF}$.*

Theorem 5 (Hirose and Nasu [14]) *There exists no universal context-sensitive grammar for $\mathcal{CS}$ with respect to the class of k-derivation bounded context-free languages (refer to [10]).*

3   Dyck Reductions

Let k be a positive integer, and let $X_k = \{x_i, x_i' | 1 \leq i \leq k\}$ be an alphabet of k pairs of matching parentheses, where x_i is the opening and x_i' is the closing parenthesis.

A Dyck language consists of the words over X_k which are "well-formed" in the usual sense.

A Dyck reduction over X_k is an operation to reduce a string $x_i x_i'$ ($1 \leq i \leq k$) within any string $u x_i x_i' v \in \Gamma^*$ to the empty word λ, where Γ is any alphabet including X_k. This reduction is denoted by $u x_i x_i' v \propto uv$. For a string $w \in \Gamma^*$, Dyck reductions are applied repeatedly until they can no longer be applied and the result of the reductions is w'. Then, we denote $w \propto^* w'$. For a language L over Γ, let $Red(L) = \{w'|w \propto^* w'\ for\ some\ w\ in\ L\}$.

Dyck languages are among the most frequently cited and the most typical context-free languages, and Dyck reductions have also been studied.

For a language class $\mathcal{L}$, the characterization with Dyck reductions of some language class $\mathcal{L}'$ is defined as follows.

Definition 2 *A language class $\mathcal{L}$ is said to be characterized with Dyck reductions of some language class $\mathcal{L}'$ when the following statement holds: for any $L \in \mathcal{L}$ over Σ, there exists an alphabet X_k, Dyck reductions Red over X_k, and a language $L' \in \mathcal{L}'$ over $\Sigma \cup X_k$ such that $L = Red(L') \cap \Sigma^*$.*

There have been many characterizations with Dyck reductions. In [12, 29, 28], Griffiths, Stanat and Savitch showed that $\mathcal{RE}$ is characterized with Dyck reductions of $\mathcal{CF}$. This result is very famous. Brandenburg has shown in [4] that even with Dyck reductions of $\mathcal{LCF}$ two pairs of matching parentheses are sufficient, and the same result has been shown by Frougny, Sakarovitch and Schupp in [7].

Theorem 6 (Griffiths, Stanat and Savitch) *$\mathcal{RE}$ can be characterized with Dyck reductions of $\mathcal{CF}$.*

Theorem 7 (Brandenburg) *$\mathcal{RE}$ can be characterized with Dyck reductions of $\mathcal{LCF}$.*

Recently, in [8], Geffert has shown that every recursively enumerable language L over Σ can be expressed in the form $L = \{h(x)^{-1}g(x)|x\ in\ \Delta^+\} \cap \Sigma^*$, where Δ is an alphabet and g, h is a pair of morphisms. In [23], Latteux and Turakainen have given a simple proof of Geffert's result and refined it so that both morphisms are nonerasing. They have also shown that one of the applications of the results is that each recursively enumerable language can be obtained from a minimal linear language with Dyck reductions. Both the class of minimal linear languages and the class of regular languages are proper subclasses of linear languages, but they are incomparable. It is also known that the class of regular languages is closed with respect to Dyck reductions [3], [21]. These facts emphasize the value of their result.

Theorem 8 (Latteux and Turakainen) *$\mathcal{RE}$ can be characterized with Dyck reductions of $\mathcal{ML}$.*

In [16], Hirose and Okawa gave direct and constructive proof of Latteux and Turakainen's result.

4 Homomorphisms

For a language class $\mathcal{L}$, there are two typical types of homomorphic characterizations. One is defined as follows.

Definition 3 *A language class $\mathcal{L}$ is said to be homomorphically characterized with two language classes $\mathcal{L}_1$ and $\mathcal{L}_2$ when the following statement holds: for any language $L \in \mathcal{L}$ over Σ, there exists an alphabet Γ, a homomorphism $h : \Gamma^* \to \Sigma^*$, a language $L_1 \in \mathcal{L}_1$, and a language $L_2 \in \mathcal{L}_2$ such that $L = h(L_1 \cap L_2)$.*

In the case that $\mathcal{L}_2$ is equal to $\mathcal{L}_1$, $\mathcal{L}$ is said to be homomorphically characterized with $\mathcal{L}_1$.

Theorem 9 (Ginsburg, Greibach and Harrison [9]) *$\mathcal{RE}$ can be homomorphically characterized with $\mathcal{DCF}$.*[1]

Theorem 10 (Baker and Book [1]) *$\mathcal{RE}$ can be homomorphically characterized with $\mathcal{LCF}$.*

Theorem 11 (Okawa and Hirose [24]) *For any suffix pair $(\mathcal{X}, \mathcal{Y}) \in \{<, = , >\} \times \{<, =, >\} - \{(=, =)\}$, $\mathcal{RE}$ can be homomorphically characterized with $\mathcal{LCF}_{\mathcal{X}}$ and $\mathcal{LCF}_{\mathcal{Y}}$.*[2]

Another type of homomorphic characterizations is as follows.

Definition 4 *A language class $\mathcal{L}$ is said to be homomorphically characterized with some language L_0 and some language class $\mathcal{L}'$ when the following statement holds: there exists an alphabet Γ, a language L_0 over Γ, and a homomorphism $h : \Gamma^* \to \Sigma^*$, and for any language $L \in \mathcal{L}$ over Σ there exists a language $L' \in \mathcal{L}'$ over Γ such that $L = h(L' \cap L_0)$.*

A Dyck language D is well used as the language L_0.

This type of characterization for $\mathcal{CF}$ by Chomsky and Stanley is very famous.

Theorem 12 (Chomsky [5] and Stanley [30]) *$\mathcal{CF}$ can be homomorphically characterized with a Dyck language and $\mathcal{R}$.*

In [20], Hirose and Yoneda refined the above theorem.

Theorem 13 (Hirose and Yoneda) *$\mathcal{CF}$ can be homomorphically characterized with a Dyck language and the class of minimal linear and regular languages.*

---

[1] $\mathcal{DCF}$ is a class of languages accepted by deterministic pushdown automata.
[2] A suffix $<$ means that for every production rule except terminating rules $A \to \omega_1 B \omega_2$, $|\omega_1| < |\omega_2|$, and other cases are defined similarly.

For $\mathcal{CS}$, under some additional condition to the definition of the homomorphic characterization, the following result was known.

Theorem 14 (Okawa, Hirose and Yoneda [25]) *$\mathcal{CS}$ cannot be homomorphically characterized with a Dyck language and $\mathcal{L}$, which is closed under λ-free homomorphisms.*

There have also been many homomorphic characterizations of this type for $\mathcal{RE}$. In [15], Hirose and Nasu showed that any recursively enumerable language is the homomorphic image of the intersection of a Dyck language and linear context-free languages. And in [19], Hirose, Okawa and Yoneda refined Hirose and Nasu's result, and directly showed that any recursively enumerable language is the homomorphic image of the intersection of a Dyck language and a minimal linear language.

Theorem 15 (Hirose and Nasu) *$\mathcal{RE}$ can be homomorphically characterized with a Dyck language and $\mathcal{LCF}$.*

Theorem 16 (Hirose, Okawa and Yoneda) *$\mathcal{RE}$ can be homomorphically characterized with a Dyck language and $\mathcal{ML}$.*

5   Relation Between Characterizations With Universal Grammars and Homomorphisms

The relation between characterizations with universal grammars and homomorphisms is proved in [15].

Theorem 17 (Hirose and Nasu) *Let $\mathcal{L}$ be a class of languages and let $\mathcal{L}'$ be some class of languages which is closed under homomorphisms and inverse homomorphisms. Then, the statements (1) and (2), below, are equivalent.*

(1) There exists a left universal context-free grammar for $\mathcal{L}$ with respect to $\mathcal{L}'$.

(2) $\mathcal{L}$ can be homomorphically characterized with a context-free language L_0 and $\mathcal{L}'$.

Some applications of the above theorem can also be found in [15].

The class of regular languages is closed under homomorphisms and inverse homomorphisms, so it follows from Theorem 12 that we have another proof of Theorem 1, since we choose a Dyck language D, which is context-free, as L_0 in Theorem 17. But, in the reverse direction, we proved the following result:

Corollary 1 *$\mathcal{CF}$ can be homomorphically characterized with a context-free language L_0 and $\mathcal{R}$.*

This result is a little weaker than Theorem 12.

Using Theorem 15, we have further proof of Theorem 4. On the other hand, using Theorem 4, we have a result that is weaker than Theorem 15 and parallel to Corollary 1.

6   Relation Between Characterizations With Homomorphisms and Dyck Reductions

The relation between characterizations with homomorphisms and Dyck reductions is discussed in [18, 17].

Theorem 18 (Hirose, Okawa and Kimura) *Let $\mathcal{L}$ be any class of languages, and $\mathcal{L}'$ be some class of languages which is closed under λ-free homomorphisms. If statement (1) below holds, then statement (2) holds:*

(1) *$\mathcal{L}$ can be characterized with Dyck reductions of $\mathcal{L}'$.*

(2) *$\mathcal{L}$ can be homomorphically characterized with a Dyck language and $\mathcal{L}'$.*

In [17], the converse of the above theorem is examined and the following almost converse result is obtained, that is, it needs some restriction on a language class $\mathcal{L}'$.

Theorem 19 (Hirose, Okawa and Kimura) *Let $\mathcal{L}$ be any class of languages, $\mathcal{L}'$ be one of $\mathcal{CF}$, $\mathcal{CS}$ and $\mathcal{RE}$. If statement (1) below holds, then statement (2) holds:*

(1) *$\mathcal{L}$ can be homomorphically characterized with a Dyck language and $\mathcal{L}'$.*

(2) *$\mathcal{L}$ can be characterized with Dyck reductions of $\mathcal{L}'$.*

Some applications of these theorems are also stated in [18, 17].

As $\mathcal{ML}$ is closed under λ-free homomorphisms, we have further proof of Theorem 16 from Theorem 8 as a corollary to Theorem 18. And though it is a weaker result than Theorem 8, we have further proof of Theorem 15 by using Theorem 7.

Although we would like to conclude Theorem 7 and 8 from Theorem 15 and 16 as a corollary to Theorem, respectively, we cannot because of the weakness of Theorem 19. However, since $\mathcal{ML}$ ($\mathcal{LCF}$) is a subclass of $\mathcal{CF}$, we can apply Theorem 19 to Theorem 15 (16) and, as a result, we have further proof of Theorem 6.

References

[1] B. Baker and R. Book, Reversal-bounded multipushdown machines, *Journal of Computer and System Sciences*, 8 (1974), 315–332.

[2] J. Berstel, *Transductions and Context-Free Languages.* Teubner, Stuttgart, 1979.

[3] R.V. Book, M. Jantzen and C. Wrathall, Monadic Thue systems, *Theoretical Computer Science*, 19 (1982), 231–251.

[4] F.J. Brandenburg, Cancellations in Linear Context-Free Languages, Technical Report MIP-8904, University of Passau, 1989.

[5] N. Chomsky, Context-free grammars and pushdown storage, *MIT Res. Lab. Electron. Quart. Prog. Reports*, 65 (1962), 187–194.

[6] N. Chomsky and M.P. Schützenberger, The algebraic theory of context-free languages. In P. Braffort and D. Hirschberg (eds.), *Computer Programming and Formal Systems*. North-Holland, Amsterdam, 1962, 118–161.

[7] C. Frougny, J. Sakarovitch and P. Schupp, Finiteness conditions on subgroups and formal language theory, *Proceedings of the London Mathematical Society*, 58 (1989), 74–88.

[8] V. Geffert, A representation of recursively enumerable languages by two homomorphisms and a quotient, *Theoretical Computer Science*, 62 (1988), 235–249.

[9] S. Ginsburg, S.A. Greibach and M.A. Harrison, One-way stack automaton, *Journal of the ACM*, 14 (1967), 389–418.

[10] S. Ginsburg and E. Spanier, Derivation-bounded languages, *Journal of Computer and System Sciences*, 2 (1968), 228.

[11] S.A. Greibach, Comments on universal and left universal grammars, context-sensitive languages, and context-free grammar forms, *Information and Control*, 39 (1978), 135–142.

[12] T.V. Griffiths, Some remarks on derivations in general rewriting systems, *Information and Control*, 12 (1968), 27–54.

[13] J.M. Hart, Two Extensions of Kasai's Universal Context-Free Grammar, Technical Report 22-75, Department of Computer Science, University of Kentucky, 1976.

[14] S. Hirose and M. Nasu, Some notes on universal grammars, *Transactions IECE*, Japan 63-D (1980), 287–294 (in Japanese).

[15] S. Hirose and M. Nasu, Left universal context-free grammars and homomorphic characterizations of languages, *Information and Control*, 50 (1981), 110–118.

[16] S. Hirose and S. Okawa, Dyck reductions of minimal linear languages yield the full class of recursively enumerable languages, *IEICE Transactions on Information and Systems*, E79-D/2 (1996), 161–164.

[17] S. Hirose, S. Okawa and H. Kimura, Dyck reductions are more powerful than homomorphic characterizations, *IEICE Transactions on Information and Systems*, E80-D/9 (1997), 958–961.

[18] S. Hirose, S. Okawa and H. Kimura, Homomorphic characterizations are more powerful than Dyck reductions, *IEICE Transactions on Information and Systems*, E80-D/3 (1997), 390–392.

[19] S. Hirose, S. Okawa and M. Yoneda, A homomorphic characterization of recursively enumerable languages, *Theoretical Computer Science*, 35 (1985), 261–269.

[20] S. Hirose and M. Yoneda, On the Chomsky and Stanley's homomorphic characterization of context-free languages, *Theoretical Computer Science*, 36 (1985), 109–112.

[21] M. Jantzen, *Confluent String Rewriting*. Springer, Berlin, 1988.

[22] T. Kasai, A universal context-free grammar, *Information and Control*, 28 (1975), 30–34.

[23] M. Latteux and P. Turakainen, On characterizations of recursively enumerable languages, *Acta Informatica*, 28 (1990), 179–186.

[24] S. Okawa and S. Hirose, Subclasses of linear context-free languages and homomorphic characterizations of languages, *Systems and Computers in Japan*, 22.5 (1991), 30–38.

[25] S. Okawa, S. Hirose and M. Yoneda, On the impossibility of the homomorphic characterization of context-sensitive languages, *Theoretical Computer Science*, 44 (1986), 225–228.

[26] G. Rozenberg, A note on universal grammars, *Information and Control*, 34 (1977), 172–175.

[27] A. Salomaa, *Jewels of Formal Language Theory*. Computer Science Press, Potomac, Md., 1981.

[28] W.J. Savitch, How to make arbitrary grammars look like context-free grammars, *SIAM Journal of Computing*, 2 (1973), 174–182.

[29] D.F. Stanat, Formal languages and power series. In *Proceedings of the 3rd ACM Symposium on Theory of Computing*, 1971, 1–11.

[30] R.J. Stanley, Finite state representations of context-free languages, *MIT Res. Lab. Electron. Quart. Prog. Reports*, 76 (1965), 276–279.

On D0L Power Series over Various Semirings[1]

Juha Honkala

Department of Mathematics
University of Turku
Finland
juha.honkala@utu.fi

Abstract. We study Fatou properties of D0L power series and related decision problems.

1 Introduction

The study of formal power series is an important area of theoretical computer science (see Eilenberg [2] and Kuich and Salomaa [8]). The formal power series that are most often considered in connection with automata and languages are the rational and algebraic series. In this paper, we discuss the D0L power series defined in Honkala [4] and studied in detail in Honkala [5-7]. The study of these series is an interesting counterpart to the customary theory of D0L languages.

In this paper, we investigate the Fatou properties of D0L power series. By definition, a semiring A is a Fatou extension of its subsemiring B with respect to D0L power series if r is an D0L power over B, whenever r is an D0L power series over A with its coefficients in B. When A is not a Fatou extension of B with respect to D0L power series, we consider the problem of deciding whether or not an D0L power series r over A is an D0L power series over B. Similar problems are a well known research area in the study

---

[1]Research supported by the Academy of Finland.

of rational series (see Salomaa and Soittola [11] and Berstel and Reutenauer [1]). In our study of D0L power series, rational series are essential tools.

For further background and motivation we refer to Honkala [3-7] and the references given therein.

The reader is assumed to be familiar with the basics of the theories of semirings, formal power series and L systems (see Kuich and Salomaa [8], Rozenberg and Salomaa [9,10]). Notions and notations that are not defined are taken from these references.

2   Definitions and Earlier Results

Suppose A is a commutative semiring and X is a finite alphabet. The set of *formal power series* (resp. *polynomials*) *with noncommuting variables* in X and coefficients in A is denoted by $A \ll X^* \gg$ (resp. $A < X^* >$).

Assume that X and Y are finite alphabets. A semialgebra morphism $h : A < X^* > \longrightarrow A < Y^* >$ is called a *monomial morphism* if for each $x \in X$ there exists a nonzero $a \in A$ and $w \in Y^*$ such that $h(x) = aw$. If $h : A < X^* > \longrightarrow A < Y^* >$ is a monomial morphism, the *underlying monoid morphism* $\overline{h} : X^* \longrightarrow Y^*$ is defined by $\overline{h}(x) = \mathrm{supp}(h(x))$ for $x \in X$. A series $r \in A \ll X^* \gg$ is called an D0L *power series* over A if there exists a nonzero $a \in A$, a word $w \in X^*$ and a monomial morphism $h : A < X^* > \longrightarrow A < X^* >$ such that:

$$r = \sum_{n=0}^{\infty} ah^n(w) \tag{1}$$

and, furthermore:

$$\mathrm{supp}(ah^i(w)) \neq \mathrm{supp}(ah^j(w)) \text{ whenever } 0 \leq i < j.$$

(Note that, if A is a field, this condition is equivalent to the local finiteness of the family $\{h^n(w)\}_{n \geq 0}$.)

Consider the series r given in (1) and denote:

$$ah^n(w) = c_n w_n,$$

where $c_n \in A$ and $w_n \in X^*$ for $n \geq 0$. Then we have:

$$r = \sum_{n=0}^{\infty} c_n w_n. \tag{2}$$

In what follows, the righthand side of (2) is called the *normal form* of r. A sequence $(c_n)_{n \geq 0}$ of elements of A is called an D0L *multiplicity sequence* over A if there exists an D0L power series r over A such that (2) is the normal form of r.

Now, suppose A and B are commutative semirings and $B \subseteq A$. We say that A is a *Fatou extension of B with respect to D0L power series* if whenever $r \in A \ll X^* \gg$ is an D0L power series over A such that $r \in B \ll X^* \gg$, r is an D0L power series over B. Similarly, A is a *Fatou extension of B with respect to D0L multiplicity sequences* if whenever $(c_n)_{n\geq 0}$ is an D0L multiplicity sequence over A such that $c_n \in B$ for all $n \geq 0$, $(c_n)_{n\geq 0}$ is an D0L multiplicity sequence over B.

For later use we recall the characterization of D0L multiplicity sequences from Honkala [7].

A sequence $(a_n)_{n\geq 0}$ of nonnegative integers is called a *modified PD0L length sequence* if there exists a nonnegative integer t such that $a_0 = a_1 = \ldots = a_{t-1} = 0$ and $(a_{n+t})_{n\geq 0}$ is an PD0L length sequence. A sequence $(a_n)_{n\geq 0}$ of nonnegative integers is a modified PD0L length sequence if and only if the sequence $(a_{n+1}-a_n)_{n\geq 0}$ is **N**-rational (see Rozenberg and Salomaa [9]).

Theorem 1 *Suppose A is a commutative semiring. A sequence $(c_n)_{n\geq 0}$ of nonzero elements of A is an D0L multiplicity sequence over A if and only if there exists a positive integer k, nonzero $a_1, \ldots, a_k \in A$ and modified PD0L length sequences $(s_{in})_{n\geq 0}$ for $1 \leq i \leq k$ such that:*

$$c_n = \prod_{i=1}^{k} a_i^{s_{in}}, \qquad (3)$$

for all $n \geq 0$.

We will also need the following result from Honkala [7].

Theorem 2 *Suppose A is a field. A sequence $(c_n)_{n\geq 0}$ of nonzero elements of A is an D0L multiplicity sequence over A if and only if there exists a positive integer t and integers $\beta_1, \ldots, \beta_t$ such that:*

$$c_{n+t} = c_{n+t-1}^{\beta_1} c_{n+t-2}^{\beta_2} \cdots c_n^{\beta_t}, \qquad (4)$$

for $n \geq 0$.

3 Fatou Properties

The positive Fatou properties given below follow easily from the definitions and Theorem 2. A direct proof of Theorem 4 would be much more difficult.

Theorem 3 *Suppose $A \subseteq \mathbf{R}$ is a semiring. Then A is a Fatou extension of A_+ with respect to D0L multiplicity sequences and D0L power series. In particular, $\mathbf{Z}$ is a Fatou extension of $\mathbf{N}$ with respect to D0L multiplicity sequences and D0L power series.*

Proof. Suppose $r \in A_+ \ll X^* \gg$ is an D0L power series over A given by:

$$r = \sum_{n=0}^{\infty} ah^n(w),$$

where $h : A < X^* > \longrightarrow A < X^* >$ is a monomial morphism, $a \in A$ and $w \in X^*$. Define the monomial morphism $h_1 : A_+ < X^* > \longrightarrow A_+ < X^* >$ by $h_1(x) = h(x)$ if the coefficient of $h(x)$ is positive and $h_1(x) = -h(x)$ if the coefficient of $h(x)$ is negative, $x \in X$. Then:

$$r = \sum_{n=0}^{\infty} ah_1^n(w)$$

is an D0L power series over A_+. Hence, A is a Fatou extension of A_+ with respect to D0L power series. It follows that A is a Fatou extension of A_+ with respect to D0L multiplicity sequences. $\square$

Theorem 4 *Suppose E and F are fields and $F \subseteq E$. Then E is a Fatou extension of F with respect to D0L multiplicity sequences.*

Proof. Suppose $(c_n)_{n \geq 0}$ is an D0L multiplicity sequence over E such that $c_n \in F$ for all $n \geq 0$. Then the only if-part of Theorem 2 implies the existence of a positive integer t and integers $\beta_1, \ldots, \beta_t$ such that:

$$c_{n+t} = c_{n+t-1}^{\beta_1} c_{n+t-2}^{\beta_2} \cdots c_n^{\beta_t}, \tag{5}$$

for $n \geq 0$. Now (5) implies by the if-part of Theorem 2 that $(c_n)_{n \geq 0}$ is an D0L multiplicity sequence over F. $\square$

Theorem 5 $\mathbf{R}_+$ *is a Fatou extension of $\mathbf{Q}_+$ with respect to D0L multiplicity sequences.*

Proof. Suppose $(c_n)_{n \geq 0}$ is an D0L multiplicity sequence over $\mathbf{R}_+$ such that $c_n \in \mathbf{Q}_+$ for all $n \geq 0$. Then $(c_n)_{n \geq 0}$ is an D0L multiplicity sequence over $\mathbf{R}$. By Theorem 4, $(c_n)_{n \geq 0}$ is an D0L multiplicity sequence over $\mathbf{Q}$. Hence, by Theorem 3, $(c_n)_{n \geq 0}$ is an D0L multiplicity sequence over $\mathbf{Q}_+$. $\square$

In general, a field is not a Fatou extension of its subfield with respect to D0L power series.

Example 1 *Let $E = \mathbf{Q}(\sqrt{2})$, $X = \{b\}$ and define the monomial morphism $h : E < X^* > \longrightarrow E < X^* >$ by $h(b) = \sqrt{2}b^2$. Then:*

$$r = \sum_{n=0}^{\infty} h^n(b^2)$$

is an D0L power series over E. *Because:*

$$r = \sum_{n=0}^{\infty} 2^{2^n-1} b^{2^{n+1}},$$

we have $r \in \mathbf{Q} \ll X^* \gg$. *However,* r *is not an D0L power series over* $\mathbf{Q}$. *Indeed, if* $g : \mathbf{Q} < X^* > \longrightarrow \mathbf{Q} < X^* >$ *were a monomial morphism such that:*

$$r = \sum_{n=0}^{\infty} g^n(b^2),$$

we would have $g(b^2) = 2b^4$ *which is not possible. Hence* $\mathbf{Q}(\sqrt{2})$ *is not a Fatou extension of* $\mathbf{Q}$ *with respect to D0L power series.*

Example 1 also shows that $\mathbf{R}_+$ is not a Fatou extension of $\mathbf{Q}_+$ with respect to D0L power series.

Example 2 *Let* $A = \mathbf{Q}_+$ *and* $X = \{b, c, d, e\}$. *Define the monomial morphism* $h : A < X^* > \longrightarrow A < X^* >$ *by:*

$$h(b) = d, \quad h(c) = \frac{1}{2}e^2, \quad h(d) = 2b^2, \quad h(e) = c.$$

Then:

$$r = 2 \sum_{n=0}^{\infty} h^n(b^2 c)$$

is an D0L power series over $\mathbf{Q}_+$. *The associated D0L multiplicity sequence* $(c_n)_{n \geq 0}$ *is given by:*

$$c_n = \begin{cases} 2^{2^{\frac{n}{2}}} & \text{if } n \text{ is even,} \\ 1 & \text{if } n \text{ is odd} \end{cases}$$

(see Example 2.4 in Honkala [6]). Hence, $c_n \in \mathbf{N}$ *for all* $n \geq 0$. *However, by Theorem 1,* $(c_n)_{n \geq 0}$ *is not an D0L multiplicity sequence over* $\mathbf{N}$ *(resp.* $\mathbf{Z}$*). It follows that* $\mathbf{Q}$ *is not a Fatou extension of* $\mathbf{N}$ *or* $\mathbf{Z}$ *and* $\mathbf{Q}_+$ *is not a Fatou extension of* $\mathbf{N}$ *with respect to D0L multiplicity sequences or D0L power series.*

4 Decidability Questions

In this section, we discuss various decidability questions which are closely related to Fatou properties of D0L multiplicity sequences and D0L power series.

Theorem 6 *Suppose* $A \subseteq \mathbf{R}$ *is a semiring. It is decidable whether or not* $c_n \in A_+$ *for all* $n \geq 0$ *if* $(c_n)_{n \geq 0}$ *is an D0L multiplicity sequence over* A.

Proof. By Theorem 2 there exists a positive integer t and integers $\beta_1, \ldots, \beta_t$ such that:

$$c_{n+t} = c_{n+t-1}^{\beta_1} c_{n+t-2}^{\beta_2} \cdots c_n^{\beta_t}, \tag{6}$$

for $n \geq 0$. Therefore, $c_n \in A_+$ for all $n \geq 0$ if and only if $c_n \in A_+$ for all $0 \leq n < t$. $\qquad\square$

Theorem 6 holds true with the same proof also if A and A_+ are replaced by E and F, respectively, where $F \subseteq E$ are fields, provided that we can check whether given elements of E belong to F.

Theorem 7 *Suppose $A \subseteq \mathbf{R}$ is a semiring. It is decidable whether or not a D0L multiplicity sequence $(c_n)_{n \geq 0}$ over A is an D0L multiplicity sequence over A_+. Similarly, it is decidable whether or not an D0L power series r over A is an D0L power series over A_+.*

Proof. The claims follow by Theorems 3 and 6. $\qquad\square$

In what follows, we tacitly assume that D0L power series are given in the normal form. In particular, an D0L power series r over A with the normal form (2) is said to be an D0L power series over a subsemiring $B \subseteq A$ only if r as an D0L power series over B also has the normal form (2). We are going to discuss the following three problems.

Problem 1 *Suppose $(c_n)_{n \geq 0}$ is an D0L multiplicity sequence over $\mathbf{Q}$. Is $(c_n)_{n \geq 0}$ an D0L multiplicity sequence over $\mathbf{Z}$ (resp. $\mathbf{N}$)?*

Problem 2 *Suppose r is an D0L power series over $\mathbf{Q}$. Is r a D0L power series over $\mathbf{Z}$ (resp. $\mathbf{N}$)?*

Problem 3 *Suppose r is an D0L power series over a computable field E of characteristic zero. Is r an D0L power series over $\mathbf{Q}$?*

We will show that Problems 1-3 are decidable. This should be contrasted with the fact that an algorithm to decide whether or not the elements of an D0L multiplicity sequence over $\mathbf{Q}$ are integers would solve Skolem's problem. Indeed, suppose $(a_n)_{n \geq 0}$ is a $\mathbf{Z}$-rational sequence and define the $\mathbf{Z}$-rational sequence $(b_n)_{n \geq 0}$ by $b_n = a_n^2 - 1$ for $n \geq 0$. Then the sequence $(2^{b_n})_{n \geq 0}$ is an D0L multiplicity sequence over $\mathbf{Q}$ which has its entries in $\mathbf{Z}$ if and only if there is no $n \geq 0$ such that $a_n = 0$.

Theorem 8 *It is decidable whether or not an D0L multiplicity sequence $(c_n)_{n \geq 0}$ over $\mathbf{Q}$ is an D0L multiplicity sequence over $\mathbf{N}$.*

Proof. Suppose $(c_n)_{n \geq 0}$ is an D0L multiplicity sequence over $\mathbf{Q}$. Then there exists a positive integer k, prime numbers $p_1, p_2, \ldots, p_k \in \mathbf{N}$, an ultimately periodic sequence $(a_{0n})_{n \geq 0}$ consisting of 0s and 1s and $\mathbf{Z}$-rational sequences $(a_{in})_{n \geq 0}$ for $1 \leq i \leq k$ such that:

$$c_n = (-1)^{a_{0n}} \prod_{i=1}^{k} p_i^{a_{in}},$$

for $n \geq 0$. (This follows easily from Theorem 1.) If there is an integer n such that $a_{0n} = 1$, $(c_n)_{n \geq 0}$ is not an D0L multiplicity sequence over $\mathbf{N}$. We proceed with the assumption that $a_{0n} = 0$ for all $n \geq 0$. Next we decide whether or not the sequences $(a_{in})_{n \geq 0}$, $1 \leq i \leq k$, are modified PD0L length sequences. This decision is made in two steps. First, decide whether or not (a_{in}) is $\mathbf{N}$-rational. If it is, decide whether or not $(a_{i,n+1} - a_{i,n})$ is $\mathbf{N}$-rational (see Salomaa and Soittola [11]).

If $(a_{in})_{n \geq 0}$ is a modified PD0L length sequence for $1 \leq i \leq k$, Theorem 1 implies that $(c_n)_{n \geq 0}$ is an D0L multiplicity sequence over $\mathbf{N}$. On the other hand, if $(c_n)_{n \geq 0}$ is an D0L multiplicity sequence over $\mathbf{N}$, Theorem 1 implies that there exist modified PD0L length sequences $(s_{in})_{n \geq 0}$, for $1 \leq i \leq k$, such that:

$$c_n = \prod_{i=1}^{k} p_i^{s_{in}}.$$

(Here we use the fact that modified PD0L length sequences are closed under addition.) Hence, by the fundamental theorem of arithmetics, we have:

$$s_{in} = a_{in}$$

for $1 \leq i \leq k$, $n \geq 0$, implying that (a_{in}) is a modified PD0L length sequence for $1 \leq i \leq k$. Consequently, we have seen that $(c_n)_{n \geq 0}$ is an D0L multiplicity sequence over $\mathbf{N}$ if and only if the sequences $(a_{in})_{n \geq 0}$, $1 \leq i \leq k$, are modified PD0L length sequences. This concludes the proof. $\square$

The solution of Problem 1 is similar if $\mathbf{N}$ is replaced by $\mathbf{Z}$.

The solution of Problems 2 and 3 requires two lemmas. Suppose $(c_n)_{n \geq 0}$ is a sequence of nonzero elements of a commutative semiring A and $(w_n)_{n \geq 0}$ is an D0L sequence over X such that $\{w_n \mid n \geq 0\}$ is an infinite language. We say that $(c_n)_{n \geq 0}$ and $(w_n)_{n \geq 0}$ are *compatible* if:

$$\sum_{n=0}^{\infty} c_n w_n$$

is the normal form of an D0L power series over A. Let $(w_n)_{n \geq 0}$ be as above and define the sequences $(s_{x,n})_{n \geq 0}$ for $x \in X$ by:

$$s_{x,n} = |w_0|_x + \ldots + |w_{n-1}|_x,$$

$n \geq 0$, where $|w|_x$ is the number of the occurrences of the letter x in w. We say that $(s_{x,n})_{n\geq0}$, $x \in X$, are the *cumulative sequences* of $(w_n)_{n\geq0}$.

Lemma 1 *Let $(c_n)_{n\geq0}$ and $(w_n)_{n\geq0}$ be as above and let $(s_{x,n})_{n\geq0}$, $x \in X$, be the cumulative sequences of $(w_n)_{n\geq0}$. Then, $(c_n)_{n\geq0}$ and $(w_n)_{n\geq0}$ are compatible if and only if there exist nonzero $a_x \in A$, $x \in X$, and nonzero $a \in A$ such that:*

$$c_n = a \prod_{x \in X} a_x^{s_{x,n}}, \tag{7}$$

for $n \geq 0$.

Proof. Suppose first that $(c_n)_{n\geq0}$ and $(w_n)_{n\geq0}$ are compatible. Then:

$$r = \sum_{n=0}^{\infty} c_n w_n$$

is an D0L power series over A. Let:

$$c_n w_n = a h^n(w_0),$$

$n \geq 0$, where $h : A < X^* > \longrightarrow A < X^* >$ is a monomial morphism. Denote $h(x) = a_x \bar{h}(x)$ for $x \in X$. We claim that (7) holds for all $n \geq 0$. First, (7) is clear if $n = 0$. Then, if (7) holds for $n \geq 0$, we have:

$$c_{n+1} w_{n+1} = a h^{n+1}(w_0) = h(a h^n(w_0)) = h(c_n w_n) = c_n h(w_n) =$$

$$c_n \prod_{x \in X} a_x^{|w_n|_x} w_{n+1} = a \prod_{x \in X} a_x^{s_{x,n}+|w_n|_x} w_{n+1} = a \prod_{x \in X} a_x^{s_{x,n+1}} w_{n+1}.$$

Hence (7) holds for all $n \geq 0$. This concludes the proof of Lemma 1 in one direction.

Suppose then that (7) holds. Let $w_n = g^n(w_0)$, $n \geq 0$, where $g : X^* \longrightarrow X^*$ is a morphism. Define the monomial morphism $h : A < X^* > \longrightarrow A < X^* >$ by $h(x) = a_x g(x)$, $x \in X$. Then it follows that:

$$\sum_{n=0}^{\infty} c_n w_n = \sum_{n=0}^{\infty} a h^n(w_0)$$

is an D0L power series over A. Hence $(c_n)_{n\geq0}$ and $(w_n)_{n\geq0}$ are compatible. $\square$

Lemma 2 *Suppose $(t_{in})_{n\geq0}$, $0 \leq i \leq m$, are $\mathbf{Z}$-rational (resp. $\mathbf{N}$-rational) sequences. It is decidable whether or not there exist $a_1, \ldots, a_m \in \mathbf{Z}$ (resp. $a_1, \ldots, a_m \in \mathbf{N}$) such that:*

$$t_{0n} - \sum_{i=1}^{m} a_i t_{in}, \tag{8}$$

for all $n \geq 0$.

Proof. Suppose $(t_{in})_{n \geq 0}$, $0 \leq i \leq m$, are **Z**-rational sequences. Then there exist a positive integer k and integers $\beta_1, \ldots, \beta_k$ such that:

$$t_{j,n+k} = \sum_{i=1}^{k} \beta_i t_{j,n+k-i}, \tag{9}$$

for all $0 \leq j \leq m$, $n \geq 0$. Next, decide whether or not there exist integers $a_1, \ldots, a_m$ such that (8) holds for $0 \leq n < k$. This decision is possible by methods from linear algebra. If such $a_1, \ldots, a_m$ do not exist, it is clearly impossible to find $a_1, \ldots, a_m \in \mathbf{Z}$ such that (8) holds for all $n \geq 0$. On the other hand, if (8) holds for $0 \leq n < k$, (9) implies (8) for all $n \geq 0$. Indeed, if (8) holds for $0 \leq n < q$, $q \geq k$, we have:

$$t_{0,q} = \sum_{i=1}^{k} \beta_i t_{0,q-i} = \sum_{i=1}^{k} \beta_i \sum_{j=1}^{m} a_j t_{j,q-i} = \sum_{j=1}^{m} a_j \sum_{i=1}^{k} \beta_i t_{j,q-i} = \sum_{j=1}^{m} a_j t_{j,q}.$$

The proof for **N**-rational sequences is similar but easier. $\qquad \square$

Theorem 9 *It is decidable whether or not an D0L power series r over* **Q** *is an D0L power series over* **Z** *(resp.* **N***).*

Proof. Suppose:

$$r = \sum_{n=0}^{\infty} c_n w_n.$$

First decide whether or not $(c_n)_{n \geq 0}$ is an D0L multiplicity sequence over **Z**. If not, r is not an D0L power series over **Z**. Suppose $(c_n)_{n \geq 0}$ is an D0L multiplicity sequence over **Z**. Then, by Theorem 1, there exists a positive integer k, primes $p_1, \ldots, p_k \in \mathbf{N}$, an ultimately periodic sequence $(a_{0n})_{n \geq 0}$ consisting of 0s and 1s and **N**-rational sequences $(a_{in})_{n \geq 0}$ for $1 \leq i \leq k$ such that:

$$c_n = c_0(-1)^{a_{0n}} \prod_{i=1}^{k} p_i^{a_{in}},$$

for $n \geq 0$. It suffices to decide whether or not $(c_n)_{n \geq 0}$ and $(w_n)_{n \geq 0}$ are compatible over **Z**. For this purpose let $(s_{x,n})_{n \geq 0}$, $x \in X$, be the cumulative sequences of $(w_n)_{n \geq 0}$.

By Lemma 1, $(c_n)_{n \geq 0}$ and $(w_n)_{n \geq 0}$ are compatible over **Z** if and only if $(a_{0n})_{n \geq 0}$ is an **N**-linear combination of $(s_{x,n})_{n \geq 0}$, $x \in X$, modulo 2 and, furthermore, for all $1 \leq i \leq k$, the sequence $(a_{in})_{n \geq 0}$ is an **N**-linear combination of $(s_{x,n})_{n \geq 0}$, $x \in X$. By Lemma 2, these conditions are decidable.

Next, if r is not an D0L power series over **Z**, r is a not an D0L power series over **N**. Otherwise, the second claim follows by Theorem 7. $\qquad \square$

Theorem 10 *Suppose E is a computable field of characteristic zero. It is decidable whether or not an D0L power series r over E is an D0L power series over* **Q**.

Proof. Suppose:

$$r = \sum_{n=0}^{\infty} c_n w_n.$$

Decide by the method explained above whether or not $c_n \in \mathbf{Q}$ for all $n \geq 0$. If so, it suffices to decide whether or not $(c_n)_{n \geq 0}$ and $(w_n)_{n \geq 0}$ are compatible over **Q**. Let:

$$c_n = c_0(-1)^{a_{0n}} \prod_{i=1}^{k} p_i^{a_{in}},$$

for $n \geq 0$, where $k \geq 1$, $p_1, \ldots, p_k \in \mathbf{N}$ are primes, $(a_{0n})_{n \geq 0}$ is an ultimately periodic sequence consisting of 0s and 1s and $(a_{in})_{n \geq 0}$, for $1 \leq i \leq k$, are **Z**-rational sequences. Furthermore, let $(s_{x,n})_{n \geq 0}$, $x \in X$, be the cumulative sequences of $(w_n)_{n \geq 0}$. By Lemma 1, $(c_n)_{n \geq 0}$ and $(w_n)_{n \geq 0}$ are compatible over **Q** if and only if $(a_{0n})_{n \geq 0}$ is an **Z**-linear combination of $(s_{x,n})_{n \geq 0}$, $x \in X$, modulo 2 and, furthermore, for all $1 \leq i \leq k$, the sequence $(a_{in})_{n \geq 0}$ is an **Z**-linear combination of $(s_{x,n})_{n \geq 0}$, $x \in X$. These conditions are decidable by Lemma 2. $\square$

References

[1]   J. Berstel and C. Reutenauer, *Rational Series and Their Languages.* Springer, Berlin, 1988.

[2]   S. Eilenberg, *Automata, Languages and Machines.* Academic Press, New York, 1974, vol. A.

[3]   J. Honkala, On morphically generated formal power series, *RAIRO, Theoretical Informatics and Applications*, 29 (1995), 105–127.

[4]   J. Honkala, On the decidability of some equivalence problems for L algebraic series, *International Journal of Algebra and Computation*, 7 (1997), 339–351.

[5]   J. Honkala, On algebraicness of D0L power series, *Journal of Universal Computer Science*, 5 (1999), 11–19.

[7]   J. Honkala, On sequences defined by D0L power series, *RAIRO, Theoretical Informatics and Applications*, 33 (1999), 125–132.

[6]   J. Honkala, On D0L power series, *Theoretical Computer Science*, to appear.

[8]   W. Kuich and A. Salomaa, *Semirings, Automata, Languages.* Springer, Berlin, 1986.

[9] G. Rozenberg and A. Salomaa, *The Mathematical Theory of L Systems*. Academic Press, New York, 1980.

[10] G. Rozenberg and A. Salomaa (eds.), *Handbook of Formal Languages*. Springer, Berlin, 1997.

[11] A. Salomaa and M. Soittola, *Automata-Theoretic Aspects of Formal Power Series*. Springer, Berlin, 1978.

The World of Unary Languages: A Quick Tour[1]

Carlo Mereghetti

Department of Informatics, Systems Science and Communication
University of Milano–Bicocca
Italy
`mereghetti@disco.unimib.it`

Giovanni Pighizzini

Department of Information Sciences
University of Milan
Italy
`pighizzi@dsi.unimi.it`

Abstract. We give two flashes from the *world of unary languages* related to the study of tight computational lower bounds for nonregular language acceptance. They show both an interesting dissymmetry with the general case, and the flavor and some typical number theoretic tools of unary computations.

1   Introduction

In mathematics, "simplifications" often lead to meaningful and interesting results. This is the case even for theoretical computer science and, particularly, for language and complexity theory. In such realms, one of the most

---

[1]Partially supported by MURST, under the project "Modelli di calcolo innovativi: metodi sintattici e combinatori".

investigated simplifications is that provided by *unary languages*. A unary language is simply a language built over a single–letter input alphabet. Several results in the literature show the relevance of dealing with unary inputs, often emphasizing sharp dissymmetries with the general case of languages on alphabets with two or more symbols.

However, it is worth remarking that simplifications do not necessarily make it easier to cope with problems, and this is the case even when considering unary languages. The total absence of structure in input strings forces our computation techniques to be much more sophisticated and sometimes tricky. In particular, as may be expected, it is interesting to record the constant use of Number Theory, which provides a valuable source of nontrivial tools and properties.

In this paper, we shall briefly survey some results showing the importance and the typical flavor of computing with unary languages. In particular, we focus on *the world of Turing machines that work within a sublogarithmic amount of space*. Here, several contributions emphasize the importance of investigating unary computations. It is well–known, for instance, that the truth of both Immerman–Szelepcsényi and Savitch's Theorems remains an open problem when considering sublogarithmic work space. In contrast, Viliam Geffert [5] proves that these two theorems hold true for *unary languages processed in sublogarithmic space*. Beside this, other examples can be found that show the relevance of considering the unary version of problems, e.g. in [3, 6, 13].

Indeed, it must be said that computing within sublogarithmic space turns out to be quite difficult since a lot of traditional programming tools are missing. For instance, in such a restricted amount of tape, we cannot store input length, or fix positions on the input tape. However, the difficulties are even greater when considering inputs that are completely structureless, such as strings in unary languages. As a result, algorithms and proofs are often highly involved, and often rely on number theoretical issues (see e.g. [6]).

In what follows, we are to deal with two instances of a lower bound problem that is closely connected to the sublogarithmic space world, and for which considering unary languages turns out to be very fruitful. Namely, we discuss the question that was first posed in 1965 by Hartmanis, Stearns, and Lewis [9, 11]: *"what is the minimal amount of space used by a Turing machine — equipped with a read-only input tape and a read/write work tape — that accepts a nonregular language?"*.

This question has been extensively answered for deterministic, nondeterministic, and alternating Turing machines, and for *different ways of measuring space* (see [3, 13]). Furthermore, a generalization requesting lower bounds on both space and number of input head reversals has also been studied [2, 3]. Here, we concentrate on two cases.

The first one, presented in Section 3 and studied in [10], describes a situation in which investigation into unary and general languages gives two different results. It concerns one–way (i.e. input is scanned once, from left to right) alternating Turing machines that work in middle space $s(n)$: all computations on all accepted inputs of length n use no more than $s(n)$ work tape cells. What is the lower bound on $s(n)$ when the recognized language is *nonregular*? For nonregular languages built over *general alphabets*, the optimal lower bound is known to be $s(n) \notin o(\log \log n)$ [12]. In contrast, for *unary* nonregular languages, we show that the optimal lower bound is strictly higher, namely $s(n) \notin o(\log n)$.

The second case, presented in Section 4 and studied in [2], enables us to appreciate both the hardness of accepting unary languages within a very restricted number of computational resources, and the great help that can be provided by Number Theory. It deals with the optimality proof of the lower bound on the product space×number of input head reversals, written $s(n) \cdot i(n)$, for nondeterministic Turing machines that accept nonregular languages in strong $s(n)$ space and $i(n)$ input head reversals: all computations on inputs of length n use no more than $s(n)$ work tape cells, and no more than $i(n)$ input head reversals. In [2], it is proved that $s(n) \cdot i(n) \notin o(\log n)$. It should be stressed that, without imposing restrictions on $i(n)$, so as to have general two–way nondeterministic Turing machines, a lower bound of strong $s(n) \notin o(\log \log n)$ is given in [8]. Thus, it would be interesting to give the optimality of $s(n) \cdot i(n) \notin o(\log n)$ in the "best possible way", i.e. by exhibiting a unary nonregular language accepted by a nondeterministic Turing machine in strong $O(\log \log n)$ space and $O(\log n / \log \log n)$ input head reversals. This is exactly what we are going to do by considering a unary nonregular language proposed by Alt and Mehlhorn in [1], namely, $\mathcal{L}_{AM} = \{1^n \mid q(n) \text{ is a power of 2}\}$, where $q(n)$ denotes the smallest integer not dividing n. We will show how a very efficient recognizing algorithm for $\mathcal{L}_{AM}$ can be designed by studying some number theoretic properties of $q(n)$.

Before taking a look at the unary world, we need some technical details which are provided in the next section.

2   Preliminary Notions

Let Σ^* be the set of all strings (with the empty string ε) on an alphabet Σ. We let $|x|$ be the length of $x \in \Sigma^*$. A language $L \subseteq \Sigma^*$ is *unary (or tally)*, whenever Σ is a single–letter alphabet. In this case, we let $\Sigma = \{1\}$.

We consider the standard Turing machine model (see [9, 11, 13]) with a finite state control, a two–way read–only input tape (with input enclosed between a left and a right endmarker symbol, '¢' and '\$', respectively), and a separate semi–infinite two–way read–write work tape. A *memory state*[2]

---

[2]Sometimes called *internal configuration* (see e.g. [12, 13]).

of a Turing machine M is an ordered triple (q, w, i), where q is a finite control state, the string w is the nonblank content of the work tape, and the integer $1 \leq i \leq |w| + 1$ denotes the head position on the work tape. A *configuration* of M on a given input string x is an ordered pair (j, m), where the integer $0 \leq j \leq |x| + 1$ represents the input head position, while m is a memory state. The *initial configuration* is the pair $(0, (q_0, \varepsilon, 1))$, where q_0 is the *initial state*. An *accepting configuration* is any configuration containing a *final state*. It is well–known that the computation of a deterministic or nondeterministic Turing machine on a given input string can be seen as a finite or infinite sequence $c_0, c_1, \ldots, c_i, \ldots$ of configurations beginning in the initial configuration c_0 and such that c_{i+1} is an *immediate successor* of c_i on the appropriate input symbol. A computation is *accepting* whenever it ends in an accepting configuration.

The *space* used in a computation is the maximum number of work tape cells taken up by the configurations in that computation. An *input head reversal* in a computation is a sequence $\alpha, \beta_1, \beta_2, \ldots, \beta_s, \gamma$ ($s \geq 1$) of configurations each one of which is the immediate successor of the previous one, and which has the form $\alpha = (j-1, m')$, $\beta_i = (j, m_i'')$, $\gamma = (j-1, m''')$, or also $\alpha = (j+1, m')$, $\beta_i = (j, m_i'')$, $\gamma = (j+1, m''')$. A Turing machine is said to be *one–way* whenever its computations do not present input head reversals; otherwise, it is *two–way*.

The reader is assumed to be familiar with the notion of *alternating Turing machine* [4]. We only recall that, in an alternating Turing machine, the finite control states — and hence configurations and memory states as well — are partitioned into *existential* and *universal*. A computation is described by a tree whose nodes are labelled by configurations. The root is labelled by the initial configuration. Any internal node labelled by an existential configuration e has a unique son labelled by an immediate successor of e. Any internal node labelled by a universal configuration u has a son for each immediate successor of u. A computation is accepting if the corresponding tree is finite and all its leaves are labelled by accepting configurations.

The *space* used in the computation of an alternating Turing machine is the maximum number of work tape cells taken up by the configurations that label the corresponding tree. The *number of input head reversals* used in that computation is the maximum number of input head reversals along computation paths from the root.

Several notions of space complexity have been defined in the literature [3, 13]. Here, we are interested in strong [8, 9, 11] and middle [12] space. A Turing machine works in strong (middle) $s(n)$ space if any computation on each (accepted) input of length n uses no more than $s(n)$ space. These two space notions (and the others proposed in the literature) coincide for *fully space constructible*[3] bounds, e.g. "normal" functions above $\log n$. We are

---

[3]A function $s(n)$ is said to be *fully space constructible* if there exists a deterministic

also interested in Turing machines *simultaneously bounded on both space and input head reversals* [2, 3]. More precisely, we say that a Turing machine works in **strong** (middle) $s(n)$ space and $i(n)$ input head reversals if any computation on each (accepted) input of length n uses no more than $s(n)$ space and $i(n)$ input head reversals.

3 A Gap Result for One–Way middle Alternation: When Unary Differs from General

What is the minimal middle space complexity $s(n)$ for a *one–way alternating Turing machine* that accepts a nonregular language? The answer to this question has been given by Szepietowski [12]: $s(n) \notin o(\log \log n)$. This lower bound is tight. In fact, [12] discusses a nonregular language that is recognized by a one–way alternating Turing machine in middle $O(\log \log n)$ space. Such a language is defined as:

$$\mathcal{L}_S = \{a^k b^m \mid m \text{ is a common multiple of all } r \leq k\}.$$

It is interesting to have some idea of how a one–way alternating Turing machine for $\mathcal{L}_S$ works. We shall have the chance to see both a very space–inexpensive computation, and Number Theory at work!

We begin by studying the structure of the strings in $\mathcal{L}_S$. To this end, we need a well–known result proved by P. Čebyšev in 1851. Let the function $\pi(x) = $ number of primes not exceeding x. Čebyšev's Theorem (see e.g. [7]) states that $c_1 x / \ln x \leq \pi(x) \leq c_2 x / \ln x$, for two positive constants c_1 and c_2, and with "ln" denoting the natural logarithm.

This enables us to give a lower bound for the function $P(x) - \prod_{p \leq x} p$, where p denotes a prime number. In fact, by Čebyšev's Theorem, we can write $P(x) \geq (c_1 x / \ln x)! \geq 2^{\frac{x}{4}}$, where the last inequality holds for sufficiently large x. This shows that:

Lemma 1 *There exists a positive constant d such that $P(x) \geq d^x$.*

At this point, we can state the following:

Theorem 1 *Let $a^k b^m$ belong to $\mathcal{L}_S$. Then $k \in O(\log m)$.*

Proof. Since any integer less than or equal to k must divide m, then any prime not exceeding k divides m as well. This clearly implies that $m \geq P(k)$. By applying Lemma 1, the claimed result follows. □

With these numerical properties at hand, we are now ready to exhibit a one–way alternating Turing machine M accepting $\mathcal{L}_S$ in middle $O(\log \log n)$

Turing machine which, on any input of length n, uses exactly $s(n)$ space.

space. On the input of string x, our machine M first writes on its work tape, in binary notation, the number k of a's at the beginning of x. Then, it *universally* branches on each $r \leq k$, checking whether r divides the number m of b's at the end of x. This latter operation can be accomplished by simply counting m modulo r.

That M is one–way is straightforward. For space requirement, we observe that, if x belongs to $\mathcal{L}_S$, then the space used by M is $O(\log k)$, namely, the amount of tape needed to store the number of a's. But Theorem 1 states that $k \in O(\log m)$. Thus, we conclude that $s(|x|) \in O(\log \log m) \in O(\log \log |x|)$. Furthermore, we notice that if x was not in $\mathcal{L}_S$, then M might use more than $\log \log |x|$ tape before rejecting. All this shows that the space is middle.

Needless to say, the language $\mathcal{L}_S$ is built over a binary alphabet. So the question arises of whether the middle space lower bound $s(n) \notin o(\log \log n)$ can be "certified" even in the *unary world*. In other words: are we able to exhibit a *unary nonregular language* accepted in middle $O(\log \log n)$ space by a one–way alternating Turing machine? A negative answer to this question is contained in the following theorem, which states that there is an exponential gap between the general and the unary case:

Theorem 2 *Let M be a one–way alternating Turing machine accepting a unary nonregular language L in middle $s(n)$ space. Then $s(n) \notin o(\log n)$. This bound is tight.*

Proof. In contradiction, we assume $s(n) \in o(\log n)$. It is easy to see that, for any integer n such that the string 1^n belongs to L, the number of memory states of M on input 1^n is bounded above by $2^{h \cdot s(n)}$, for a suitable positive constant h. Moreover, because of our initial assumption on $s(n)$ and since L is nonregular, there must exist an integer n' such that the string $1^{n'}$ belongs to L and $2^{h \cdot s(n')} < n'$. Let us prove the following:

Claim 1 *Any memory state reachable by M without scanning the right end-marker symbol can be reached in less than n' moves.*

Proof. It is enough to show that every memory state m reachable by M in $k \geq n'$ moves by reading 1's on the input tape can be reached in $k' < k$ moves as well. Thus, let $m_0, m_1, \ldots, m_k$ be a sequence of memory states of M such that m_0 is the initial memory state, $m_k = m$, and m_{i+1} is an immediate successor of m_i, for each $0 \leq i \leq k - 1$.

Consider the initial segment $\mathcal{S} = m_0, m_1, \ldots, m_{n'}$. Each memory state in $\mathcal{S}$ can be reached on input $1^{n'}$. Moreover, since $1^{n'}$ belongs to L and M is middle space bounded, then each memory state in $\mathcal{S}$ takes at most $s(n')$ work tape cells. Recall that $2^{h \cdot s(n')} < n'$. Hence, a simple pigeonhole argument shows the existence of $0 \leq i < j \leq n'$ satisfying $m_i - m_j$. Thus, the sequence $m_0, m_1 \ldots, m_i, m_{j+1}, \ldots, m_k = m$ leads to m in $k' = k + i - j$ moves.   □

An easy consequence of this Claim is that the computations of M on 1's of any input must go only through memory states requiring no more than $s(n')$ work tape cells. This enables us to exhibit a *one–way alternating automaton*[4] A that accepts the language L.

Informally, A has the set $\mathcal{M}$ of all the memory states of M using at most $s(n')$ work tape cells as the set of states. Its transition function δ is the "immediate successor" relation on the memory states assumed by M on reading 1's. Our Claim ensures that $\delta(c, 1)$ is well defined for any m in $\mathcal{M}$. It is not hard to verify that both A and M behave in the same way on 1's of any input string. Furthermore, those computations of M taking place on the right endmarker symbol may be correctly resumed in A by choosing the final states as follows: each $m \in \mathcal{M}$ is a final state if and only if M accepts by starting from m and scanning only the symbol '$\$$'.

A well–known result in [4] states that one–way alternating automata exactly characterize the class of regular languages. This proves that L is regular, and contradicts our initial assumption.

The optimality of our new lower bound for unary nonregular languages is easily provable since we are given a "wide" (logarithmic) amount of work space in which to store, e.g. the length of the input by one input scan. Thus, for instance, the language $\mathcal{L}_{\text{AM}}$ mentioned in Section 1 — for which we will exhibit very space–efficient recognizing algorithms in Section 4 — is easily seen to be recognized in $O(\log n)$ space even by a one–way strong deterministic Turing machine. $\qquad\square$

4   Number Theory for Optimality

We now come to our second lower bound investigation. The question is now: what is the minimal amount of strong $s(n)$ space and $i(n)$ input head reversals for a nondeterministic Turing machine that accepts a nonregular language? The answer is given in [2]: $s(n) \cdot i(n) \notin o(\log n)$. This lower bound has to be compared with the bound proved in [8], which says that a *two–way* nondeterministic Turing machine that accepts a nonregular language in strong $s(n)$ space must satisfy $s(n) \notin o(\log \log n)$.

The comparison of these two lower bounds challenges us to prove the optimality of $s(n) \cdot i(n) \notin o(\log n)$ in the "best possible way", namely: can we exhibit a nonregular language which is *unary*, and is accepted by a nondeterministic Turing machine performing $O(\log n / \log \log n)$ input head reversals in strong $O(\log \log n)$ space? The answer is positive. Actually we will do even better, since the unary nonregular language we are going to consider will be accepted within the prescribed resource bounds by a *deterministic* machine.

---

[4]We recall that a one–way alternating automaton is a one–way finite state automaton where we distinguish between *existential* and *universal* states [4].

So, let us take a look at the unary nonregular language $\mathcal{L}_{\text{AM}}$ proposed by Alt and Mehlhorn in [1]. For any positive integer n, let $q(n)$ denote the smallest integer not dividing n. The language is defined as:

$$\mathcal{L}_{\text{AM}} = \{1^n \mid q(n) \text{ is a power of } 2\}.$$

As for the language $\mathcal{L}_S$ in the previous section, we need some number theoretic tools to analyze the structure of the strings in $\mathcal{L}_{\text{AM}}$, and to evaluate the complexity of the algorithms for $\mathcal{L}_{\text{AM}}$. We begin with:

Theorem 3 $q(n) \in O(\log n)$.

Proof. Since $q(n)$ is the smallest integer not dividing n, then each prime less than $q(n)$ must divide n. This clearly implies that $n \geq P(q(n) - 1)$, whence the result follows from Lemma 1. □

Theorem 3 enables us to estimate the space and input head reversals of the following algorithm for $\mathcal{L}_{\text{AM}}$ whose correctness follows trivially:

input(1^n)
$k := 2$
while $n \equiv 0 \pmod{k}$ **do**   /* at the end of this loop k will contain $q(n)$ */
    $k := k + 1$
if k is a power of 2 **then** ACCEPT **else** REJECT

Such an algorithm can be easily implemented on a *deterministic* Turing machine M whose space requirement, on any input 1^n, is basically that for storing the counter k. Since the greatest value assumed by k is exactly $q(n)$, we get that M uses strong $s(n) = O(\log q(n))$ space which, in the light of Theorem 3, becomes $s(n) \in O(\log \log n)$. So, as far as the space is concerned, we have reached our goal.

Unfortunately, problems come with the input head reversals. In fact, M performs an input head reversal for each new value of k, up to $q(n)$. On each input traversal, M checks the loop condition "$n \equiv 0 \pmod{k}$" by simply counting n modulo k (no added space is needed). Hence, we get $i(n) = q(n)$ which, by Theorem 3, reads as $i(n) \in O(\log n)$. To save some head reversals, we must improve our knowledge of $q(n)$:

Lemma 2 *For any positive integer n, $q(n)$ is a prime power.*

Proof. Assume that $q(n)$ is not a prime power. Thus, $q(n) = a \cdot b$, for two coprime integers $1 < a < b < q(n)$. Since $q(n)$ is the smallest integer not dividing n, then both a and b must divide n, and since they are coprime, their product $a \cdot b = q(n)$ divides n as well, a contradiction. □

This result immediately leads to the following improved algorithm for $\mathcal{L}_{\text{AM}}$, where $pp(x)$ denotes the predicate "x is a prime power":

```
input(1ⁿ)
k := 2
while n ≡ 0 (mod k) do
    begin
        k := k + 1
        while not(pp(k)) do k := k + 1        /* skip non–prime powers */
    end
if k is a power of 2 then ACCEPT else REJECT
```

The space requirement is still seen to be $O(\log \log n)$ (in particular, it is not hard to argue that testing $pp(x)$ needs no extra space). What has changed is that *now $i(n)$ equals the number of prime powers not exceeding $q(n)$*. To compute this new value of $i(n)$, let $\Pi(x)$ = number of prime powers not exceeding x. We now estimate the growth of $\Pi(x)$. For this purpose, we use the celebrated Prime Number Theorem of J. Hadamard and C. de la Vallée Poussin, 1896 (see e.g. [7]). This theorem refines Čebyšev's by stating that $\pi(x) \sim x/\ln x$.

Theorem 4 $\Pi(x) \sim \frac{x}{\ln x}$.

Proof. For $1 \leq i \leq \pi(x)$, let N_i be the number of powers of the i-th prime p_i that do not exceed x. Hence, $\Pi(x) = \sum_{i=1}^{\pi(x)} N_i$. It is easy to see that $x \geq p_i^{N_i}$, and hence $N_i \leq \ln x/\ln p_i$. With these facts, and using the Prime Number Theorem, we can write:

$$\frac{x}{\ln x} \sim \pi(x) \leq \Pi(x) \leq \sum_{i=2}^{\pi(x)} \frac{\ln x}{\ln p_i} \leq \sum_{i=2}^{\pi(x)} \frac{\ln x}{\ln i} \sim \ln x \int_2^{\pi(x)} \frac{1}{\ln t}\, dt \sim \frac{x}{\ln x}.$$

□

Summing up, we made the improved algorithm for $\mathcal{L}_{AM}$ perform $i(n) = \Pi(q(n))$ input head reversals. By both Theorem 4 and Theorem 3, we obtain $i(n) \in O(\log n/\log \log n)$, which concludes the story:

Theorem 5 *$\mathcal{L}_{AM}$ is recognized by a deterministic Turing machine in* strong $O(\log \log n)$ *space with* $O(\log n/\log \log n)$ *input head reversals.*

References

[1] H. Alt and K. Mehlhorn, A language over a one symbol alphabet requiring only $O(\log \log n)$ space, *SIGACT News*, 7 (1975), 31–33.

[2] A. Bertoni, C. Mereghetti and G. Pighizzini, An optimal lower bound for nonregular languages, *Information Processing Letters*, 50 (1994), 289–292. Corrigendum: *ibid.*, 52 (1994), 339.

[3] A. Bertoni, C. Mereghetti and G. Pighizzini, Space and Reversals Complexity of Nonregular Languages, Technical Report 224-98, Dipartimento di Scienze dell'Informazione, Università di Milano, 1998. Submitted for publication.

[4] A. Chandra, D. Kozen and L. Stockmeyer, Alternation, *Journal of the ACM*, 28 (1981), 114–133.

[5] V. Geffert, Tally version of the Savitch and Immerman–Szelepcsényi theorems for sublogarithmic space, *SIAM Journal of Computing*, 22 (1993), 102–113.

[6] V. Geffert, C. Mereghetti and G. Pighizzini, Sublogarithmic bounds on space and reversals, *SIAM Journal of Computing*, 28 (1999), 325–340.

[7] G. Hardy and E. Wright, *An Introduction to the Theory of Numbers*. Oxford University Press, Oxford, 1979.

[8] J.E. Hopcroft and J.D. Ullman, Some results on tape-bounded Turing machines, *Journal of the ACM*, 16 (1969), 168–177.

[9] P. Lewis, R. Stearns and J. Hartmanis, Memory bounds for the recognition of context free and context sensitive languages. In *Proceedings of the IEEE Conference Record on Switching Circuit Theory and Logical Design*, 1965, 191–202.

[10] C. Mereghetti and G. Pighizzini, A remark on middle space bounded alternating Turing machines, *Information Processing Letters*, 56 (1995), 229–232.

[11] R. Stearns, J. Hartmanis and P. Lewis, Hierarchies of memory limited computations. In *Proceedings of the IEEE Conference Record on Switching Circuit Theory and Logical Design*, 1965, 179–190.

[12] A. Szepietowski, Remarks on languages acceptable in $\log \log n$ space, *Information Processing Letters*, 27 (1988), 201–203.

[13] A. Szepietowski, *Turing Machines with Sublogarithmic Space*. Springer, Berlin, 1994.

A New Universal Logic Element for Reversible Computing

Kenichi Morita

Faculty of Engineering
Hiroshima University
Higashi-Hiroshima, Japan
morita@ke.sys.hiroshima-u.ac.jp

Abstract. A "rotary memory" (RM) is a simple logic element with both reversibility and a bit-conserving property. In this paper, we show that this element is logically universal by using only RMs to construct a Fredkin gate, which is already known to be universal. We then describe a method for concisely constructing a reversible sequential machine out of RMs.

1 Introduction

Recently, various computing models that directly reflect laws of nature have been proposed and investigated: quantum computing (e.g. [2]), DNA computing (e.g. [6]), reversible computing (e.g. [1, 2, 3]), and so on. Reversible computing is a model that reflects physical reversibility, and it has played an important role in the study of inevitable power dissipation in a computing process. Until now, such reversible systems as reversible Turing machines, reversible cellular automata, and reversible logic circuits have been proposed and investigated.

A logic gate is called reversible if its logical function is one-to-one. A reversible logic circuit is one constructed only of reversible logic gates. Some reversible gates are 'universal' in the sense that any logic circuit (even if it is irreversible) can be embedded in a circuit composed only of this type of gate. A Fredkin gate [3] and a Toffoli gate [7] are typical universal reversible gates with 3 inputs and 3 outputs. In particular, it has been shown that logical functions of AND and NOT can be realized by a single Fredkin or Toffoli gate [3, 7].

A rotary memory (RM), which is also a reversible logic element, was introduced by Morita, Tojima, and Imai [5]. They proposed a simple model of 2-D reversible cellular automaton called P_4 in which any reversible two-counter machine can be embedded in a finite configuration. They showed that an RM and a blinking marker (BM) as well as several kinds of signal routing (wiring) elements can be implemented in this cellular space, and they described a method for constructing a reversible counter machine out of these elements. An RM is a 4-input 4-output reversible element, which has two states (hence it is not a pure logic gate). It is a kind of switching element that changes the path of an input signal depending on its state. A BM is a pebble-like object that can be placed in a 2-D cellular space, and its position can be shifted by giving an appropriate signal. In [5], RMs are used to build a finite-state control of a counter machine, while BMs are used to realize counters.

In this paper, we prove the universality of an RM by showing that a Fredkin gate can be embedded in a circuit composed only of RMs. We then describe a simple method for constructing a reversible sequential machine (i.e. a reversible finite-state transducer) out of RMs.

2   A Rotary Memory

A *rotary memory* (RM) is the logic element depicted in Fig. 1. It has four input lines $\{n, e, s, w\}$, four output lines $\{n', e', s', w'\}$, and has two states called H-state and V-state. All the values of the inputs and outputs are either 0 or 1. To explain how it works, we use the following intuitive interpretation. Signals 1 and 0 are interpreted as existence and non-existence of a particle. An RM is a square-shaped object, and it has a "rotating bar" which controls the direction of movement of a particle. We assume that input is at most

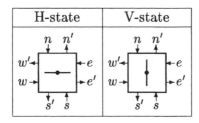

Figure 1: Two states of a rotary memory.

one particle at a time (the operation of an RM is undefined for those cases in which particles are given to two or more input lines). When no particle exists, nothing happens in the RM. If a particle comes from a direction parallel to the rotating bar, then it leaves the output line from the opposite side (i.e. it goes straight ahead) without affecting the direction of the bar (Fig. 2 (a)). On the other hand, if a particle comes from a direction orthogonal to the bar, then it makes a right turn, and rotates the bar by 90 degrees counterclockwise (Fig. 2 (b)). We can see that the operation of an RM is reversible, because, from the present state and the output, the previous state and the input are determined uniquely. Since a particle is neither annihilated nor newly created, an RM also has a bit-conserving property, i.e. the number of 1's is conserved between inputs and outputs.

Since one particle at most appears as an input and an output, the sets of lines $\{n, e, s, w\}$, and $\{n', e', s', w'\}$ can be regarded as input and output alphabets, respectively. Thus, the state-transition function and the output function can be described as shown in Table 1 (for instance, if the present state is H and a particle comes from the input line n, then the state becomes V and a particle goes out from w').

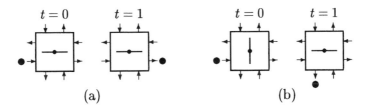

Figure 2: Operations of a rotary memory: (a) the parallel case (i.e. the particle's direction of approach is parallel to the rotating bar), and (b) the orthogonal case.

Present state	Input			
	n	e	s	w
H-state: ⊡	⊡ w'	⊡ w'	⊡ e'	⊡ e'
V-state: ⊡	⊡ s'	⊡ n'	⊡ n'	⊡ s'

Table 1: The state-transition function and the output function of a rotary memory.

In the following construction of a reversible logic circuit, we also use delay elements. A *unit-time delay* is a one-input one-output element that simply delays the input signal by one unit of time (Fig. 3 (a)). An *n-time delay* is a series composition of n unit-time delays (Fig. 3 (b)). Note that, since an RM has by definition a delay of one unit of time between inputs and outputs, the RM itself can be used as a unit-time delay by setting the state appropriately (hence no special element for delay is needed).

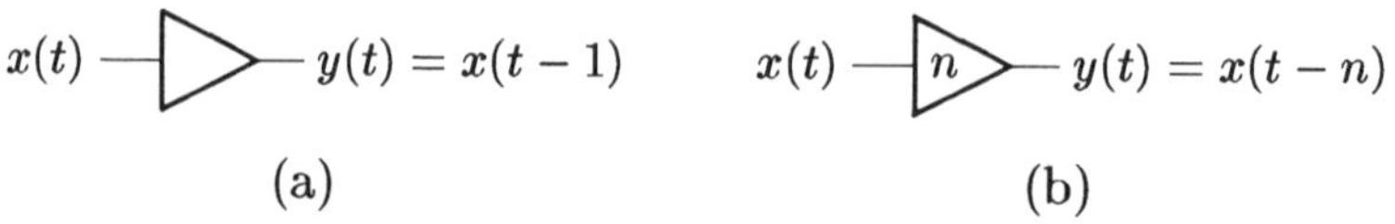

$$x(t) \longrightarrow\!\!\!\triangleright\!\!\!- y(t) = x(t-1) \qquad x(t) \longrightarrow\!\!\!\boxed{n}\!\!\!\triangleright\!\!\!- y(t) = x(t-n)$$

$$\text{(a)} \hspace{6cm} \text{(b)}$$

Figure 3: (a) A unit-time delay, and (b) an n-time delay.

3 Logical Universality of a Rotary Memory

In this section, we prove that an RM is a universal logic element by giving an RM circuit (i.e. a circuit composed of RMs and delays) that simulates a Fredkin gate (F-gate). An F-gate [3] is a reversible and bit-conserving logic gate (see Fig. 4). It is known that any combinational logic element can be realized only with F-gates [3]. Thus, any sequential circuit can be composed of F-gates and delay elements.

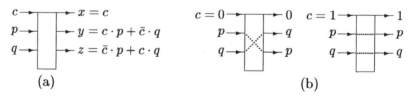

$$\begin{aligned}
c &\longrightarrow \quad\quad\quad \longrightarrow x = c \\
p &\longrightarrow \quad\quad\quad \longrightarrow y = c \cdot p + \bar{c} \cdot q \\
q &\longrightarrow \quad\quad\quad \longrightarrow z = \bar{c} \cdot p + c \cdot q
\end{aligned}$$

$$\text{(a)} \hspace{6cm} \text{(b)}$$

Figure 4: (a) A Fredkin gate (F-gate), and (b) its function.

It is also known that an F-gate can be constructed from much simpler gates called switch gates (S-gate) and inverse S-gate [3]. An S-gate is a 2-input 3-output gate, and it is also reversible and bit-conserving (Fig. 5 (a)). An inverse S-gate is one that realizes inverse logical function of an S-gate (Fig. 5 (b)).

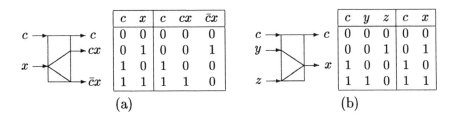

c	x	c	cx	$\bar{c}x$
0	0	0	0	0
0	1	0	0	1
1	0	1	0	0
1	1	1	1	0

(a)

c	y	z	c	x
0	0	0	0	0
0	0	1	0	1
1	0	0	1	0
1	1	0	1	1

(b)

Figure 5: (a) A Switch gate (S-gate), and (b) an inverse S-gate (note that the logical function of an inverse S-gate is not totally defined on $\{0,1\}^3$ as shown in the above table).

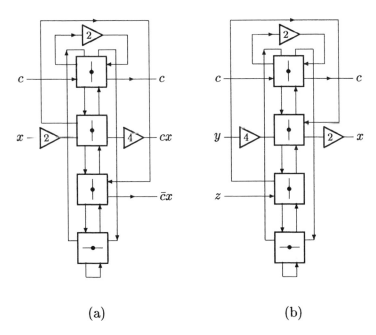

(a)   (b)

Figure 6: Realization of (a) an S-gate, and (b) an inverse S-gate by RMs.

An S-gate and an inverse S-gate are realized by RM circuits shown in Fig. 6. We can verify their correctness by testing all the cases of inputs. For example, Fig. 7 shows the simulating process of an S-gate in the case $c = x = 1$. The time delay of the circuits is 12.

Based on the method given in [3] and using two copies of the circuits which realize an S-gate and an inverse S-gate, we can obtain a circuit that simulates an F-gate, as shown in Fig. 8 (its time delay is 48). From above, we can conclude that an RM is a universal.

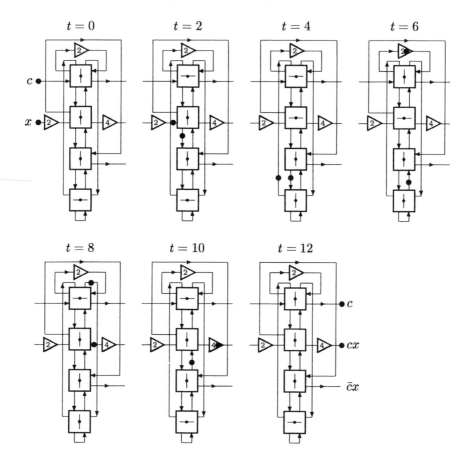

Figure 7: Simulating an S-gate by an RM circuit (the case $c = x = 1$).

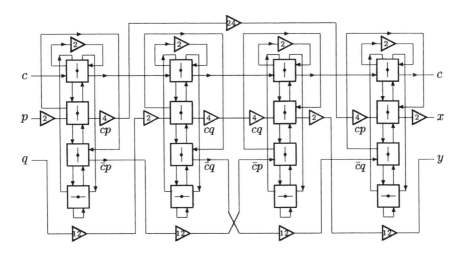

Figure 8: An RM circuit that simulates an F-gate.

4 Constructing a Reversible Sequential Machine

From Fredkin and Toffoli's result [3], we can see that any sequential machine (even if it is irreversible) can be embedded in an F-gate circuit if we supply constant signals and produce garbage (i.e. useless) signals as well as true inputs and outputs. As shown in the section above, RMs can simulate F-gate, so we can construct any sequential machine only of RMs under the same conditions as above. On the other hand, a reversible sequential machine can be embedded in a closed F-gate circuit [4], where "closed" means that there is neither a supply of constant signals nor a production of garbage signals. Thus, we can also construct a closed RM circuit for it. However, when constructing a reversible sequential machine out of RMs, there is a direct and much simpler method, as we shall show below.

A *reversible sequential machine* (RSM) is a system defined by:

$$M = (Q, \Sigma, \Gamma, q_1, \delta),$$

where Q is a finite non-empty set of states, Σ and Γ are finite non-empty sets of input and output symbols, respectively, and $q_1 \in Q$ is an initial state. $\delta : Q \times \Sigma \rightarrow Q \times \Gamma$ is a one-to-one mapping called a move function. Note that, since δ is one-to-one, $|\Sigma| \leq |\Gamma|$ must hold. Also note that an RM itself is an RSM with the move function shown in Table 1.

Assume $Q = \{q_1, \cdots, q_n\}$, $\Sigma = \{a_1, \cdots, a_h\}$, $\Gamma = \{b_1, \cdots, b_k\}$, $(k \geq h)$. To construct an RM circuit that simulates M, we use a q_i-*module* as shown in Fig. 9 (a). It consists of $k+1$ RMs, and has $2k$ input lines and $2k$ output lines. We assume that all RMs except the bottom one are initially set to V-states (the bottom RM may be either H- or V-state). Further assume

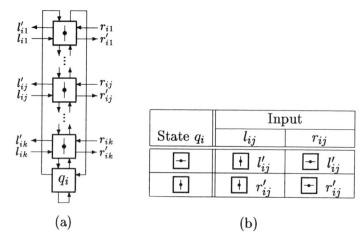

Figure 9: (a) A q_i-module, and (b) its move function ($j \in \{1, \cdots, k\}$).

that a particle is given to at most one input line. Then, a q_i-module acts like the RSM shown in Fig. 9 (b), where the input and output alphabets are $\{l_{i1}, \cdots, l_{ik}, r_{i1}, \cdots, r_{ik}\}$ and $\{l'_{i1}, \cdots, l'_{ik}, r'_{i1}, \cdots, r'_{ik}\}$, respectively, and the state set is $\{\boxed{\leftarrow}, \boxed{\uparrow}\}$ which matches that of the bottom RM. Though there are $k+1$ RMs, we can consider the q_i-module as a two-state machine, because all the RMs except the bottom one are reset to V-states when a particle leaves some output line. For example, if it is in the state $\boxed{\leftarrow}$ and the input is l_{ij}, then after *some* time steps the state becomes $\boxed{\uparrow}$ and gives an output l'_{ij}.

We put n q_i-modules in a row, and connect r'_{ij} to $l_{i+1,j}$ for each $i \in \{1, \cdots, n-1\}$ and $j \in \{1, \cdots, k\}$. Next, for each $i_0 \in \{1, \cdots, n\}$ and $j_0 \in \{1, \cdots, h\}$, if $\delta(q_{i_0}, a_{j_0}) = (q_{i_1}, b_{j_1})$ then connect $l'_{i_0 j_0}$ to $r_{i_1 j_1}$. If M's present state is q_{i_0}, we assume that the q_{i_0}-module is in the state $\boxed{\leftarrow}$, and that all the other q_i-modules are in $\boxed{\uparrow}$. We correspond each input a_j of M to the lines l_{ij}'s and each output b_j of M to the lines r'_{ij}'s.

Consider that the case M is in q_{i_0} and the input is a_{j_0}. Then a particle is given to the line l_{1j_0}. From the move function of a q_i-module (Fig. 9 (b)), we can see that a particle eventually leaves $l'_{i_0 j_0}$ making all the q_i-modules in the state $\boxed{\uparrow}$. As explained above, if $\delta(q_{i_0}, a_{j_0}) = (q_{i_1}, b_{j_1})$ then $l'_{i_0 j_0}$ is connected to $r_{i_1 j_1}$. Hence, a particle will be put into the line $r_{i_1 j_1}$. Again from the move function of a q_i-module, we can see that the q_{i_1}-module will be set to the state $\boxed{\leftarrow}$, and a particle will leave $r'_{n j_1}$. In this way, $\delta(q_{i_0}, a_{j_0}) = (q_{i_1}, b_{j_1})$ is simulated.

We give a simple example. Let $M_1 = (\{q_1, q_2, q_3\}, \{a_1, a_2\}, \{b_1, b_2\}, q_1, \delta_1)$ be an RSM, where δ_1 is given in Table 2. The RM circuit for M_1 is shown in

Fig. 10. Note that when connecting $l'_{i_0 j_0}$ to $r_{i_1 j_1}$ by the relation $\delta(q_{i_0}, a_{j_0}) = (q_{i_1}, b_{j_1})$, a delay is inserted appropriately so that the total delay between inputs and outputs is constant in all the cases.

State	Input	
	a_1	a_2
q_1	$q_2 b_1$	$q_3 b_2$
q_2	$q_2 b_2$	$q_1 b_1$
q_3	$q_1 b_2$	$q_3 b_1$

Table 2: δ_1 of a reversible sequential machine M_1 (an example).

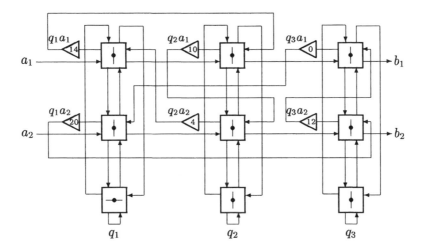

Figure 10: An RM circuit for a reversible sequential machine M_1 (in the state q_1). The delay between inputs and outputs is 28.

5   Concluding Remarks

In this paper, we have shown that a rotary memory is a universal reversible logic element. We have also given a method for constructing a reversible sequential machine concisely as a closed circuit consisting only of rotary memories.

It is known that a reversible Turing machine is computation-universal [1]. In principle, it is possible -though complicated- to formulate both (1) a finite-state control of a reversible Turing machine, and (2) a tape cell of it, as reversible sequential machines. Hence, we shall be able to realize a universal reversible computing machine as a closed circuit of rotary memories (but the entire circuit acting as a tape unit becomes infinite). Its detailed design is left for the future study. We are unable to say whether there is a simpler circuit of rotary memories that simulates a Fredkin gate simpler than the one given in this paper.

References

[1] C.H. Bennett, Logical reversibility of computation, *IBM Journal of Research and Development*, 17 (1973), 525–532.

[2] R.P. Feynman, *Feynman Lectures on Computation*, A.J.G. Hey and R.W. Allen (eds.). Perseus, Reading, Mass., 1996.

[3] E. Fredkin and T. Toffoli, Conservative logic, *International Journal of Theoretical Physics*, 21 (1982), 219–253.

[4] K. Morita, A simple construction method of a reversible finite automaton out of Fredkin gates, and its related problem, *Transactions of IEICE Japan*, E-73 (1990), 978–984.

[5] K. Morita, Y. Tojima and K. Imai, A simple computer embedded in a reversible and number-conserving two-dimensional cellular space, *Multiple-Valued Logic*, to appear.

[6] Gh. Păun, G. Rozenberg and A. Salomaa, *DNA Computing: New Computing Paradigms*. Springer, Berlin, 1998.

[7] T. Toffoli, Reversible computing. In *Automata, Languages and Programming*. Springer, Berlin, 1980, 632–644.

Church-Rosser Languages and Their Relationship to Other Language Classes

Gundula Niemann

Friedrich Otto

Department of Mathematics and Computer Science
University of Kassel
Germany
{niemann,otto}@theory.informatik.uni-kassel.de

Abstract. Church-Rosser languages were introduced by McNaughton et al (1988) by using finite, length-reducing, and confluent string-rewriting systems as a kind of accepting device. Here we give a survey of some recent results concerning characterizations of the Church-Rosser languages through certain machine models and their relationship to other language classes.

1 Introduction

For a finite length-reducing string-rewriting system R there exists an algorithm that, given a string w as input, computes an irreducible descendant $\hat{w}$ of w with respect to R in linear time [2]. If R is confluent, then this irreducible descendant is uniquely determined. Accordingly, two strings u and v are congruent with respect to the Thue congruence generated by R if and only if their irreducible descendants $\hat{u}$ and $\hat{v}$ coincide, and hence the *word problem* for R is solvable in linear time.

Motivated by this observation, McNaughton et al. [14] used the finite, length-reducing, and confluent string-rewriting systems to define two classes of languages: the class CRL of *Church-Rosser languages* and the class CRCL of *Church-Rosser congruential languages* (see Section 2 for the definitions), which is a proper subclass of CRL. The membership problem for these languages is solvable in linear time and, hence, CRL is contained in the class CSL of context-sensitive languages. In addition, it was shown in [14] that CRL contains the class DCFL of deterministic context-free languages, and that it contains some languages that are not context-free. However, the exact relationship between CRL and the class CFL of context-free languages remained open at the time.

Context-sensitive languages are generated by phrase-structure grammars that contain only length-increasing and length-preserving productions. Dahlhaus and Warmuth [5] investigated the subclass GCSL of growing context-sensitive languages, which are generated solely by length-increasing productions. GCSL lies properly between CFL and CSL. The membership problem for these languages is solvable in polynomial time, and it turned out that GCSL is an abstract family of languages [3].

In this survey we describe some recent results that establish machine characterizations for CRL and for GCSL, respectively, leading to the observation that CRL can be interpreted as the *deterministic variant* of GCSL. Using the machine characterization obtained for CRL, various closure and non-closure properties have been derived for CRL.

Also we will describe another characterization of this class by a machine model, which is a specific type of the *restarting automata with rewriting* introduced by Jančar et al [11, 12]. Further, we will present some typical examples of Church-Rosser languages that nicely illustrate the power of this concept, and we will compare CRL as well as CRCL to some other language classes.

The paper closes with a discussion of some open problems and several ideas for future work.

2   CRL and GCSL

A *string-rewriting system* R on some alphabet Σ is a subset of $\Sigma^* \times \Sigma^*$. An element $(\ell, r) \in R$ is called a *(rewrite) rule*, and it will be denoted as $(\ell \to r)$. Here we will only be dealing with finite string-rewriting systems.

The string-rewriting system R induces a *reduction relation* $\to_R^*$ on Σ^*, which is the reflexive, transitive closure of the single-step reduction relation $\to_R = \{(x\ell y, xry) \mid x, y \in \Sigma^*, (\ell \to r) \in R\}$. If $u \to_R^* v$, then u is an *ancestor* of v, and v is a *descendant* of u. If there is no $v \in \Sigma^*$ such that $u \to_R v$ holds, then the string u is called *irreducible* (mod R). By IRR(R) we denote the set of all such irreducible strings.

The smallest equivalence relation $\leftrightarrow_R^*$ containing $\to_R$ is called the *Thue congruence* generated by R. For $u \in \Sigma^*$, $[u]_R$ denotes the congruence class $\{v \in \Sigma^* \mid v \leftrightarrow_R^* u\}$ of u.

A string-rewriting system R is *length-reducing* if $|\ell| > |r|$ holds for each rule $(\ell \to r) \in R$, and it is *confluent* if, for all $u, v, w \in \Sigma^*$, $u \to_R^* v$ and $u \to_R^* w$ imply that v and w have a common descendant.

Using string-rewriting systems of this restricted form the following two language classes have been defined in [14].

Definition 1 *(a) A language $L \subseteq \Sigma^*$ is a* Church-Rosser language *if there exist an alphabet $\Gamma \supsetneq \Sigma$, a finite, length-reducing, and confluent string-rewriting system R on Γ, two irreducible strings $t_1, t_2 \in (\Gamma \smallsetminus \Sigma)^*$, and an irreducible letter $Y \in \Gamma \smallsetminus \Sigma$ such that, for all $w \in \Sigma^*$, $t_1 w t_2 \to_R^* Y$ if and only if $w \in L$.*

(b) A language $L \subseteq \Sigma^*$ is a Church-Rosser congruential language *if there exist a finite, length-reducing, and confluent string-rewriting system R on Σ and finitely many strings $w_1, \ldots, w_n \in \mathrm{IRR}(R)$ such that*
$$L = \bigcup_{i=1}^{n} [w_i]_R.$$

By CRL we denote the class of Church-Rosser languages, while CRCL denotes the class of Church-Rosser congruential languages. CRCL is obviously contained in CRL, and CRCL contains non-regular languages. For example, if $R = \{aabb \to ab\}$ and $L_1 = [ab]_R$, then we see that $L_1 \in$ CRCL is the non-regular language $L_1 = \{a^n b^n \mid n \geq 1\}$. Further, CRL contains DCFL [14], the class of deterministic context-free languages. CRCL is incomparable to DCFL, because DCFL contains the language $L_2 = L_1 \cup b^+$, which is not even congruential, while CRCL contains some languages that are not even context-free, as shown by the following example.

Example 1 [14] Consider the non-context-free language $L_3 = \{a^{2^n} \mid n \geq 0\}$, and let $R = \{\text{¢}aaaa \to \text{¢}aaF, Faa \to aF, F\$ \to \$, \text{¢}aa\$ \to Y, \text{¢}a\$ \to Y\}$. Then R is length-reducing and confluent, and it is easily verified that, for all $n \geq 0$, $a^n \in L_3$ iff $\text{¢}a^n\$ \to_R^* Y$ holds. Thus, $L_3 \in$ CRL.

Now let $\Sigma = \{a, F, \text{¢}, \$, Y\}$, and let $L_4 = [Y]_R$. Then $L_4 \in$ CRCL. On the other hand, $L_4 \notin$ CFL, as $\text{¢} \cdot L_3 \cdot \$ = L_4 \cap \text{¢} \cdot a^* \cdot \$$, and CFL is closed under intersection with regular sets and left- and right-derivatives.

Example 2 Let $L_5 = \{a^n b^n c^n \mid n \geq 0\}$, and let R be the string-rewriting system containing the following rules:

$\text{¢}aaaa$	$\to$	$\text{¢}aaF,$		$O_1 bcc$	$\to$	$cO_2,$
Faa	$\to$	$aF,$		$E_2 cc$	$\to$	$cE_2,$
Fbb	$\to$	$bE_1,$		$O_2 cc$	$\to$	$cO_2,$
$Fabb$	$\to$	$bO_1,$		$E_2\$$	$\to$	$\$,$
$E_1 bb$	$\to$	$bE_1,$		$O_2 c\$$	$\to$	$\$,$
$O_1 bb$	$\to$	$bO_1,$		$\text{¢}\$$	$\to$	$Y,$
$E_1 cc$	$\to$	$cE_2,$		$\text{¢}a^i b^i c^i \$$	$\to$	$\text{¢}\$$ for $i = 1, 2, 3.$

Then R is length-reducing and confluent, and for all $w \in \{a, b, c\}^*$, $\text{¢} w\text{\$} \to_R^* Y$ iff $w \in L_5$, that is, $L_5 \in \text{CRL}$.

In [4] the so-called *shrinking two-pushdown automata* (sTPDA) are introduced. A two-pushdown automaton $M = (Q, \Sigma, \Gamma, \delta, q_0, \perp, F)$ has two pushdown stores and a finite control. Based on the actual state and the topmost symbols seen on both pushdowns, M enters a new state and replaces the topmost symbols on both pushdowns by strings from the pushdown alphabet. A *configuration* of M is described as uqv, where $q \in Q$ is the actual state, $u \in \Gamma^*$ is the contents of the first pushdown store with the first letter of u at the bottom and the last letter of u at the top, and $v \in \Gamma^*$ is the contents of the second pushdown store with the last letter of v at the bottom and the first letter of v at the top. For an input string $w \in \Sigma^*$, the corresponding initial configuration is $\perp q_0 w \perp$, where $\perp$ is the bottom marker of the pushdowns, and $q_0 \in Q$ is the initial state of M. M induces a computation relation $\vdash_M^*$ on the set of configurations, which is the reflexive, transitive closure of the single-step computation relation $\vdash_M$ (see e.g. [9]). M accepts with an empty right pushdown store, that is:

$$L(M) = \{w \in \Sigma^* \mid \perp q_0 w \perp \vdash_M^* \alpha q \text{ for some } q \in F \text{ and } \alpha \in \Gamma^*\}.$$

Here $F \subset Q$ is the set of final states of M. Finally, M is called *shrinking* if there exists a *weight function* $\varphi : Q \cup \Gamma \to \mathbb{N}$ such that each step of M reduces the combined weight of the actual configuration, that is, whenever $uqv \vdash_M u'q'v'$, then $\varphi(u'q'v') < \varphi(uqv)$. By sTPDA and sDTPDA we denote the class of shrinking two-pushdown automata and the class of shrinking deterministic two-pushdown automata, respectively.

Instead of requiring that the weight of the actual configuration decrease, the combined length of the contents of the two pushdown stores can be required to decrease in each step. However, in order to make this variant of the TPDA sufficiently expressive, we allow it to see and to replace the topmost k symbols of each pushdown store for some fixed constant $k > 1$. In this way we obtain the classes of *length-reducing* deterministic and nondeterministic TPDAs.

Concerning these classes of two-pushdown automata the following characterizations have been established.

Proposition 1 [4, 17]

(a) *A language is accepted by some shrinking TPDA if and only if it is a growing context-sensitive language.*

(b) *A language is accepted by some shrinking DTPDA if and only if it is a Church-Rosser language.*

In [15] it is shown that the same results hold for the length-reducing TPDA.

Using the strategy from Example 2 a length-reducing DTPDA can be constructed for the language L_5. Actually, this strategy applies to more general situations and, for example, it can be used to show that the following languages belong to class CRL:

$$L_6 = \{a^n b^{2n} c^{4n} \mid n \geq 0\}, \quad L_7 = \{a^i b^j \mid 0 \leq i \leq j \leq 2i\},$$
$$L_8 = \{a^{2^n} b^{3^n} \mid n \geq 0\}, \quad L_9 = \{a^i b^j c^k \mid i, j, k \geq 0, i = j \text{ or } j = k\}.$$

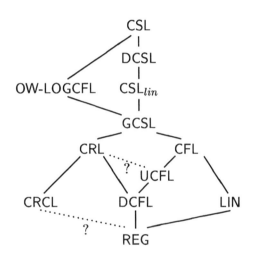

In [1, 8] it is shown that the language $L_{10} = \{wcw^R cw \mid w \in \{a, b\}^*\}$ is in CSL $\smallsetminus$ CSL$_{lin}$, where CSL$_{lin}$ denotes the class of those languages that are generated by context-sensitive grammars with linear bounds on the length of derivation sequences. The complement $L_{10}^c = \{a, b, c\}^* \smallsetminus L_{10}$ is context-free (in fact, linear), and hence, it is in GCSL. As CRL is characterized by a deterministic machine model (Proposition 1 (b)), it is closed under complement. Thus, the language L_{10}^c belongs to the difference GCSL $\smallsetminus$ CRL. In particular, this language shows that CFL is not contained in CRL, thus proving a conjecture of McNaughton et al [14].

The figure above depicts the various inclusion results concerning classes CRL and CRCL and the classes in the Chomsky hierarchy. Here OW-LOGCFL denotes the class of languages that are *one-way log-space reducible* to a context-free language, DCSL is the class of *deterministic context-sensitive languages*, and UCFL denotes *unambiguous context-free languages*. These results are proved in [14, 4, 17]. The question marks indicate possible inclusions that are still unproven.

In addition it has been observed that CRL is closed under complement, inverse morphisms, intersection with regular languages, left- and right derivatives, reversal, marked concatenation, marked iteration, and union with the empty string, and so CRL is a pre-AFL. On the other hand, CRL is not closed under union, intersection, concatenation, iteration, ε-free morphisms, quotients with regular sets, or substitution by regular languages. In fact, CRL is a *basis* for the recursively enumerable languages, that is, its closure under arbitrary morphisms yields the set of recursively enumerable languages [3, 20].

3   CRL and Restarting Automata

In [10], Jančar et al. presented the *restarting automaton*, which is a non-deterministic machine model processing strings that are stored in lists or 'rubber tapes'. These automata model certain elementary aspects of the syntactical analysis of natural languages.

A restarting automaton, or R-automaton for short, has a finite control, and it has a read/write-head with a finite look-ahead working on a list of symbols (or a tape). As defined in [10], it can perform two kinds of operations: a *move-right step*, which shifts the read/write-window one position to the right and possibly changes the actual state, and a *restart step*, which deletes some letters from the read/write-window (and therewith shortens the list or tape), places this window over the left end of the list (or tape), and puts the automaton back into its initial state. Given an input string w, it starts with the read/write-window placed on the left end of the tape, and it accepts if and when an accepting state is reached.

In subsequent papers, Jančar and his co-workers extended the restarting automaton by introducing rewrite steps that instead of simply deleting some letters replace the contents of the read/write-window by some shorter string from the input alphabet [12]. This is the so-called RW-automaton.

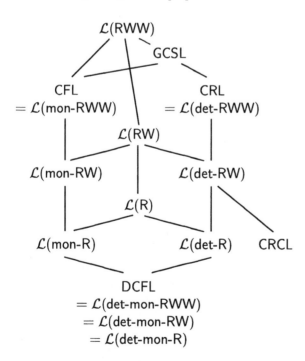

Later the restarting automaton was allowed to use auxiliary symbols, so-called nonterminals, in the replacement operation [13, 18], which lead to the so-called RWW-automaton. Finally the restarting operation was separated from the rewriting operation [13], which yielded the RRW-automaton. Obviously, the later variations can be combined, giving the so-called RRWW-automaton.

Since various additional restrictions can be put on each of these variants of the restarting automaton, a potentially large family of automata and language classes is obtained. For example, various notions of monotonicity have been defined, and it has been shown that monotonous deterministic RW-automata accept deterministic context-free languages [12], and that monotonous RWW-automata accept context-free languages [13].

To denote the class of RW-automata (R-,RWW-,RRWW-automata) that are monotonous or deterministic, respectively, we use the prefixes mon- and det-, respectively, which can also be combined. Further, by $\mathcal{L}(\mathcal{A})$ we denote the class of languages that are accepted by the automata of type $\mathcal{A}$.

Concerning the relationship of CRL to the various classes of RW-automata, the following results have been obtained.

Proposition 2 [18, 19]

(a) GCSL *is contained in the class* $\mathcal{L}$(RWW).

(b) CRL $= \mathcal{L}$(det-RWW) $= \mathcal{L}$(det-RRWW).

(c) CRCL *is a proper subclass of* $\mathcal{L}$(det-RW).

Hence, we have the hierarchy depicted in the figure above. In particular, for the Church-Rosser languages L_7 and L_9 it is known that $L_7 \in \mathcal{L}$(mon-R)$\smallsetminus$ $\mathcal{L}$(det-RW), while $L_9 \notin \mathcal{L}$(RW) [12].

4   Some Open Problems

It is an open problem whether or not GCSL coincides with the class $\mathcal{L}$(RWW). The difficulty in proving equality between these two classes stems from the fact that it is not clear how a shrinking TPDA may simulate an RWW-automaton M. It is straightforward to simulate the tape of M by two push-down stores, but each time M performs a rewrite, it also makes a restart, thus putting its read/write-window over the left end of the actual tape. However, a shrinking TPDA cannot shift the contents of its first pushdown onto its second pushdown without reducing the weight of the actual configuration. It appears that such a simulation should be possible whenever M satisfies a certain monotonicity condition, that is, if $\mathdollar xquy\mathdollar \vdash_M q_0 \mathdollar xvy\mathdollar$ is a rewrite step of M, where u is replaced by v and q_0 denotes the initial state, then the next rewrite step is not applicable before the read/write-window has at least seen the first letter of v. Thus, the above question reduces to the question of whether each RWW-automaton is equivalent to an RWW-automaton satisfying this condition.

In [14] it has been conjectured that the language $L_{11} = \{ww^R \mid w \in \{a,b\}^*\}$ is not a Church-Rosser language. This conjecture remains open. Also it is open whether each regular language belongs to CRCL, although some progress has been made recently by showing that at least each regular language of polynomial density is in CRCL [16].

Finally, the exact relationship of CRL to some other classes of languages remains to be determined. For example, it is easily seen that CRL is not contained in the class $\mathcal{L}(\lambda P)$ of languages that are generated by the *programmed context-free grammars with erasing rules*, as L_3 is in CRL, while each unary language contained in $\mathcal{L}(\lambda \mathcal{P})$ is necessarily regular [6]. However, it is not known whether CRL is contained in the class $\mathcal{L}(P_{ac})$ of languages generated by the *programmed context-free grammars without erasing rules* and with *appearance checking*.

Further, CRL contains the language $L_{12} = \{(a^n b)^n \mid n \geq 0\}$, which is not an *indexed language* (see [7]), while the indexed language $L_{13} = \{ww \mid w \in \{a, b\}^*\}$ is not Church-Rosser. To prove that L_{12} is indeed a Church-Rosser language, we can apply the same strategy as in Example 2, but this time it is not only used for comparing the length of different blocks of a's, but also for comparing the length of these blocks to the number of blocks. However, a block is not simply deleted, but it is marked as 'deleted' and kept, since it is still needed for comparing its length to the length of the other blocks. Hence, CRL is incomparable to the class of indexed languages, and the question of determining those indexed languages that are Church-Rosser still remains. A corresponding question can be asked for the OL-languages, for which a preliminary result can be found in [14].

References

[1] R.V. Book, Grammars with Time Functions, PhD dissertation, Harvard University, Cambridge, Mass., 1969.

[2] R.V. Book and F. Otto, *String-Rewriting Systems*. Springer, New York, 1993.

[3] G. Buntrock, Wachsende kontext-sensitive Sprachen. Habilitationsschrift, Fakultät für Mathematik und Informatik, Universität Würzburg, 1996.

[4] G. Buntrock and F. Otto, Growing context-sensitive languages and Church-Rosser languages, *Information and Computation*, 141 (1998), 1–36.

[5] E. Dahlhaus and M. Warmuth, Membership for growing context-sensitive grammars is polynomial, *Journal of Computer and System Sciences*, 33 (1986), 456–472.

[6] J. Dassow, Gh. Păun and A. Salomaa, Grammars with controlled derivations. In G. Rozenberg and A. Salomaa (eds.), *Handbook of Formal Languages*. Springer, Berlin, 1996, vol. 2, 101–154.

[7] R.H. Gilman, A shrinking lemma for indexed languages, *Theoretical Computer Science*, 163 (1996), 277–281.

[8] A.V. Gladkij, On the complexity of derivations for context-sensitive grammars, *Algebri i Logika Sem.*, 3 (1964), 29–44 (in Russian).

[9] J.E. Hopcroft and J.D. Ullman, *Introduction to Automata Theory, Languages, and Computation.* Addison-Wesley, Reading, Mass., 1979.

[10] P. Jančar, F. Mráz, M. Plátek and J. Vogel, Restarting automata. In H. Reichel (ed.), *Fundamentals of Computation Theory*, Proceedings of FCT'95. Springer, Berlin, 1995, 283–292.

[11] P. Jančar, F. Mráz, M. Plátek and J. Vogel, Monotonic rewriting automata with a restart operation. In F. Plášil and K.G. Jeffery (eds.), *Proceedings of SOFSEM'97: Theory and Practise of Informatics.* Springer, Berlin, 1997, 505–512.

[12] P. Jančar, F. Mráz, M. Plátek and J. Vogel, On restarting automata with rewriting. In Gh. Păun and A. Salomaa (eds.), *New Trends in Formal Languages.* Springer, Berlin, 1997, 119–136.

[13] P. Jančar, F. Mráz, M. Plátek and J. Vogel, Different types of monotonicity for restarting automata. In V. Arvind and R. Ramanujam (eds.), *Proceedings of Foundations of Software Technology and Theoretical Computer Science 18th Conference.* Springer, Berlin, 1998, 343–354.

[14] R. McNaughton, P. Narendran and F. Otto, Church-Rosser Thue systems and formal languages, *Journal of the ACM*, 35 (1998), 324–344.

[15] G. Niemann, CRL, CRDL und verkürzende Zweikellerautomaten, unpublished note, June 1997.

[16] G. Niemann, The regular languages of polynomial density are Church-Rosser congruential, unpublished note, November 1999.

[17] G. Niemann and F. Otto, The Church-Rosser languages are the deterministic variants of the growing context-sensitive languages. In M. Nivat (ed.), *Proceedings of Foundations of Software Science and Computation Structures, FoSSaCS'98.* Springer, Berlin, 1998, 243–257.

[18] G. Niemann and F. Otto, Restarting automata, Church-Rosser languages, and representations of r.e. languages. In W. Thomas (ed.), *Preproceedings of Developments in Language Theory, DLT 99*, Aachener Informatik-Berichte 99-5, RWTH Aachen, 1999, 49–62.

[19] G. Niemann and F. Otto, Some results on deterministic restarting automata, unpublished note, September 1999.

[20] F. Otto, M. Katsura and Y. Kobayashi, Infinite convergent string-rewriting systems and cross-sections for finitely presented monoids. *Journal of Symbolic Computation*, 26 (1998), 621–648.

Hiding Regular Languages

Valtteri Niemi

Nokia Research Center
Nokia Group
Helsinki, Finland
`valtteri.niemi@nokia.com`

Abstract. A cryptosystem based on formal language theory is proposed. We prove that the problem of finding a system compatible decryption key is NP-complete.

1 Introduction

During the last twenty years many public-key cryptosystems based on different mathematical fields have been published. The security of a few systems, like RSA or those based on elliptic curves, has been systematically studied while many systems are simply announced to look hard to break down. We propose another public-key system and we present some arguments that are in favor of the security of the system.

More specifically, we prove that a certain cryptanalytic approach leads to an NP-complete problem. The approach searches for a decryption key that is compatible with the system; this means that the cryptanalyst tries to find a decryption key that corresponds to the publicized encryption key via the known key generation mechanism. When this search succeeds, the cryptanalyst does not necessarily find the original decryption key but, since the key matches the original one, the two keys are equivalent in decryption. Note that Shamir [5] broke the Merkle-Hellman knapsack system using this approach.

On the other hand, the impact of a result dealing with worst-case complexity measures is always restricted in cryptography. See [1] for some discussion on this issue.

Our system is based on the idea of hiding significant letters in a considerable number of dummy letters (see e.g. [4] for the first public key cryptosystem in which this idea was introduced). The encryption is performed bit-by-bit. Afterwards, it can be decided which of the two bit values has been encrypted, by checking whether the word consisting of all significant letters in the cryptotext belongs to a specific regular language. Of course, both the set of significant letters and the regular language must be known by the receiver who decrypts the words. The system has been briefly described in [2].

Bit-by-bit encryption has many advantages, one of which is the possibility of generating cryptowords in advance. In order to encrypt an actual message you simply pick up words from two lists. An obvious disadvantage is the length of cryptowords. To exclude exhaustive search attacks, the total amount of cryptotext must be at least a hundred times more than the amount of plaintext. However, in the case of a hybrid system (where only a key to a conventional one-key cipher is encrypted by the public key system) this amount of data expansion may be acceptable for some applications.

We use the standard terminology of cryptography (see e.g. [3]). The section below contains the detailed construction of our system and the third section provides proof of its results.

2   Definition of the System

The public encryption key is constructed by the following steps:

(1) Choose two arbitrary grammars G_0 and G_1 over some terminal alphabet Σ.

(2) Choose a deterministic finite automaton A_0 over the same terminal alphabet Σ.

(3) Replace the final state set of A_0 by its complement to obtain another DFA A_1. Now it is clear that $L(A_0) \cap L(A_1) = \emptyset$ and, on the other hand, $L(A_0) \cup L(A_1) = \Sigma^*$. Moreover, all languages in this construction should be nonempty. Hence, we assume that $L(A_i) \neq \emptyset$ for $i = 0, 1$. If this is not originally the case we change the final state set of A_0.

(4) Apply the standard triple construction (see end of this section for details) to obtain grammars G_i' over Σ such that $L(G_i) \cap L(A_i) = L(G_i')$ for $i = 0, 1$.

(5) Choose another alphabet Γ and a morphism h from Γ to Σ such that each letter in Γ is mapped either to a letter in Σ or to the empty word λ. This is the division between significant terminal letters and dummy terminal letters.

(6) Construct grammars G_i'' over the terminal alphabet Γ such that $h(G_i'') \subseteq G_i'$, for $i = 0, 1$, where the morphism h is enlarged to cover non-terminals as well. A morphism from one grammar to another is defined in a natural fashion. Note that G_0'' and G_1'' may have totally distinct nonterminal alphabets. This last step introduces a division between significant nonterminal letters and dummy nonterminals.

Now the pair (G_0'', G_1'') constitutes the public encryption key. A bit 0 is encrypted by generating a word $w_0 \in L(G_0'')$. Similarly, 1 is encrypted by some $w_1 \in L(G_1'')$. In the sequel, the words w_i are called *cryptowords*.

Steps (3) and (4) guarantee that $L(G_0') \cap L(G_1') = \emptyset$ and from (6) it follows that also $L(G_0'') \cap L(G_1'') = \emptyset$. Hence 0 and 1 can never be encrypted by the same cryptoword.

The pair (h, A_0) constitutes the secret decryption key. The receiver first applies the morphism h to each cryptoword, and then checks whether the automaton A_0 accepts the resulting word (in which case the decrypted bit is 0).

We conclude this section by describing the structure of the triple construction used in step (4).

First we replace each terminal a with a new nonterminal A and a new production $A \to a$ is added. In the second phase, we build a totally new nonterminal alphabet that consists of all triples $[p, C, q]$ where p and q are states of the underlying automaton and C is an old nonterminal letter.

Then we add a new start symbol S_0 and all productions $S_0 \to [s, S, f]$, where S is the original start symbol, s is the initial state of the underlying automaton and f is one of its final states.

Each old production is replaced with a whole bunch of new ones by simply converting all nonterminals to corresponding triples. There are two restrictions:

- the state sequence of the left-hand-side and that of the right-hand side must begin and end similarly,

- the intermediate states in the sequences (in both sides) must "chain" properly.

For example, a production $AB \to CDC$ is replaced with all productions of the type $[p, A, q][q, B, r] \to [p, C, t][t, D, u][u, C, r]$.

In the case of terminating productions $A \to a$ the left-hand side is converted to the production $[p, A, q] \to a$ if and only if the transition from the state p to the state q by the input symbol a is valid in the automaton.

Finally, each erasing production $C_1 \cdots C_k \to \lambda$ is replaced with all productions of the type $[p, C_1, r] \cdots [t, C_k, p] \to \lambda$.

3 Cryptanalytic Views

The cryptanalytic task is related to the membership problem of recursively enumerable languages, which is an undecidable problem. Indeed, it is easy to see that the problem of whether a given word is a cryptoword or not is undecidable. Unfortunately, the problem is as hard from the legitimate receiver's point of view.

The next result is more relevant.

Theorem 1 *Given an automaton A_0 and a grammar G_0'', it is an NP-complete problem to decide whether there exist construction steps (1)-(6) that lead from A_0 to G_0''.*

Proof. We give a reduction of the well known NP-complete *Hamiltonian circuit problem for digraphs*: "given a directed graph D, is there a circuit in D that meets every vertex of D exactly once?".

Let us demonstrate the reduction by means of a concrete example. The general construction can be derived from it in a straightforward manner.

Let D be the following digraph:

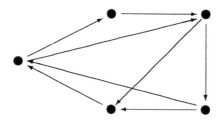

(i) First fix the start (and end) point of a (possibly existing) Hamiltonian circuit (HC, for short). It can be done, since the circuit meets every vertex anyway.

(ii) Name the vertices (the start vertex is s).

(iii) Name each arc according to the initial vertex in such a way that the terminal vertex occurs as a subscript (e.g. s_a, a_c).

Now D looks like the following:

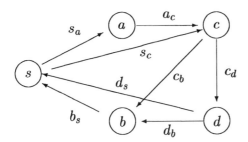

(iv) We obtain a deterministic finite automaton A_0 after adding a "dead end" state: all transitions from a "wrong" initial state lead to the dead end (for instance, the transition from the state a by the letter c_b). The correspondence between elements of the digraph D and elements of the DFA A_0 is given below:

digraph D	DFA A_0
vertices	states
arcs	transitions
start point of the HC	initial state
= end point of the HC	= final state
names of arcs	terminal letters

(v) Define a grammar G_0'' as follows. The start symbol is S_0. We have one long production that refers exactly once to each arc with s as the initial vertex and exactly $|D| - 1$ times to every other arc (here $|D| = 5$):

$$S_0 \to S_{a1} S_{c1} A_{c1} B_{s1} C_{b1} C_{d1} D_{s1} D_{b1} A_{c2} \cdots A_{c4} B_{s4} C_{b4} C_{d4} D_{s4} D_{b4}.$$

Moreover, we have productions:

$M_S \to S_{a1} S_{c1},$
$M_A \to A_{c1} A_{c2} A_{c3} A_{c4},$
$M_B \to B_{s1} B_{s2} B_{s3} B_{s4},$
$M_C \to C_{b1} C_{d1} C_{b2} C_{d2} C_{b3} C_{d3} C_{b4} C_{d4}$ and
$M_D \to D_{s1} D_{b1} D_{s2} D_{b2} D_{s3} D_{b3} D_{s4} D_{b4}.$

The right-hand sides of these productions are formed by picking up all the letters in the long production such that the corresponding arcs have the same initial vertex. The M-letters are new nonterminals.

Finally, we have terminating productions: $S_{a1} \to s_{a1}, S_{c1} \to s_{ci}, A_{ci} \to a_{ci}, B_{si} \to b_{si}, C_{bi} \to c_{bi}, C_{di} \to c_{di}, D_{si} \to d_{si}, D_{bi} \to d_{bi} (i = 1, 2, 3, 4)$.

If the grammar G_0'' and the automaton A_0 together satisfy the following two conditions, then we are able to show that the digraph D has an HC iff there are construction steps (1)-(6) that lead from A_0 to G_0''.

The crucial conditions are:

(I) Each terminal letter x_{yi} (where $x, y \in \{s, a, b, c, d\}$ and $1 \leq i \leq 4$) is mapped either to λ or to the letter x_y by the morphism h of construction step (5) (if such an h exists).

(II) For each $x \in \{s, a, b, c, d\}$ there exists at least one pair (y, i) such that $h(x_{yi}) = x_y$ (if h exists).

Let us now show how these conditions imply the desired criterion for the existence of an HC. The argument is based on properties of the triple construction.

First, assume there exist construction steps (1)-(6). Because of (II), the long production (with S_0 as the left-hand side) has at least five significant nonterminals on its right-hand side. Significant nonterminals are those that terminate in significant terminal letters, i.e. significant letters have nonempty h-images. On the other hand, the structure of the triple construction and condition (I) together imply that each significant nonterminal X_{yi} (where $X \in \{S, A, B, C, D\}$, $y \in \{s, a, b, c, d\}$ and $i \in \{1, 2, 3, 4\}$) must map either into a triple $[x, \alpha, y]$ (where $\alpha \to x_y$ is a production of the grammar G_0') or into a triple $[z, \alpha, g]$ (where z is a state different from x and g is the dead end state).

If all significant nonterminals map according to the first alternative then the productions with M-letters on the left-hand side cannot satisfy the triple construction principle unless there is at most one significant nonterminal on the right-hand side of each such production. Otherwise the triples do not chain properly. This means that there are at most five significant nonterminals on the right-hand side of the long S_0-production. Hence, there are *exactly* five significant nonterminals and, moreover, all the states of the automaton appear in the corresponding triples.

On the other hand, if some of the letters X_{yi} map into a type $[z, \alpha, g]$ triple, it follows that the h-image of the long production cannot chain properly since there are no transitions coming out from the dead end state g. Thus, this case is impossible and an HC exists in the original graph.

Secondly, assume an HC exists. In our case, indeed, the circuit $s_a a_c c_d d_b b_s$ is Hamiltonian. We can choose $S_{a1}, A_{c1}, C_{d2}, D_{b3}$ and B_{s4} to be the only significant nonterminals. They correspond to the triples $[s, \alpha_1, a]$, $[a, \alpha_2, c]$, $[c, \alpha_3, d]$, $[d, \alpha_4, b]$, $[b, \alpha_5, s]$ and they chain properly in the long production. Therefore, a valid construction procedure exists.

Unfortunately, the grammar G_0'' and the automaton A_0 do not necessarily fulfil conditions (I) and (II). However, we can modify these two objects so that the conditions are fulfilled. In fact, assuming that (I) holds, condition (II) can be inserted into the cryptosystem itself. This is due to the fact that if (II) is not satisfied then the state x is totally useless in the decryption process. We did not want to add this condition to the construction steps (1)-(6) of the previous section in order to avoid nonessential complications.

Note also that if (II) is not required then the answer to our construction problem is trivially positive: we may choose the morphism h that maps everything to λ.

As regards the condition (I), we may proceed as follows. We modify the structure of the automaton A_0: for each terminal x_y in A_0 we add a completely new cycle consisting of new states and x_y-transitions between them. All these new cycles are pairwise distinct and of different length. A "wrong" letter in between a cycle leads to the dead end state. Suitable productions can be added to the grammar G_0'' in order to force (I) to be satisfied. Of course, these modifications to the automaton and the grammar could affect the first part of our proof. Check that the addition of cycles does not make it possible to implement construction steps (1)-(6) without having an HC in the original digraph. However, we omit the details of this checking procedure.

Thus, the problem is NP-hard. On the other hand, it is clearly in NP.
□

Intuitively, the search for a decryption key (h, A_0) from the public key (G_0'', G_1'') seems to be harder than the problem of Theorem 1 that we just proved to be NP-complete. Below we give a formal proof of this conclusion.

Theorem 2 *Given a possible encryption key (G_0'', G_1''), it is an NP-complete problem to decide whether there exists a corresponding decryption key (h, A_0).*

Proof. Our argument is based on the fact that only half of the public key, i.e. G_0'', was constructed in the proof of Theorem 1. The other half, G_1'', supplies enough leeway to force the automaton A_0 to the form described in the proof of Theorem 1. After this is done, the proof of Theorem 1 also provides proof of Theorem 2.

We begin the forcing procedure by placing some restrictions on the model of our original cryptosystem. This does not look very elegant since, essentially, we start to prove a property of a problem by modifying the problem itself. However, these restrictions do not make the cryptosystem impractical. We could have introduced the restrictions in the construction of section 2 but, for the sake of clarity, we did not.

We assume that our cryptosystem satisfies the following conditions:

(a) The grammars G_i of construction step (1) do not contain any productions of the type $A \rightarrow B$ where both A and B are nonterminals.

(b) If there is a production $A \rightarrow \lambda$ in G_i then A cannot occur either as the leftmost or as the rightmost letter on the right-hand side of any production.

Now we are ready to describe the grammar G_1''. The grammar G_0'' is the same as the one that was constructed in the proof of Theorem 1. We follow construction steps (1)-(6) starting with G_1, which is chosen to be the right-linear grammatical version of the automaton A_1. In turn, A_1 is the complement automaton of the DFA A_0 constructed in the proof of Theorem 1.

Once again, we illustrate the general argument with our example digraph D. At the end of the previous proof, we introduced new cycles into the original automaton A_0 and corresponding changes to G_0''. Now we ignore these cycles. We can afford to do this because our forcing procedure will automatically guarantee that conditions (I) and (II) are satisfied.

According to the triple construction, we have five initial productions $S_0 \to [s, S, a], ..., S_0 \to [s, S, g]$ and five complete copies of the state set of A_1: $[s, S, a], [a, A, a], [b, B, a], ..., [g, G, a]; [s, S, b], ..., [g, G, b]; ...; [s, S, g], ..., [g, G, g]$. Here the upper case letters are the nonterminals of G_1 corresponding to the states of A_1 which are marked by lower case letters. Moreover, we have triples obtained from the terminating nonterminals $S_a, S_c, A_c, ..., D_s$. Recall that each terminal was replaced with a new nonterminal at the beginning of the triple construction. Hence, we have productions $[s, S_a, a] \to s_a$, $[a, S_a, g] \to s_a, ..., [d, D_s, s] \to d_s$ in the grammar G_1', which is the intermediate grammar in the construction of our system. For each transition of A_1 we have respective productions in G_1', for instance, $[s, S, a] \to [s, S_a, a][a, A, a]$, $[a, A, a] \to [a, S_a, g][g, G, a]$, $[a, A, d] \to [a, A_c, c][c, C, d]$ etc.

We also have "semiterminating" productions that correspond to the terminal productions of G_1: $[s, S, a] \to [s, S_a, a]$, $[a, A, g] \to [a, S_a, g]$, $[c, C, d] \to [c, C_d, d]$ etc.

The last two construction steps (5) and (6) make only minor changes to G_1': each terminal is replaced with a string of terminals. In our example, we have terminating productions $[s, S_a, a] \to s_{a1}s_{a2}, ..., [a, A_c, c] \to a_{c1}a_{c2}a_{c3}a_{c4}, ..., [d, A_c, g] \to a_{c1}a_{c2}a_{c3}a_{c4}$ etc.

After this, the construction is complete.

It is now clear that G_1'' can be obtained from the automaton A_1. We show next that G_1'' cannot be derived from any other automaton. Of course, A_1 (and, similarly, A_0) could be part of a larger automaton with transitions by such terminals that do not occur in G_1 (or G_0) at all, but these "invisible" terminals do not affect the reduction from the Hamiltonian circuit problem anyhow.

The structure of the triple construction implies that at most one of the terminals a_{ci} (and, similarly, s_{ai}, s_{ci} etc.) is significant. If they are all dummies, then it follows from restriction (b) that $[a, A_c, c]$, $[b, A_c, g]$ etc. are also dummies. This in turn would imply that, e.g. the production $[a, A, d] \to [a, A_c, c][c, C, d]$ maps under the morphism h into a type $A \to B$ production unless $[c, C, d]$ is also a dummy. Because of the restriction (a), all the $A \to B$ type productions in G_1' have either A as a start symbol or B as a terminating nonterminal. However, both cases are impossible here. Indeed, the letter $[a, A, d]$ cannot map to the start symbol of G_1' since it appears on the right-hand side of the production $[s, S, d] \to [s, S_a, a][a, A, d]$. Secondly,

$[c, C, d]$ cannot map to a terminating nonterminal in G'_1 since it appears on the left-hand side of the production $[c, C, d] \to [c, C_d, b][b, B, d]$ which cannot map to a terminating production in G'_1.

Finally, if $[c, C, d]$ is a dummy, then everything derived from it is also a dummy. Since we can derive all symbols from $[c, C, d]$, this would imply total collapse into a trivial case. Of course, the original digraph may contain vertices from which you cannot go into other vertices (which is not the case in our example) but this property is easily checked and would imply that no HC exists.

Thus, exactly one of the terminals a_{ci} (and, similarly, s_{ai}, s_{ci} etc.) is significant. By an argument similar to the one above, we see that if there is at least one dummy nonterminal in G''_1, then all other nonterminals are also dummies. More specifically, we have already seen that if a terminating nonterminal, e.g. $[a, A_c, c]$, is a dummy then the whole system collapses. On the other hand, if some other nonterminal, e.g. $[c, C, d]$ is a dummy, then all we can directly infer is that nonterminals of the form $[*, *, d]$ are dummies. However, since there exists a production $[c, C, d] \to [c, C_b, b][b, B, d]$ (and, similarly, there exist respective productions for other nonterminals), $[c, C_b, b]$ is also a dummy and we return to the first case. Hence, there are no dummy nonterminals in G''_1.

We can now analyze the transition structure of the underlying automaton. Let us denote an arbitrary underlying automaton by A'_1 while A_1 is still the automaton of our example. The useful terminals of A'_1 are $h(s_{a1}s_{a2})$, $h(s_{c1}s_{c2})$, $h(a_{c1}a_{c2}a_{c3}a_{c4})$,..., $h(d_{s1}d_{s2}d_{s3}d_{s4})$, some of which may be identical. Next we want to show that each pair of transition arcs which have a common terminal vertex in A_1 correspond to a similar pair in A'_1 after the morphism h is applied. This means, for instance, that there exists a pair of transition arcs labeled by the letters $h(c_{b1} \cdots c_{b4})$ and $h(d_{b1} \cdots d_{b4})$ in A'_1 which have equal terminal vertices. Of course, in the case $h(c_{b1} \cdots c_{b4}) = h(d_{b1} \cdots d_{b4})$ the two arcs may be identical. We know that the terminating nonterminal $h([c, C_b, b])$ corresponds to some transition by terminal $h(c_{b1} \cdots c_{b4})$ in A'_1. Similarly, $h([d, D_b, b])$ corresponds to some transition by $h(d_{b1} \cdots d_{b4})$. On the other hand, we have productions $h([c, C, *]) \to h([c, C_b, b])h([b, B, *])$ and $h([d, D, *]) \to h([d, D_b, b])h([b, B, *])$ in G'_1. Hence, by the triple construction rules, the two transitions must have a common terminal vertex in (the graph of) A'_1. Above, $*$ denotes any final state of the original example automaton A_1.

An analogous argument shows that if two transition arcs in our example automaton have common initial vertices then the same holds for the respective h-images in the automaton A'_1. Also, if the terminal vertex of some transition arc in the example automaton coincides with the initial vertex of some other transition arc, then the same relation holds for h-images as well.

It is now easy to see that the transition structure of A'_1 may differ from that of the example automaton A_1 in only two ways:

(i) for some distinct terminals of the example automaton (e.g. c_b and d_b) the respective terminals (e.g. $h(c_{b1} \cdots c_{b4})$ and $h(d_{b1} \cdots d_{b4})$) in A_1' are identical,

(ii) in A_1' some states of the example automaton are merged.

The example automaton is deterministic. Hence (i) implies (ii). Furthermore, (ii) clearly forces all states to be merged with the dead end state since the example automaton is connected (and we have already noted that this means no loss of generality). It is also clear that the case where the underlying automaton consists of one single state can be excluded because of restrictions in step (3) of the previous section.

What is still needed is to show that the final state set of A_1' is correct, i.e. it consists of all states other than the original start state. Fortunately, we can generate suitable words in G_1'' in order to see that this must be the case. For instance, the derivation:

$$S_0 \Rightarrow [s, S, c] \Rightarrow [s, S_a, a][a, A, c] \Rightarrow^2 s_{a1}s_{a2}a_{c1}a_{c2}a_{c3}a_{c4}$$

shows that $h(s_{a1} \cdots a_{c4}) \in L(G_1'')$. Hence the counterpart of the state c in A_1' is a final state.   □

Of course, theorem 2 also implies that the search of a decryption key is NP-complete. Hence, we may establish:

Corollary 1 *Given a decryption key (G_0'', G_1''), it is an NP-complete problem to find the corresponding decryption key (h, A_0).*

Note that our results say nothing about the complexity of cryptanalysis in general. In fact, all the systems constructed in our proofs are easy to break down since the grammars generate only small finite languages. However, there are no reasons to believe that typical instances of the system which are hard to break by searching a system compatible decryption key would be especially insecure otherwise.

Finally, we present one more point of view which supports the applicability of our system. It is possible to strengthen the system in a simple way: at any stage of the construction we may replace any grammar by an *equivalent* one. Whether the system really becomes more secure in this way depends on the method used to generate equivalent grammars.

References

[1] J. Kari, A cryptosystem based on propositional logic. In J. Dassow and J. Kelemen (eds.), *Machines, Languages, and Complexity*. Springer, Berlin, 1989, 210–219.

[2] V. Niemi, Cryptology: language-theoretic aspects. In G. Rozenberg and A. Salomaa (eds.), *Handbook of Formal Languages*. Springer, Berlin, 1997, vol. 2, 507–524.

[3] A. Salomaa, *Public-Key Cryptography*. Springer, Berlin, 1990.

[4] A. Salomaa and S. Yu, On a public-key cryptosystem based on iterated morphisms and substitutions, *Theoretical Computer Science*, 48 (1986), 283–296.

[5] A. Shamir, A polynomial time algorithm for breaking the basic Merkle-Hellman cryptosystem. In *Proceedings of the 23rd FOCS Symposium*, 1982, 145–152.

On the Difference Problem for Semilinear Power Series[1]

Ion Petre

Turku Centre for Computer Science (TUCS)
and
Department of Mathematics
University of Turku
Finland
ipetre@cs.utu.fi

Abstract. We prove in this paper that if r and s are two semilinear power series in commuting variables and s has bounded coefficients, then $r - s$ is a rational series. It may be thought that this result extends one of Eilenberg's results for series in noncommuting variables. However, our proof is combinatorial, and does not use a deeply algebraic cross-section theorem as the noncommutative case does.

1 Introduction

The notion of semilinearity for formal power series was recently introduced in Petre [7], extending the classical notion of a semilinear set. In the same paper, some basic properties of this family of series were investigated. We continue this research here by considering the difference of semilinear series.

It is proved in Eilenberg and Schützenberger [2] that the family of semilinear subsets of a commutative monoid is closed under difference. Moreover, it is proved in Ginsburg [3] that the family of a semilinear set of vectors with

---

[1]The work of the author is supported by the Academy of Finland, Project 44087.

nonnegative integers as components is closed under difference, and the result is effectively computable.

The problem seems more difficult for formal power series. If the coefficients are taken in a ring, then the problem is trivial. If they are not, the problem becomes challenging. Some results are known for rational series in noncommuting variables. Eilenberg [1] considers on $\mathbb{N}$-rational series an operation of *quasidifference* $r \doteq s$ defined as $(r \doteq s, x) = \max(0, (r, x) - (s, x))$. The result is that if r and s are rational series and s has bounded multiplicities, then the quasidifference $r \doteq s$ is also rational.

In this paper, we consider the difference problem on the family of semilinear series, and establish a result similar to Eilenberg [1]: if r and s are semilinear series, s has bounded coefficients, and $r \geq s$, then $r - s$ is a rational series. However, the commutativity makes the problem completely different, and we prove our result by combinatorial means, without using a deeply algebraic tool such as the cross-section theorem in Eilenberg [1].

2   Preliminaries

We recall that for a semiring K and a commutative monoid M, a series $r \in K\langle\langle M \rangle\rangle$ is called linear if it is of the form $r = pq^*$, with p a monomial in $K\langle\langle M \rangle\rangle$, and q a proper polynomial in the same family. A series is called semilinear if it is a finite sum of some linear series.

The family of semilinear power series generally behaves like the semilinear subsets of a commutative monoid, as is proved in Petre [7]: the semilinear power series are closed under rational operations if the semiring of coefficients is idempotent. Likewise, they are closed under morphisms and Parikh's theorem holds if the coefficients are in a commutative, idempotent, ω-continuous semiring.

In this paper, we will denote by $\mathbb{N}$ the family of nonnegative integers. We will restrict ourselves to series with coefficients in $\mathbb{N}$ over the free commutative monoid $\Sigma^\oplus$, for some finite alphabet Σ. Unless otherwise stated, all the series in the following will be considered in $\mathbb{N}\langle\langle \Sigma^\oplus \rangle\rangle$.

We say that r is a *quasilinear* series if $r = p_0 p_1^* \dots p_m^*$, where p_0 is a monomial and $p_1, \dots, p_m$ are proper polynomials. The series r is said to be *elementary* if $p_0 = 1$. A series is called *quasisemilinear* if it is a finite sum of quasilinear series.

We should remark that the family of quasisemilinear series coincides with the rational series of star height 1, but we will use the name quasisemilinear to underline their close connection to semilinearity.

Let k be a positive integer, and consider on the set of vectors $\mathbb{N}^k$ the partial order $\leq$ defined as follows: for any $t, t' \in \mathbb{N}^k$, we have $t \leq t'$ if the following two conditions are satisfied:

(i) $t_i > 0$ if and only if $t_i' > 0$, for all $1 \leq i \leq k$.

(ii) $t_i \leq t_i'$, for all $1 \leq i \leq k$.

Let r be a quasilinear series: $r = p_0 p_1^* \ldots p_m^*$, where $p_0 = \alpha_0 u_0$ is a monomial, and $p_1, \ldots, p_m$ are proper polynomials, $p_i = \alpha_{i1} u_{i1} + \ldots + \alpha_{ik_i} u_{ik_i}$, with $\alpha_{ij} \in \mathbb{N}$, and $u_{ij} \in \Sigma^\oplus$, for all $1 \leq i \leq m$, and $1 \leq j \leq k_i$. We denote $k = k_1 + \ldots + k_m$. Let us collect in the vector ω, with k components from $\Sigma^\oplus$, all the monomials appearing in $p_1, \ldots, p_m$:

$$\omega = (u_{11}, \ldots, u_{1k_1}, \ldots, u_{m1}, \ldots, u_{mk_m}).$$

We will call the vector ω the base of the series r.

For any word $v \in \Sigma^\oplus$, we can associate with respect to r a finite set of vectors from $\mathbb{N}^k$ in the following way; if:

$$v = u_0 u_{11}^{n_{11}} \ldots u_{1k_1}^{n_{1k_1}} \ldots u_{m1}^{n_{m1}} \ldots u_{mk_m}^{n_{mk_m}},$$

then the vector $t = (n_{11}, \ldots, n_{1k_1}, \ldots, n_{m1}, \ldots, n_{mk_m})$ is associated to v. We denote $v = u_0 \omega^t$ and say that t is a representation of v in the base ω. Note that in general the associated vector of a given word is not unique, and that the set of associated words is nonempty if and only if $(r, v) > 0$.

Before proving several technical results, we state two lemmata needed in the sequel. The first one is a well known number theoretical result, and the second one can be found in Ginsburg [3].

Lemma 1 *Let $n_1, \ldots, n_k$ be positive integers, and let d be their greatest common divisor. For any large enough integer n, n can be written as a linear combination of $n_1 \ldots, n_k$ if and only if n is a multiple of d.*

Lemma 2 *Any semilinear set of nonnegative integers can be written as a disjoint union of linear sets of nonnegative integers.*

The following lemma describes some basic properties of the quasilinear series.

Lemma 3 *Let $p_0, p_1, \ldots p_m$ be some polynomials, and $r = p_0 p_1^* \ldots p_m^*$ be a quasilinear series. Let ω be the base of r:*

(i) For any $v \in \Sigma^\oplus$, such that $(r, v) > 0$, there is a quasisemilinear series s such that $r = s + v p_1^* \ldots p_m^*$.

(ii) Let v_1, v_2 be two words in $\Sigma^\oplus$ such that both (r, v_1) and (r, v_2) are nonzero. There are two vectors of nonnegative integers $t_1 \leq t_2$, representations in the base ω of v_1 and v_2, respectively if and only if there is a quasisemilinear series s such that $r = s + v_1 p_1^* \ldots p_m^*$, and $(r - s, v_2) > 0$.

(iii) Let $v_1, v_2, \ldots, v_n$ be some words in $\Sigma^\oplus$, and $t_1, t_2, \ldots, t_n$ representations in the base ω of $v_1, v_2, \ldots, v_n$, respectively. If $t_1, t_2, \ldots, t_n$ are incomparable vectors, then $r = r_0 + v_1 r_1 + v_2 r_2 + \ldots + v_n r_n$, for some quasisemilinear series r_0, and some elementary quasilinear series $r_1, r_2, \ldots, r_n$.

Proof. (i) Using the notations above, it is clear that since $(r, v) > 0$, there is a vector t of nonnegative integers, the representation of v in the base ω:

$$t = (t_{11}, \ldots, t_{1k_1} \ldots t_{m1} \ldots t_{mk_m}).$$

This means that v is a monomial of the polynomial $p_0 p_1^{t_1} p_2^{t_2} \ldots p_m^{t_m}$, where $t_i = t_{i1} + \ldots = t_{ik_i}$, for all $1 \leq i \leq m$. Hence:

$$
\begin{aligned}
r &= p_0 p_1^* \ldots p_m^* = \\
&= p_0(1 + p_1 + \ldots + p_1^{n_1} p_1^*) \ldots (1 + p_m + \ldots + p_m^{n_m} p_m^*) \\
&= r' + p_1^{t_1} \ldots p_m^{t_m} p_1^* \ldots p_m^* = \\
&= s + v p_1^* \ldots p_m^*,
\end{aligned}
$$

with r' and s quasisemilinear series.
(ii) The "if" part is clear. For the other implication, we use (i) and obtain that $r = s + v_1 p_1^* \ldots p_m^*$, for some quasisemilinear series s. It is not difficult to see that the vector $t_2 - t_1$ is a representation of some word w with respect to r, since both t_1 and t_2 are, and $t_1 \leq t_2$. Moreover, this word w satisfies the relation $v_1 w = v_2$. Thus, $(r - s, v_2) = (v_1 p_1^* \ldots p_m^*, v_2) > 0$.
(iii) We consider a decomposition of r similar to the case (i) above. So the fact that $t_1, \ldots, t_n$ are incomparable can be used to obtain the claim. $\quad\square$

The next two results will be the main tools which we use to obtain our main result. We prove that for any quasisemilinear series r, and any word u, the set of powers of u in the support of r is a semilinear set. This result, interesting in itself, shows once again the strong connections between the notions of semilinearity for sets and for power series.

Lemma 4 *Let* $r = p_1^* \ldots p_m^*$ *be an elementary quasilinear series and* u *a word in* $\Sigma^\oplus$. *There is a nonnegative integer* n_0 *such that the set:*

$$P_{r,u} = \{n \geq n_0 \mid (r, u^n) > 0\}$$

is a linear set of nonnegative integers.

Proof. Let us consider all the words of the form u^n, with $n \in \mathbb{N}$ and $(r, u^n) > 0$, and let $V_{r,u}$ be the set of their representations with respect to r. By König's lemma, we find that the set of minimal elements in $V_{r,u}$ is finite. We denote it by $V_{r,u}^{(0)}$. Using lemma 3(i), it can be seen that a vector of nonnegative integers t is a representation of some u^n with respect to r, if and only if t is a linear combination of the vectors in $V_{r,u}^{(0)}$. This implies that, denoting by $W_{r,u}^{(0)}$ the set of words represented by the vectors in $V_{r,u}^{(0)}$, we have $(r, u^n) > 0$, for some $n \in \mathbb{N}$, if and only if u^n is a linear combination of words in $W_{r,u}^{(0)}$: $u^n = u^{n_1} \ldots u^{n_k}$, with $u^{n_1}, \ldots, u^{n_k} \in W_{r,u}^{(0)}$.

Furthermore, let us denote $P_{r,u}^{(0)} = \{n \in \mathbb{N} \mid (r, u^n) > 0, u^n \in W_{r,u}^{(0)}\}$. From the above, we finally obtain that $(r, u^n) > 0$ if and only if n is a linear combination of integers in $P_{r,u}^{(0)}$. The claim follows now by lemma 1.   □

Theorem 1 *Let r be a quasisemilinear series and u be a word in $\Sigma^{\oplus}$. There is a nonnegative integer n_0 such that:*

$$P_{r,u} = \{n \geq n_0 \mid (r, u^n) > 0\}$$

is a semilinear set of nonnegative integers.

Proof. Clearly, $P_{r_1+r_2,u} = P_{r_1,u} \cup P_{r_2,u}$ and thus, it is enough to prove the theorem for a quasilinear series $r = p_0 p_1^* \ldots p_m^*$, where p_0 is a monomial, and $p_1, \ldots, p_m$ are proper polynomials.

As in the proof of lemma 4, we consider the words u^n, with $(r, u^n) > 0$ and $V_{r,u}$ the set of their representations with respect to r. By König's lemma, the set of minimal elements in $V_{r,u}$ is finite and, hence, by lemma 3(iii), $r = r_0 + u^{n_1} r_1 + \ldots + u^{n_k} r_k$, for some nonnegative integers $n_1, \ldots, n_k$ and some elementary quasilinear series $r_1, \ldots, r_k$. Moreover, $(r_0, u^n) = 0$, for all $n \in \mathbb{N}$. The claim follows now using lemma 4.   □

3   The Result

The first step towards the main result of this paper is to characterize those $\mathbb{N}$-semilinear power series that have bounded coefficients.

Lemma 5 *If s is a semilinear series with bounded coeficients, then s is of the form $s = m_1 u_1^* + \ldots + m_k u_k^*$, for some monomials $m_1, \ldots, m_k$ and some words $u_1, \ldots, u_k$.*

Proof. Since s is a semilinear series, then it is of the form:

$$s = m_1 p_1^* + \ldots + m_k p_k^*,$$

for some monomials $m_1, \ldots, m_k$, and some polynomials $p_1, \ldots, p_k$. Consider the polynomial $p_1 = \alpha_1 u_1 + \ldots + \alpha_i u_i$, with $\alpha_1, \ldots, \alpha_i \in \mathbb{N}$, and $u_1, \ldots, u_i \in \Sigma^{\oplus}$. All the coefficients $\alpha_1, \ldots, \alpha_i$ must be equal to 1, since otherwise s does not have bounded coefficients. Moreover, if p_1 is not a monomial, i.e. $i \geq 2$, then $(p_1, u_1^{j_1} u_2^{j_2}) \geq \binom{j_1+j_2}{j_1}$, and this is not bounded for $j_1, j_2 \geq 0$. The lemma is thus proved.   □

We are now ready for the main result of this paper.

Theorem 2 *If r and s are semilinear series such that $r \geq s$, and s has bounded multiplicities, then $r - s$ is a rational series.*

Proof. We prove slightly more generally that if r is a quasisemilinear series, $s = v'v^*$, with $v, v' \in \Sigma^{\oplus}$, and $r \geq s$, then $r - s$ is a quasisemilinear series. Using lemma 5, the theorem obviously follows from this. We begin by making several reductions of the general problem to simpler instances of it.

Let the quasisemilinear form of the series r be $r = r_1 + \ldots + r_i$, with $r_1, \ldots, r_i$ quasilinear series.

Our first claim is that we can assume $v' = 1$, without any loss of generality.

Assume that this is not the case, i.e. $v' \neq 1$, and consider the linear series r_1, $r_1 = p_0 p_1^* \ldots p_m^*$, where p_0 is monomial, and $p_1, \ldots, p_m$ are proper polynomials. Take then the representations with respect to r_1 of the words $v'v^n$, for all $n \geq 0$. By König's lemma, the set of minimal elements in this set of vectors is finite, and then, by lemma 3(iii), we obtain that:

$$r_1 = r_1' + v'v^{n_1} r_{11} + \ldots + v'v^{n_{k_1}} r_{1k_1},$$

where r_1' is a quasisemilinear series such that $(r_1', v'v^n) = 0$, for all $n \geq 0$, and $r_{11}, \ldots, r_{1k_1}$ is a quasilinear series.

We proceed similarly with $r_2, \ldots, r_i$, and obtain that:

$$r = r' + v'r'',$$

where r' and r'' are quasisemilinear series, and $(r', v'v^n) = 0$, for all $n \geq 0$. Since $r \geq s$, we then have that $v'r'' \geq s$, and hence $r'' \geq v^*$. Moreover, $r - s = r' + v'(r'' - v^*)$. It is enough now to prove that the difference $r'' - v^*$ is quasisemilinear. The first claim is thus proved.

Our second goal is to reduce the problem to the case when r is a linear series.

Let us assume that this is not the case. Due to our first reduction, we can assume that r is of the form $r = v^{n_1} r_1 + \ldots + v^{n_i} r_i$, with $r_1, \ldots, r_i$ elementary quasilinear series. By theorem 1, the set P_k of powers of v in support of each of the series $v^{n_k} r_k$, $1 \leq k \leq i$, is a linear set, modulo a finite number of elements. Hence, P_k is of the form $P_k = A_k \cup B_k$, with A_k linear and B_k finite sets of nonnegative integers, for all $1 \leq k \leq i$. Let $B = \cup_{k=1}^i B_k$.

Since $r \geq v^*$, the set $\cup_{k=1}^i A_k$ covers the entire set $\mathbb{N} - B$. By lemma 2, $\mathbb{N} - B$ can thus be covered also by a disjoint union of linear sets $\cup_{k=1}^i A_k'$, with $A_k' \subseteq A_k$, for all $1 \leq k \leq i$. From this, one can easily obtain that the series v^* can be written as:

$$v^* = s_0 + s_1 + \ldots + s_i,$$

with $s_k = \sum_{n \in A_k'} v^n$ a linear series, for all $1 \leq k \leq i$, and $s_0 = \sum_{n \in B} v^n$ a polynomial. We need not be concerned with the polynomial s_0 since clearly, by lemma 3(i), $t - s_0$ is a quasisemilinear series, for any quasisemilinear series t such that $t \geq s_0$. So, we can assume that $s_0 = 0$.

Since $A'_k \subseteq A_k$, we have that $v^{n_k} r_k \geq s_k$, for all $1 \leq k \leq i$. To obtain the claim of the theorem, it is enough to prove that $v^{n_k} r_k - s_k$ is quasisemilinear, where r_k is an elementary quasilinear series, say $r_k = q_1^* \ldots q_{m_k}^*$, and s_k is of the form $s_k = v^{j_k} (v^{d_k})^*$. This completes the second reduction.

Before computing the difference in this case, we make one last reduction by observing that it is enough to assume that $n_k = j_k = 0$. To see this, we need only point out that $n_k \leq j_k$, since otherwise $(v^{n_k} r_k, v^{j_k}) = 0 < 1 = (s_k, v^{j_k})$, which is impossible. Hence, $v^{n_k} r_k - s_k = v^{n_k}(q_1^* \ldots q_{m_k}^* - v^{j_k - n_k}(v^{d_k})^*)$, and thus, we can assume that $n_k = 0$. In this hypothesis, we now have:

$$r_k = q_1^* \ldots q_{m_k}^* \geq s_k = v^{j_k}(v^{d_k})^*.$$

By lemma 4, there is an integer d such that $(r_k, v^n) > 0$ if and only if n is a multiple of d. This and the fact that $r_k \geq s_k$ shows that both j_k and d_k must be multiples of d. In particular this implies that $r_k \geq (v^{d_k})^*$. We then have:

$$
\begin{aligned}
r_k - s_k &= q_0 q_1^* \ldots q_{m_k}^* + v^{j_k} q_1^* \ldots q_{m_k}^* - v^{j_k}(v^{d_k})^* = \\
&= q_0 q_1^* \ldots q_{m_k}^* + v^{j_k}(r_k - (v^{d_k})^*).
\end{aligned}
$$

Hence, we can indeed assume that $n_k = j_k = 0$.

In this case, the difference can be computed as follows:

$$
\begin{aligned}
r_k - s_k &= q_1 * \ldots q_{m_k}^* - (v^{d_k})^* = \\
&= q_0 q_1^* \ldots q_{m_k}^* + v^{d_k} q_1^* \ldots q_{m_k}^* - (1 + v^{d_k}(v^{d_k})^*) = \\
&= q_0' q_1^* \ldots q_{m_k}^* + v^{d_k}(r_k \quad s_k),
\end{aligned}
$$

and, thus, $r_k - s_k = q_0' q_1^* \ldots q_{m_k}^* (v^{d_k})^*$, which is quasisemilinear. The proof of the theorem is now complete. $\square$

We now give an example of how to compute the difference of two semilinear power series.

Example 1 *Let $r = (a^2 + a^3)^*$ and let $s = (a^2)^* + a^3(a^3)^*$. We first compute the difference $r' = r - (a^2)^*$ as follows:*

$$
\begin{aligned}
r - (a^2)^* &= 1 + a^2 r + a^3 r - 1 - a^2(a^2)^* = \\
&= a^2(r - (a^2)^*.
\end{aligned}
$$

Hence, $r' = a^3(a^2)^*(a^2 + a^3)^*$. We continue as follows:

$$
r' - a^3(a^3)^* = a^3\left((a^2 + a^3)^*(a^2)^* - (a^3)^*\right).
$$

Let us denote now $r'' = (a^2)^*(a^2 + a^3)^*$, and $s'' = (a^3)^*$. Since the representation of a^3 with respect to r'' is the vector $(0, 0, 1)$, we have:

$$
\begin{aligned}
r'' &= (1 + (a^2 + a^3)(a^2 + a^3)^*)(1 + a^2(a^2)^*) = \\
&= 1 + a^2(a^2)^* + (a^4 + a^5)(a^2)^*(a^2 + a^3)^* + a^2(a^2 + a^3)^* + \\
&\quad + a^3(a^2 + a^3)^*.
\end{aligned}
$$

The difference $r'' - s''$ is now:

$$
\begin{aligned}
r'' - s'' &= a^2(a^2)^* + (a^4 + a^5)(a^2)^*(a^2 + a^3)^* + a^2(a^2 + a^3)^* + \\
&\quad + a^3\left((a^2 + a^3)^* - (a^3)^*\right).
\end{aligned}
$$

All that remains to be done is to compute the difference $(a^2 + a^3)^* - (a^3)^*$, which is:

$$
\begin{aligned}
(a^2 + a^3)^* - (a^3)^* &= a^2(a^2 + a^3)^* + a^3\left((a^2 + a^3)^* - (a^3)^*\right) \\
&= a^2(a^2 + a^3)^*(a^3)^*.
\end{aligned}
$$

This proves the rationality of the difference.

4 Conclusions

In this paper, we have continued the research initiated in Petre [7] and we have shown once again that the semilinear formal power series in commuting variables behave nicely under basic operations. Namely, we have extended one of Eilenberg's results for rational series in noncommuting variables, and proved that the difference of two semilinear power series is rational, provided that one of them has bounded coefficients. It is an open problem whether this is semilinear or not, and we conjecture that the answer to this question is negative.

References

[1] S. Eilenberg, *Automata, Languages and Machines*. Academic Press, New York, 1974.

[2] S. Eilenberg and M.P. Schützenberger, Rational sets in commutative monoids, *Journal of Algebra*, 13 (1969), 173–191.

[3] S. Ginsburg, *The Mathematical Theory of Context-Free Languages*. McGraw-Hill, New York, 1966.

[4] W. Kuich, The Kleene and the Parikh theorem in complete semirings. In *Automata, Languages and Programming*. Springer, Berlin, 1987, 211–215.

[5] W. Kuich, Semirings and formal power series: their relevance to formal languages and automata. In G. Rozenberg and A. Salomaa (eds.), *Handbook of Formal Languages*. Springer, Berlin, 1997, vol. 1, 609–677.

[6] W. Kuich and A. Salomaa, *Semirings, Automata, Languages*. Springer, Berlin, 1986.

[7] I. Petre, On semilinearity in formal power series. In *Proceedings of Developments in Language Theory, DLT 1999*, to appear.

[8] A. Salomaa and M. Soittola, *Automata-Theoretic Aspects of Formal Power Series*. Springer, Berlin, 1978.

On Spatial Reasoning via Rough Mereology

Lech Polkowski

Polish-Japanese Institute of Information Technology

and

Institute of Mathematics

Warsaw University of Technology

Poland

email:polkow@pjwstk.waw.pl

Abstract. The content of this note is a demonstration that Mereotopology, one of the principal ingredients of spatial reasoning toolery, may be developed in the framework of Rough Mereology. We introduce Rough Mereology into Stanisław Leśniewski's Ontological and Mereological framework and we reveal topological structures which may be defined therefrom. These structures turn out to be closely related to those studied in the Calculus of Individuals [20].

1   Introduction

In this paper, we study some properties of reasoning devices applicable in spatial reasoning and founded on rough mereology. Rough mereology was proposed by Polkowski and Skowron [22], [21], [25] as a paradigm for reasoning under uncertainty. In particular, they discussed logics for synthesizing approximate solutions by distributed systems [21] as well as problems of design, analysis and control in those systems [25]. This theory is rooted in rough set theory proposed by Pawlak [18] and in Stanisław Leśniewski's mereological theory [15], [27], [26] and may be regarded as an extension of the latter. We underline here this aspect of rough mereology by proposing

a simple and convenient formalization of this theory within Leśniewski's ontological framework [16], [17], [13], [28]. As spatial reasoning is by its very
nature related to continuous objects, it is convenient to develop it in terms of
part-whole relations, i.e. in a mereological framework, e.g. the one indicated
above.

Spatial reasoning is an important aspect of reasoning under uncertainty
and is related to logical and philosophical investigations about the nature of
space and time [5], [23], [29], [30], [31] as well as to research in natural language processing (cf. [2], [3]) and geographic information systems (cf. [9]),
etc. Although often not explicitly mentioned, important aspects of spatial
reasoning are present in the field of mobile robotics under the guise of navi-
gation, obstacle avoidance and other techniques related to spatial reasoning
[1], [8], [10]. Our study of rough mereology as a vehicle for spatial reasoning
leads us to mereotopological structures immanent to spatial reasoning. We
begin with a formal introduction to Stanisław Leśniewski's Ontology and
Mereology and we continue with a section on Rough Mereology. Then we
discuss the mereotopological features of Rough Mereology.

2   Ontology: An Introduction

The theory of part-whole relations may be conveniently presented in an
ontological language intended by Stanisław Leśniewski [15] as a formulation
of general principles of being (cf. [26], [11], [28], [13]). This language is an
alternative to the standard language of set theory.

The only primitive notion of Leśniewski's Ontology is the copula "*is*"
denoted by the symbol ε.

The original axiom of Ontology defining the meaning of ε is as follows.

2.1   The Ontological Axiom

$$X\varepsilon Y \iff \exists Z.Z\varepsilon X \wedge \forall U,W.(U\varepsilon X \wedge W\varepsilon X \implies U\varepsilon W) \wedge \forall Z.(Z\varepsilon X \implies Z\varepsilon Y).$$

In this axiom, the defined copula ε happens to occur on both sides of the
equivalence: however, the definiendum $X\varepsilon Y$ belongs to the left side only
and we may perceive the axiom as the definition of the meaning of $X\varepsilon Y$ via
the meaning of terms of "lower level" $Z\varepsilon X$, $Z\varepsilon Y$, etc.

According to this reading of the axiom, we gather that the proposition
$X\varepsilon Y$ is true if and only if the conjunction holds for the following three
propositions:

(I) $\exists Z.Z\varepsilon X$.

This proposition asserts the existence of an object (name) Z which is X,
and so X is not an empty name.

(II) $\forall U, W.(U\varepsilon X \wedge W\varepsilon X \Longrightarrow U\varepsilon W)$.

This proposition asserts that any two objects which are X are each other ('a fortiori', they will be identified later on): X is an individual name (or X is an individual object, a singleton).

(III) $\forall Z.(Z\varepsilon X \Longrightarrow Z\varepsilon Y)$.

This proposition asserts that every object which is X is Y (or, X is contained in Y).

The meaning of $X\varepsilon Y$ can be made clear now: X is an individual and this individual is Y.

Identity of individual objects is introduced via:

$$X = Y \Longleftrightarrow X\varepsilon X \wedge Y\varepsilon Y \wedge X\varepsilon Y \wedge Y\varepsilon X.$$

The universal name V and the empty name Λ are introduced by: $X\varepsilon V \Longleftrightarrow \exists Y\, X\varepsilon Y$; $X\varepsilon\Lambda \Longleftrightarrow X\varepsilon X \wedge non(X\varepsilon X)$.

3 Mereology: An Introduction

Mereology is a theory of collective classes, unlike ontology which is a theory of distributive classes. Mereology may be based on each of a few notions such as of a *part*, an *element*, a *class* etc. Historically, it was conceived by Stanisław Leśniewski as a theory of the relation of being a part and we follow this line of development. We assume that the copula ε is given and that the Ontology Axiom holds. Under these assumptions, we introduce the notion of the name-forming functor *pt* of part.

3.1 Mereology Axioms

(A0) $X\varepsilon ptY \Longrightarrow X\varepsilon X \wedge Y\varepsilon Y$.
(A1) $X\varepsilon ptY \wedge Y\varepsilon ptZ \Longrightarrow X\varepsilon ptZ$.
(A2) $non(X\varepsilon ptX)$.

On the basis of the notion of a part, we define the notion of an element (a possible improper part) as a name-forming functor *el*.

Definition 1 $X\varepsilon elY \Longleftrightarrow X\varepsilon ptY \vee X = Y$.

We may now introduce the notion of a (collective) class via a functor *Kl*.

Definition 2 $X\varepsilon KlY \Longleftrightarrow \exists Z.Z\varepsilon Y \wedge \forall Z.(Z\varepsilon Y \Longrightarrow Z\varepsilon elX) \wedge \forall Z.(Z\varepsilon elX \Longrightarrow \exists U, W.(U\varepsilon Y \wedge W\varepsilon elU \wedge W\varepsilon elZ)$.

The class functor is subject to additional postulates.

(A4) $X \varepsilon KlY \wedge Z \varepsilon KlY \implies X \varepsilon Z$.

Hence, KlY is an individual name.

(A5) $\exists Z. Z \varepsilon Y \iff \exists Z. Z \varepsilon KlY$.

The class operator may be regarded as an aggregation operator which turns any non-empty collection of objects (a general name) into an individual object (an individual name); we show below that it may serve as an efficient neighborhood–forming operator.

3.2 Subset, Complement

We define the notions of a subset and a complement and we will look at relations and functions in a mereological context. We define first the notion of a subset as a name–forming functor sub of an individual variable.

Definition 3 $X \varepsilon subY \iff X \varepsilon X \wedge Y \varepsilon Y \wedge \forall Z (Z \varepsilon elX \implies Z \varepsilon elY)$.

We now define the notion of being external as a binary proposition-forming functor ext of individual variables.

Definition 4 $ext(X, Y) \iff X \varepsilon X \wedge Y \varepsilon Y \wedge non(\exists Z. Z \varepsilon elX \wedge Z \varepsilon elY)$.

The notion of a complement is rendered as a name-forming functor $comp$ of two individual variables.

Definition 5 $X \varepsilon comp(Y, Z) \iff Y \varepsilon subZ \wedge X \varepsilon Kl\Theta$ where $U \varepsilon \Theta \iff U \varepsilon elZ \wedge ext(U, Y)$.

We may now go on to Rough Mereology.

4 Rough Mereology: An Introduction

Rough mereology is an extension of mereology based on the predicate of being a part in a degree; this predicate is rendered here as a family of name-forming functors μ_r parameterized by a real parameter r in the interval $[0, 1]$ with the intent that $X \varepsilon \mu_r Y$ reads "X is a part of Y in degree at least r". We begin with the set of axioms and we construct the axiom system as an extension of systems for ontology and mereology.

We assume, therefore, that a functor el of an element satisfying the mereology axiom system within a given ontology of ε is given; around this, we develop a system of axioms for rough mereology.

4.1 The Axiom System

The following is the list of basic postulates:

(RM1) $X\varepsilon\mu_1 Y \Longleftrightarrow X\varepsilon el Y$.

(RM2) $X\varepsilon\mu_1 Y \Longrightarrow \forall Z.(Z\varepsilon\mu_r X \Longrightarrow Z\varepsilon\mu_r Y)$.

(RM3) $X = Y \wedge X\varepsilon\mu_r Z \Longrightarrow Y\varepsilon\mu_r Z)$.

(RM4) $X\varepsilon\mu_r Y \wedge s \leq r \Longrightarrow X\varepsilon\mu_s Y$.

It follows that the functor μ_1 coincides with the given functor el and thus establishes a link between rough mereology and mereology while functors μ_r with $r < 1$ diffuse the functor el to a hierarchy of functors expressing the fact of being an element (or, part) in various degrees.

We introduce a new name-forming functor μ_r^+ via:

Definition 6 $X\varepsilon\mu_r^+ Y \Longleftrightarrow X\varepsilon\mu_r Y \wedge \forall s.(s > r \Longrightarrow non(X\varepsilon\mu_s Y))$.

We find a usage for this functor in two new postulates.

(RM5) $ext(X, Y) \Longrightarrow X\varepsilon\mu_0^+ Y$.

This postulate expresses our intuition that objects which are external to each other should be elements of each other in no positive degree. This assumption, however, reflects a high degree of certainty of our knowledge and it will lead to models in which connection coincides with overlapping (see below). It will be more realistic to assume that our knowledge is uncertain to the extent that we may not be able to state beyond doubt that two given objects are external to each other, rather we will be pleased with the statement that in such cases they are elements of each other in a bounded degree. Hence, we introduce a weaker form of the postulate (RM5):

(RM5*) $ext(X, Y) \Longrightarrow \exists s < 1.X\varepsilon\mu_s^+ Y$.

4.2 A Certain Model: t-Norm Modifiers

We introduce now, following Polkowski & Skowron [22], a modification to our functors μ_r; it is based on an application of residuated implication [12] and a measure of containment defined within the fuzzy set theory [4]. Combining the two ideas, we achieve a formula for μ_r which allows a transitivity rule; this rule will, in turn, allow rough mereological topologies, to be introduced into our universe.

We therefore introduce a $t - norm$ $\top$ as a function of two arguments $\top : [0,1]^2 \longrightarrow [0,1]$ which satisfies the following requirements:

$$\mathsf{T}(x,y) = \mathsf{T}(y,x);$$

$$\mathsf{T}(x, \mathsf{T}(y,z)) = \mathsf{T}(\mathsf{T}(x,y),z);$$

$$\mathsf{T}(x,1) = x;$$

$$x' \geq x \wedge y' \geq y \Longrightarrow \mathsf{T}(x',y') \geq \mathsf{T}(x,y).$$

We also invoke a notion of fuzzy containment $\subset_r$ [4]; it relies on a many-valued implication Υ, i.e. on a function $\Upsilon : [0,1]^2 \longrightarrow [0,1]$ according to the formula: $X \subset_r Y \iff \forall Z.(\Upsilon(\mu_X(Z), \mu_Y(Z)) \geq r)$, where μ_A is the fuzzy membership function [12] of the fuzzy set A.

We replace Υ with a specific implication viz. the residuated implication $\overrightarrow{\mathsf{T}}$ induced by T and defined by the following prescription.

Definition 7 $\overrightarrow{\mathsf{T}}(r,s) \geq t \iff \mathsf{T}(t,r) \leq s.$

We define a functor $\mu_{\mathsf{T},r}$ where $r \in [0,1]$, according to the recipe.

Definition 8 $X\varepsilon\mu_{\mathsf{T},r}Y \iff \forall Z.(\exists t, w.Z\varepsilon\mu_t X \wedge Z\varepsilon\mu_w Y \wedge \overrightarrow{\mathsf{T}}(t,w) \geq r).$

It turns out, as first proved in a different context in [22], that $\mu_{\mathsf{T},r}$ satisfies axioms (RM1-RM5*).

Proposition 1 *Functors $\mu_{\mathsf{T},r}$ satisfy (RM1)-(RM5*).*

Yet another important feature of this model is that the following rule (DR) is observed [22].

Proposition 2 *(DR) holds in the form :* $\dfrac{X\varepsilon\mu_{\mathsf{T},r}Y, Y\varepsilon\mu_{\mathsf{T},s}Z}{X\varepsilon\mu_{\mathsf{T},\mathsf{T}(r,s)}Z}.$

5 Mereotopology

We shall now discuss the topological structures arising in the mereological universe endowed with the functors μ_r. These structures provide a mereotopological environment in which it is possible to carry out spatial reasoning in the part concerning spatial relationships (cf. e.g. [2], [3]).

5.1 Mereotopology: Čech Topologies

It has been demonstrated that in a mereological setting a Čech quasi-topology may be defined (cf. [7]), which under additional artificial assumptions (op.cit.) may be made into a quasi-topology. Here, we induce a Čech topology in any rough mereological universe. To this end, we define the class $Kl_r X$ for any object X and $r < 1$ by:

Definition 9 $Z\varepsilon Kl_r X \iff Z\varepsilon Kl\mu_r X.$

Thus $Kl_r X$ is the class of objects with the property $\mu_r X$. From the definition of a class, it follows that

Proposition 3 $Kl_r X = B_r X$ where $Z\varepsilon el B_r X \iff \exists T.Z\varepsilon el T \wedge T\varepsilon \mu_r X.$

Proof. Assume that $Z\varepsilon el Kl_r X$; then for some U, W, we have $U\varepsilon el Z$, $U\varepsilon el W$, $W\varepsilon \mu_r X$. Hence $U\varepsilon el B_r X$ and thus $Kl_r X\varepsilon el(B_r X)$.

Conversely, $Z\varepsilon el(B_r X)$ implies $\exists T.Z\varepsilon el T \wedge T\varepsilon \mu_r X$. Hence $T\varepsilon el(Kl_r X)$ and $Z\varepsilon el(Kl_r X)$ implying $B_r X\varepsilon el(Kl_r X)$. $\square$

From this, the corollary below follows.

Corollary 1 *For* $s \leq r : Kl_r X\varepsilon el(Kl_s X).$

We may therefore accept the name B defined as follows as a base for open sets which will make our universe V into a topological space.

Definition 10 $Z\varepsilon B \iff \exists X, r < 1.Z\varepsilon Kl_r X.$

Along the same lines, we may define a new functor Int of one argument.

Definition 11 $Z\varepsilon Int(X) \iff Z\varepsilon Klint X$ where $T\varepsilon int X \iff$ $\exists U, s < 1.Kl_s U\varepsilon el X \wedge T\varepsilon el Kl_s U.$

We may state the properties of Int.

Proposition 4 *For any* $X, Y :$
  (i) $Int(X)\varepsilon el X;$
  (ii) $Int(Int(X))\varepsilon Int(X);$
  (iii) $X\varepsilon el Y \implies Int(X)\varepsilon el Int(Y);$
  (iv) $Int(Kl V) = Kl V;$
  (v) $Int(\Lambda)\varepsilon Int(\Lambda) \implies \Lambda\varepsilon \Lambda.$

Proof. For (i): assume that $Z\varepsilon el(Int(X))$; there are U, W with $U\varepsilon el Z$, $U\varepsilon el W$, $Kl_s W\varepsilon el X$ for some $s < 1$;hence, $U\varepsilon el X$ and $Int(X)\varepsilon el X$. For (ii): $Int(Int(X))$ $\varepsilon el(Int(X))$ by (i). Assume that $Z\varepsilon el(Int(X))$; then, there are U, W with $U\varepsilon el Z$, $U\varepsilon el W$, $Kl_s W\varepsilon el X$ for some $s < 1$ and thus $Kl_s W\varepsilon el(Int(X))$ which implies that $Z\varepsilon el(Int(Int(X)))$. (iii) and (iv) are obvious, and (v) expresses the fact that $Int(\Lambda)$ is the empty name. $\square$

Properties (i)-(v) show that the topology introduced by B is a Čech *topology* [6], when restricted to non–empty objects. We denote this by the symbol τ_μ.

Proposition 5 *Rough mereotopology* τ_μ *induced by functors* μ_r *is a Čech topology.*

We now study the case of mereotopology under functors $\mu_{T,r}$; in this case, the Čech topology τ_μ turns out to be a topology.

5.2   Mereotopology: The Case of $\mu_\top$

We begin with an application of our deduction rule (DR). We denote by the symbol $Kl_{\top,r}X$ the set $Kl_r X$ when $\mu_\top$. We assume that $\top(r,s) < 1$ when $rs < 1$.

Proposition 6 $Z\varepsilon el(Kl_{\top,r}X) \iff Z\varepsilon\mu_{\top,r}X$.

Proof. By proposition 3, $Z\varepsilon el(Kl_{\top,r}X)$ means that $Z\varepsilon T$, $T\varepsilon\mu_{\top,r}X$ for some T. Hence $Z\varepsilon\mu_{\top,1}(T)$, $T\varepsilon\mu_{\top,r}X$ imply by (DR) that $Z\varepsilon\mu_{\top,\top(1,r)}X$, i.e. $Z\varepsilon\mu_{\top,r}X$.   □

This proposition means that $Kl_{\top,r}X$ may be regarded as *"an open ball of radius r centered at X"*.

Proposition 7 *For* $Z\varepsilon el(Kl_{\top,r}X$,
   if $s_0 = \arg\_\min\{s : \top(r,s) \geq r\}$ *then* $Kl_{\top,s_0}Z\varepsilon el(Kl_{\top,r}X)$.

Proof. Let $s \geq s_0$; consider $T\varepsilon el(Kl_{\top,s}Z$ so $T\varepsilon\mu_{\top,s}Z$. Then $T\varepsilon\mu_{\top,\top(s,r)}X$. Hence $T\varepsilon\mu_{\top,r}X$ so $T\varepsilon el(Kl_{\top,r}X)$ implying finally that $Kl_{\top,s_0}Z\varepsilon el(Kl_{\top,r}X$.   □

We define a proposition–forming functor of two nominal individual variables AND.

Definition 12

$$Z\varepsilon el(AND(X,Y)) \iff Ov(X,Y) \wedge Z\varepsilon el X \wedge Z\varepsilon el Y.$$

Proposition 8 *The rough mereotopology* $\tau_{\mu_\top}$ *is a topology when restricted to non–empty objects, i.e. the property:*

$$AND(Int(X), Int(Y)) = Int(AND(X,Y))$$

holds whenever $AND(Int(X), Int(Y))$ *is non–empty.*

Proof. The intersection of two open basic classes may be described effectively by means of proposition 7 : assume that $Z\varepsilon el(Kl_{\top,r}X)$ and $Z\varepsilon el(Kl_{\top,s}Y)$ for some $r, s < 1$. Then, for $t_0 = \arg\_\min\{t : \top(r,t) \geq r, \top(s,t) \geq s\}$ and $1 > t \geq t_0$, we have $Kl_{\top,t}Z\varepsilon el(Kl_{\top,r}X)$ and $Kl_{\top,t}Z\varepsilon el(Kl_{\top,s}Y)$. The general case follows easily.   □

Acknowledgement

This work was carried out under Grant no 8T 11C 024 17 from the State Committee for Scientific Research (KBN) of the Republic of Poland.

References

[1] R.C. Arkin, *Behavior-Based Robotics*. MIT Press, Cambridge, 1998.

[2] N. Asher and L. Vieu, Toward a geometry of commonsense: a semantics and a complete axiomatization of mereotopology. In *Proceedings of IJCAI'95*, Morgan Kaufmann, San Mateo, Ca., 1995, 846–852.

[3] M. Aurnague and L. Vieu, A theory of space-time for natural language semantics. In K. Korta and J.M. Larrazábal (eds.), *Semantics and Pragmatics of Natural Language: Logical and Computational Aspects*, ILCLI Series I, University of País Vasco, San Sebastián, 1995, 69–126.

[4] W. Bandler and L. J. Kohout, Fuzzy power sets and fuzzy implication operators. *Fuzzy Sets and Systems*, 4 (1980), 13–30.

[5] J. van Benthem, *The Logic of Time*. Reidel, Dordrecht, 1983.

[6] E. Čech, *Topological Spaces*. Academia, Praha, 1966.

[7] B.L. Clarke, A calculus of individuals based on connection. *Notre Dame Journal of Formal Logic*, 22/2 (1981), 204–218.

[8] M. Dorigo and M. Colombetti, *Robot Shaping. An Experiment in Behavior Engineering*. MIT Press, Cambridge, 1998.

[9] A. Frank, I. Campari and U. Formentini (eds.), *Theories and Methods of Spatial Reasoning in Geographic Space*, Proceedings of the International Conference GIS: From Space to Territory. Springer, Berlin, 1992.

[10] D. Kortenkamp, R.P. Bonasso and R. Murphy (eds.), *Artificial Intelligence and Mobile Robots*. AAAI Press/MIT Press, Cambridge, 1998.

[11] T. Kotarbiński, *Elements of the Theory of Knowledge, Formal Logic and Methodology of Science*. Polish Scientific Publishers, Warsaw, 1966.

[12] R. Kruse, J. Gebhardt and F. Klawonn, *Foundations of Fuzzy Systems*. John Wiley, Chichester, 1984.

[13] Cz. Lejewski, On Leśniewski's Ontology. *Ratio*, I/2 (1958), 150–176.

[14] H. Leonard and N. Goodman, The calculus of individuals and its uses. *Journal of Symbolic Logic*, 5 (1940), 45–55.

[15] St. Leśniewski, Grundzüge eines neuen Systems der Grundlagen der Mathematik. *Fundamenta Mathematicae*, 24 (1929), 242–251.

[16] St. Leśniewski, Über die Grundlagen der Ontologie. *Comptes Rendus Soc. Sci. Lettr. Varsovie*, III (1930), 111–132.

[17] St. Leśniewski, On the Foundations of Mathematics. *Przegląd Filozoficzny*, 30 (1927), 164-206; 31 (1928), 261-291; 32 (1929), 60-101; 33 (1930), 77-105; 34 (1931), 142-170 (in Polish).

[18] Z. Pawlak, *Rough Sets: Theoretical Aspects of Reasoning about Data*. Kluwer, Dordrecht, 1992.

[19] Z. Pawlak and A. Skowron, Rough membership functions. In R.R. Yager, M. Fedrizi and J. Kacprzyk (eds.) *Advances in the Dempster-Schafer Theory of Evidence*. John Wiley, New York, 1994, 251–271.

[20] L. Polkowski, On synthesis of constructs for spatial reasoning via rough mereology. *Fundamenta Informaticae*, to appear.

[21] L. Polkowski and A. Skowron, Adaptive decision-making by systems of cooperative intelligent agents organized on rough mereological principles. *Intelligent Automation and Soft Computing. An International Journal*, 2/2 (1996), 123–132.

[22] L. Polkowski and A. Skowron, Rough mereology: a new paradigm for approximate reasoning. *International Journal of Approximate Reasoning*, 15/4 (1997), 333–365.

[23] H. Reichenbach, *The Philosophy of Space and Time*. Dover, New York, 1957.

[24] P. Simons, *Parts - A Study in Ontology*. Clarendon, Oxford, 1987.

[25] A. Skowron and L. Polkowski, Rough mereological foundations for design, analysis, synthesis and control in distributed systems. *Information Sciences*, 104 (1998), 129–156.

[26] B. Sobociński, Studies in Leśniewski's Mereology. *Yearbook for 1954-55 of the Polish Society of Art and Sciences Abroad*, 5 (1954), 34–43.

[27] J. Srzednicki, S.J. Surma, D. Barnett and V.F. Rickey (eds.), *Collected Works of Stanisław Leśniewski*. Kluwer, Dordrecht, 1992.

[28] J. Słupecki and S. Lesniewski, Calculus of Names. *Studia Logica*, 3 (1955), 7–72.

[29] A. Tarski, Les fondements de la géométrie des corps. In *Księga Pamiątkowa I Polskiego Zjazdu Matematycznego (Memorial Book of the Ist Polish Mathematical Congress)*, a supplement to *Annales de la Société Polonaise de Mathématique*, Cracow, 1929, 29–33.

[30] A. Tarski, Appendix E. In J.H. Woodger, *The Axiomatic Method in Biology*. Cambridge University Press, Cambridge, 1937.

[31] A. Tarski, What is elementary geometry? In L. Henkin, P. Suppes and A. Tarski (eds.) *The Axiomatic Method, with Special Reference to Geometry and Physics*. North-Holland, Amsterdam, 1959, 16–29.

Languages and Problem Specification

Loutfi Soufi

University Henri Poincaré Nancy I
LORIA-INRIA
Vandoeuvre-lès-Nancy, France
`soufi@loria.fr`

Abstract. Generally, programming problems are formally described as function computation problems. In this paper we want to see programming problems as language recognition problems. More precisely, we suggest specifying the former problems as systems of languages, i.e. recursive languages built from the concatenation and transformation of recursive languages. To understand our grammatical approach, we illustrate it by means of typical examples of programming.

1 Introduction

This paper tries to show how programming problems can be described as language recognition problems. Generally, formal methods of program development describe programming problems as function computation problems (see for example [1, 7, 13, 9]). As mentioned, for example, in [8], a function $f : N \to N$ (N being the natural numbers) can be specified by a grammar G. More generally, if f is a function $f : T \to S$, where T, S are any *type*, denumerable domains, then the elements of T and S can be put in a bijective correspondence with the words of Σ^*, i.e. words (strings) over Σ (Σ being some alphabet).

Example 1 *An integer type.*
*The usual integer type can be specified by $L(G) = \{1^n \mid n \geq 0\}$, where x^i
denotes the word obtained by concatenation of x with itself i times, and $x^0 =
\epsilon$ (the emtpy word). Thus, one can imagine a programming environment (a
mechanical device) in which 1^n corresponds to the integer n.*

Example 2 *A function type $f : N \to N$.*
*For instance $f(n) = n^2$ can be specified by the context-sensitive language
$L(G) = \{1^n m 1^{n^2} \mid n \geq 0\}$. Given 1^n interpreted as an integer, the outcome
of f is 1^{n^2} (the marker m separates the input from the output).*

Our goal is to specify programming problems (problems for short) using
formal languages. In our grammatical approach a problem is formally described as a "system of languages", i.e. a recursive language structured in
a hierarchical way. This hierarchy of languages is only realized by using the
concatenation and transformations of recursive languages and forms what
we call a *language of concatenation of level n* depending on the number of
transformations applied to a language. Thus, we describe a problem P as a
recursive set S. Since S is recursive, then, by definition [8], its characteristic
function is computable. As a consequence, we can translate the language
S into another formalism for solving P using mechanical devices such as
programming (specification) languages, or formal methods of program construction, or the like. Here we do not tackle this translation of formalisms.
We concentrate, rather, on the specification of problems using generative devices. Chomsky grammars and the Lindenmayer systems are such devices.
Besides these two basic devices we can mention [10, 3, 4, 6, 11]. In this paper
we will see how one can specify problems using D0L and *recurrent* sytems by
means of typical examples of programming. The use of generative devices
liberates us of any particular syntax which appears in programming (specification) languages. For example, there is no syntax of types in a generative
device. Since we are liberated of any syntax, then we augment our freedom
of expressions for specifying problems. Such a freedom of specification (one
can define an "exotic" generative device which can be suitably described
our problem), and the non-definition of a particular syntax for describing a
problem are the main positive aspects of our grammatical approach.

The rest of the paper is organized as follows. In Section 2, we introduce
some notations and the D0L and recurrent systems. Section 3 deals with
hierarchy of languages. Section 4 gives examples of problem specification.
Finally, in Section 5, a short summary is given.

1.1   Related Work

Firstly, we would include the present work in the general topic of the relationship between grammar systems and Programming. Examples of grammar systems are [3, 4, 6]. Our particular field is the use of a kind of hierarchy languages for specifying programming problems in a structured and modular way. Finally, our work has nothing to do with work on tree grammars and related topics as discussed in [2]. In our approach a word over an alphabet is not viewed as a term in the algebraic sense.

2   Preliminaries

We use the following general notation: Σ^* (possibly subscripted) is the set of all finite words (sentences, strings) over the alphabet Σ, ϵ is the empty word, and $\Sigma^+ = \Sigma^* - \epsilon$. If w and v are words, $w, v \in \Sigma^*$, then so is their contatenation, written wv or $w.v$. For further elements of formal language theory we refer to [12, 11]. We suppose that the reader is familiar with the notions of regular expressions, and type-0, type-1, type-2, type-3 languages called respectively recursively enumerable, context-sensitive, context-free, and right-linear (regular) languages. We recall that a grammar is termed *right-linear* if all productions have the form $A \to x$, $A \to xB$, with A, B non-terminals and x a terminal string. As showed in [5], every right-linear grammar can be transformed into a regular grammar generating the same language. This paper does not distinguish between the regular and right-linear languages.

2.1   D0L System

Now, we define a D0L system as given in [11]. First we recall that a mapping $h : \Sigma^* \to \Sigma_1^*$ is a morphism if:

$$h(vw) = h(v)h(w), \quad \text{for all words } v \text{ and } w.$$

A D0L system is a triple $D = (\Sigma,\ h,\ w)$, where $h : \Sigma^* \to \Sigma^*$ is a morphism and $w \in \Sigma^*$. The system D^i, where i is the number of word to be generated, defines the following sequence $S(D)$ of words:

$$w = h^0(w), h(w) = h^1(w), h(h(w)) = h^2, h(h(h(w))) = h^3, h^4, ...$$

The word h^0 is referred to as the first element of $S(D)$, denoted by D_1, and h^1 as the second, denoted by D_2, and h^2 as the third, D_3, and so on.

The system D defines the following language:

$$L(D) = \{h^i(w) \mid i \geq 0\}.$$

Example 3

$D = (\{a, b\}, h, ab)$, *where* $h(a) = (ab)^2$, $h(b) = \epsilon$:
$L(D) = \{(ab)^{2^n} \mid n \geq 0\}$.
$S(D) = ab$, *abab*, *abababab*, ...

2.2   Word Transformation

By a transformation $\boldsymbol{T}$ defined on Σ and Σ_1 we shall understand a single-valued correspondence between Σ^* and Σ_1^* (where possibly $\Sigma^* = \Sigma_1^*$). It is represented by:

$$[w1/v1, w2/v2, ..., wi/vi],$$

where $wi \in \Sigma^*$ and $vi \in \Sigma_1^*$ and the wis are all distinct and the same holds for the vis and such as we have $[\epsilon/\epsilon]$. Now, its effect on a word $\alpha \in \Sigma^*$ is a word belonging to $\Sigma^* \bigcup \Sigma_1^*$, written $\boldsymbol{T}\alpha$, that is the same as α but with all wis occuring in α replaced by the vis. This simultaneous replacement in α of subwords of α by words is called *word transformation*.

Note. In general $[w1/v1, ..., wi/vi]\alpha$ is different from $[w1/v1]...[wi/vi]\alpha$, as illustrated by the following two word transformations.

Example 4 $[a/b, b/c]a = b$, $[b/c][a/b]a = c$.

2.3   Recurrent Systems

We introduce a special symbol @, called the *recurrent word*, in order to characterize regularities occurring in the words of a language. For instance, the words of the precedent language are generated from the word ab: it is a recurrent word. The word $abab$ is also a recurrent word and so on. The idea is that when we derive a word w from a word v it means that v must become an occurence (a new subword) of w. The symbol @ is used to denote the derived word to be occured in the next word to be derived. Formally, a recurrent system is a triple of the form: $S = (V, R, P)$, where $V = \Sigma \bigcup \{@\}$, R is a finite language over the alphabet Σ, and P is a finite language whose the words are of the form $u{:}v$, with $u, v \in V^*$. These words drive the derivation of the words of S. We define the relation $\Longrightarrow$ on V^* by:

$$w \Longrightarrow z \quad \text{iff} \quad w = u, \ z = [@/w]v \text{ for } u{:}v \in P.$$

Then, the language generated by S is defined by:

$$L(S) - \{z \in \Sigma^* \mid w \longrightarrow^* z, \ w \subset R\}.$$

A language generated by a recurrent system is called a *recurrent language*.

Example 5 *The language* $L = \{(ab)^{2^n} \mid n \geq 0\}$ *can be defined by the recurrent system* $S = \{ \{a, b, @\}, \{ab\}, P \}$ *with* $P = \{ab:@, @:@@, @@:@@\}$:

$$ab \Longrightarrow [@/ab]@ = ab, \; now \; @ = ab$$
$$ab \Longrightarrow @ \Longrightarrow [@/ab]@@ = abab, \; now \; @ = abab$$
$$ab \Longrightarrow @ \Longrightarrow @@ \Longrightarrow [@/abab]@@ = abababab$$
$$ab \Longrightarrow @ \Longrightarrow @@ \Longrightarrow @@ \Longrightarrow [@/abababab]@@$$

The sequence of words generated by S^i, where i is the number of words to generated, is: ab, the first element, $abab$ the second one, ... We write S_i for the ith word generated by S. We can put many recurrent systems together in order to form a *complex recurrent system* SS (a *complex* for short) as illustrated in section 4. In this case the ith element of SS is a recurrent system $S(k)^i$. Consequently $S(k)^i_n$ denotes the nth word of the recurrent system $S(k)^i$ of the complex SS. More formally, a complex is a construct of the form $SS(i) = \{S(1), ..., S(i)\}$, where $S^j = \{V_j, R_j, P_j\}$, for $j = 1, 2, ..i$, are recurrent systems such that $S^k = \{V_k, \boldsymbol{T}kR_{k-1}, \boldsymbol{U}kP_{k-1}\}$ for $1 < k \leq i$ and $i > 1$. The language generated by $SS(i)$ is defined by:

$$L(SS(i)) = \{w \mid w \in L(S^j) \; for \; j = 1, 2, .., i\}.$$

Theorem 1 *A language defined by a D0L system is recurrent.*

The proof of this theorem and those given in the next section can be found in [16].

3   Hierarchy of Languages

The goal of this section is to define the languages of concatenation of level n (languages of level n for short). Such languages are used for building languages of type $0 \leq i \leq 3$ from languages of type 3. Before defining what we mean by languages of level n, we come back to the notion of transformation on which these languages are based.

3.1   Transformations

We have already defined what we mean by word transformations. Now, suppose that we apply $\boldsymbol{T} = [w/v]$ to a word α. It could happen that α contains overlapping occurences of w. We decide that a transformation is to be applied to the leftmost occurence in α, reading α from the left to right. Such transformations are called *leftmost transformations*. From now we will only concern ourselves with leftmost transformations.

We define the *product* of two transformations (not to be confused with concatenation) $\boldsymbol{ST}$ by $(\boldsymbol{ST})\alpha = \boldsymbol{S}(\boldsymbol{T}\alpha)$, and the *sum* $\boldsymbol{S+T}$ by $(\boldsymbol{T+S})\alpha =$

$T\alpha + S\alpha$ denoting either the word $T\alpha$ or the word $S\alpha$. Thus, from such a sum we can construct the set $SUM = \{T\alpha\} \bigcup \{S\alpha\}$. If $T=S$ we can write that $T(\alpha + \alpha) = T\alpha$. More generally we can state that $T(\alpha + \beta) = T\alpha + T\beta$, where $\beta \in \Sigma^*$. Obviously sum is associative and commutative. But the product is not commutative. Furthermore sum and product are not mutually distributive. We have only: $T(U + S) = (TU) + (TS)$. The two operations, product and sum, are referred to as *transformation operations*.

These operations are extended in a natural way to languages as follows. Let T be a transformation on Σ and $\Sigma 1$. For a language L over Σ we define:

$$TL = \{\alpha' \mid \alpha' = T\alpha \text{ for all } \alpha \in L\}.$$

In fact some α may remain unchanged, which means $T\alpha = \alpha$. Such α words also belong to TL:

$$[w/L1]L = \{z \mid z = [w/v]x, \text{ for all } x = ywz \in L, \text{ and } v \in L1\}$$

$$[L1/L2] = \{[w/L2] \text{ for all } w \in L1\}$$

$$(S + T)L = \{TL\} \bigcup \{SL\}$$

Theorem 2 *For a given transformation T if L is a regular language then TL is a regular language.*

3.2   Generalized Transformations

Generalized transformations are transformations with D0L and recurrent systems, i.e. transformations of the form either:

$[K_i/K_j]$, where K is a D0L or recurrent system,
or $[K_i/K^j]$ (corresponding to j transformations),
or $[K^i/K^j]$ (corresponding to $i * j$ transformations).

Theorem 3 *If L is a regular language and T a generalized transformation then L can be transformed into another language:*

$$TL = \{\alpha' \mid \alpha' = [sk/si]\alpha \text{ for all } \alpha \in L\} \text{ of type } 0 \leq i \leq 3.$$

Example 6 *The language $L1 = \{a^{n^2} \mid n \geq 1\}$ can be constructed from the regular language $L = \{a^n \mid n \geq 2\}$ using the transformations:*

$T= [b/a]$,
$S= [a/(\{a,b\}, a, h_n)]$, *where $h(a) = ab$, $h(b) = b$, and*
$U= [a/ab]$.

We can prove that $L1 = T(SL + U\{a\})$.
Given n, $R1 = SL = \{(ab^n)^n \mid n \geq 2\}$, so we have:

$$TR1 = \{a^{n^2} \mid n > 2\} \text{ and } T\{ab\} = \{a^2\}.$$

In the sequel the term "transformation" will also denote generalized transformations.

3.3   Structure

We recall that our objective is to express complex languages in terms of simpler ones as illustrated in the above example. Theorem 3 asserts that some languages can be obtained by applying transformations to regular languages. Now we are going to strengthen that theorem. This strengthening is our main result (theorem 4) and is based on a method of classifying languages. We construct hierarchies by considering the concatenation product and the transformation operations only.

A regular expression over Σ^* is termed a *regular expression of level* 0. The regular languages over Σ^* and the empty language are regular expressions of level 0, denoted by $RE(0)$ (more shortly $RE(0)$ languages). Obviously, the concatenation of languages $RE(0)$ is a $RE(0)$ language. If E is a $RE(0)$ language and $\boldsymbol{T}$ a transformation then the language $\boldsymbol{T}E$ is a $RE(1)$ language. The concatenation of $RE(0)$ languages and $RE(1)$ languages is a $RE(1)$ language (1 is the higher level) in this product of concatenation. If L is a $RE(1)$ language then $\boldsymbol{S}L$ is a language $RE(2)$ and so on. More generally, in a product of languages, the higher level of n is determined by a finite combination of n transformation operations. A $RE(n)$ language is of the form:

$$L1L2...Lk,$$

where $L1$, $L2$, ..., LK are $RE(m)$ languages, for $0 \leq m \leq n$, $k \geq 1$, and there exists at least a language Li, $1 \leq i \leq k$, of level n.

Languages transformed into $RE(n)$ languages, $n \geq 0$, without using intersection, difference, union, and the complementation operations, are called *languages of concatenation of level n*.

Note. One can use transformation operations in order to get the same effect as the set operations (union,...). For example let's take the difference operation:

$$\{a1, a2, ..., an, b1, ..., bn\} - \{b1, b2, ..., bn\} = \{[b1/\epsilon, ..., bi/\epsilon]\{a1, a2, ..., an, b1, ..., bn\}.$$

Theorem 4 *The languages of type $0 \leq i \leq 3$ are languages of concatenation of level n, $n > 0$.*

We have not yet an algorithm which transforms a language of type i into a $RE(n)$ language. However, this theorem suggests a bottom-up method. Given a language L, we decompose L into $RE(0)$ languages, $R1$, $R2$, ..., Rk, for $k \geq 1$, such that the words of Rj are subwords of words of L and there is a bijection between Rj and L, for $j = 1, 2, ..., k$. Then we construct a language of concatenation of level 1. After that, we construct a language of concatenation of level 1, and so on until we obtain the words of L. So at each step i of our construction process of L we build a $RE(i)$ language.

Example 7 *The context-sensitive language $L = \{a^n b^n c^n \mid n \geq 1\}$ can be obtained as follows:*

> *Step 0 from L we choose to extract $L0 = \{a^n \mid n \geq 1\}$.*
> *Step 1 $L1 = [a/(\{a, b\}, h_n, a)]L0$ where $h(a) = ab$.*
> *Step 2 $L2 = [a/b]L1$ and $L2_1 = [b/(\{b, c\}, h_n, b)]L2$ where $h(b) = bc$.*
> *Step 3 $L3 = [b/c]L2_1$ and $M = L0.L2.L3$.*

By construction we have $M = L$:
From $L0$ we obtain $L1 = \{ab^{n-1}\}$. Then we construct $L2 = \{b^n\}$. From $L2$ we obtain $L2_1 = \{bc^{n-1}\}$. Finally we construct $L3 = \{c^n\}$.

4   Examples of Problem Specification

This section presents four examples of specification of programming problems: the factorial, the Fibonacci numbers, the tree type, and the Hanoi towers.

Example 8 *The factorial.*
The factorial problem is specified by the language FAC using a D0L system. $S = (\{a, b, c\}, h, abc)$ with $h(a) = ab$, $h(b) = b$, $h(c) = abc$:

$$FAC = [n/\{\{a, b, c\}, h_n, abc\}]\{n\}.$$

For instance, the word $abbabc$ in our mind denotes $2 * 1$, and $abbbabbabc$ denotes $3 * 2 * 1$.

Example 9 *The Fibonacci numbers.*
This problem can be specified by the language FIN over $\{f0, f1\}$. $f1$, $f0$ are terminal symbols:

$$FIN = [f0/(\{f0, f1\}, h^i, f0)]\{f0\},$$

where $h(f0) = f1$, $h(f1) = f0.f1$. The elements of FIN are $f0, f1, f0.f1, f1.f0.f1. \ldots$. The length of a word of FIN is a Fibonacci number.

Example 10 *How to represent a tree.*
This problem of representation can be specified by using the recurrent system:

$$S = (\{a, c, e, @\}, \{a\}, ac\text{:}@a, @a\text{:}@a, @a\text{:}@e, @a\text{:}@c, @e\text{:}@a).$$

The symbols c, c (open and close parentheses respectively) play the role of levels in a tree, and the as are nodes. The node of the root is optional. The word $(a(aa)a)$ is an element of $TREE$.

Our goal is to construct $TREE$ by induction on the words. This question is not tackled in this paper. Let us give a second example.

Example 11 *The Hanoi towers.*
We are given n disks of different sizes and three pegs $0, 1, 2$. At the beginning all disks are stocked in decreasing order of size on peg 0. The objective is to move all disks from this peg to another one moving just one disk at a time and never moving a larger disk (greater number) onto a shorter one (lower number). Let us call the shorter disk 1.
We observe that the moves of the disk 1 can be characterized by the word:
$w1 = 01\ 12\ 20\ 01\ 12\ 20\ 01\ ....$ *such that* $|w1| = 2^n$.
"move 0 to 1, no move, move 1 to 2, no move,...
The moves of the disk 2 can be characterized by the word:
$w2 = 0022\ 2211\ 1100\ 0022\ 2211\ 1100\ ...$ *such that* $|w2| = 2^n$.
$w3 = 00001111\ 11112222\ 222220000\ 00001111...$ *such that* $|w3| = 2^n$.
Now, how to generate the words $w1$, $w2$, $w3$, ...?
We set $T = [0/1, 1/2, 2/0]$, and $S = [0/2, 1/0, 2/1]$. Then:
$w1 = 01T(01)S(01T(01))T(01T(01)S(01T(01)))...$
So we can define the words $w1$ by the following recurrent system:
$RR = \{\{0, 1, 2, @\}, \{01, R\}, \{(01, @T(@)), (@T(@), @S(@)), (@S(@), @T(@))\}\}$
$S^1 = \{\{0, 1, 2, @\}, \{R^1\}, \{(00, @S(@)), (@S(@), @T(@)), (@T(@), @S(@))\}\}$
with $R^1 = \{00\}$.
$S^i = \{\{0, 1, 2, @\}, \{[0/00]R^{i-1}\}, [@(S)@/@(T)@, @(T)@/@(S)@]P^{i-1}\}$.
$SS^k = (S^1, ..., S^k)$.
$HANOI = [n/RR]\{n\}.[n/SS^n]\{n\}$.

The strategy for building languages, for specifying problems is based on inductions to be invented.

5 Conclusion

We have presented a grammatical approach to problem specification based on a notion of hierarchy languages. We have said that any language of type $0 \leq i \leq 3$ can be characterized by languages of concatenation of level n. Then we have illustrated our approach by means of examples. Finally, our intention using our grammatical approach is to bridge the gap between the concepts naturally associated with a problem and those of any specific mechanical devices with which a problem can be solved.

References

[1] M. Bidoit, H. Kreowski, F. Orejas, P. Lescanne and D. Sannela, *A Comprehensive Algebraic Approach to System Specification and Development Annotated Bibliography*. Springer, Berlin, 1991.

[2] H. Common et al., Tree automata techniques and applications. Web École Normale Supérieure, 1999.

[3] E. Csuhaj-Varjú, J. Dassow, J. Kelemen and Gh. Păun, *Grammar Systems: A Grammatical Approach to Distribution and Cooperation*. Gordon and Breach, London, 1994.

[4] J. Dassow, Gh. Păun and G. Rozenberg, Grammar systems. In G. Rozenberg and A. Salomaa (eds.), *Handbook of Formal Languages*. Springer, Berlin, 1997, vol. 2, 155–213.

[5] J.M. Howie, *Automata and Languages*. Oxford University Press, Oxford, 1991.

[6] A. Kelemenová (ed.), *Proceedings of the Workshop on Grammar Systems*. Silesian University, Opava, 1997.

[7] Z. Luo, Program Specification and Data Refinement in Type Theory, Report LFCS, Edinburgh, 1995.

[8] D. Mandrioli and C. Ghezzi, *Theoretical Foundations of Computer Science*. John Wiley, New York, 1987.

[9] A. Partsch, *Specification and Transformation of Programs*. Springer, Berlin, 1990.

[10] Gh. Păun, G. Rozenberg and A. Salomaa, Grammars based on the shuffle operation, *Journal of Universal Computer Science*, 1995.

[11] A. Salomaa, *Formal Languages*. Academic Press, New York, 1973.

[12] A. Salomaa, *Jewels of Formal Language Theory*. Pitman, London, 1981.

[13] D. Sannella and A. Tarlecki, Foundations for Program Development: Basic Concepts and Motivation, Report LFCS, Edinburgh, 1995.

[14] L. Soufi, A Language of Transformations, Report Lirsia, Dijon, 1999.

[15] L. Soufi, Designing Types as Automata, BTCS15, Keele University, April 1999.

[16] L. Soufi, Formal languages and programming, draft Lirsia, 1999.

The Identities of Local Threshold Testability

Avraam N. Trakhtman

Department of Mathematics and Computer Science
Bar-Ilan University
Ramat Gan, Israel
trakht@macs.biu.ac.il

Abstract. A locally threshold testable semigroup S is a semigroup with the property that for some nonnegative integers k and l, whether or not words u and v are equal in S depends on (1) the prefix and suffix of the words of length $k - 1$ and (2) the set of intermediate substrings of length k where the number of occurrences of the substrings ($\leq l$) is taken into account. For given k and l, the semigroup is called l-threshold k-testable. We present a finite basis of identities of the variety of l-threshold k-testable semigroups.

1 Introduction

The concept of local testability was introduced by McNaughton and Papert [17] and by Brzozowski and Simon [6]. Local testability can be considered as a special case of local l-threshold testability for $l = 1$. Locally testable languages, automata and semigroups have been investigated from different points of view (see [5] - [13], [16], [20], [23] - [27]). In [18], local testability was discussed in terms of "diameter-limited perceptrons". Locally testable languages are a generalization of the definite and reverse-definite languages, which can be found, for example, in [10] and [19]. Some variations of the

concept of local testability (strictly, strongly) obtained by changing or omit-
ting prefixes and suffixes in the definition of the concept were studied in [5],
[7], [17], [20], [23].

Locally testable automata have a wide spectrum of applications. Regular
languages and picture languages can be described by strictly locally testable
languages [5], [12]. Local automata (a kind of locally testable automata)
are heavily used to construct transducers and coding schemes adapted to
constrained channels [2]. In [11], locally testable languages are used in the
study of DNA and informational macromolecules in biology.

The syntactic characterization of locally testable semigroups has been
given in [17] and [6]: for any idempotent e in locally testable semigroup S,
the semigroup eSe is commutative and idempotent. The set of k-testable
semigroups forms a variety [27] generated by a finite semigroup [23]. A
finite basis of identities of the variety of k-testable (l-threshold for $l = 1$)
semigroups is presented in [24].

The local threshold testability generalizes the concept of local testability.
The locally threshold testable languages have been studied extensively in
recent years (see [3], [4], [20], [22], [26], [25]). The set of l-threshold k-testable
semigroups forms a variety of semigroups [25]. A locally threshold testable
semigroup has only trivial subgroups [6], [3]. The syntactic characterization
of locally threshold testable languages can be found in [3].

Given the syntactic semigroup S of the language L, we form a graph
G(S) as follows. The vertices of G(S) are the idempotents of S, and the
edges from e to f are the elements of the form esf. A language L is locally
threshold testable if and only if S is aperiodic and for any two nodes e, f
and three edges p, q, r such that p and q are edges from e to f and r is an
edge from f to e we have:

$$prq = qrp.$$

In this paper, we prove that the variety of l-threshold k-testable semigroups is finitely based and present a finite basis of identities of the variety.

2   Notation and Definitions

Let Σ be an alphabet and let Σ^+ denote the free semigroup on Σ. If $w \in \Sigma^+$,
let $|w|$ denote the length of w.

Let $h_k(w)$ $[t_k(w)]$ denote the prefix [suffix] of w of length k or w if $|w| < k$.

Let $F_{k,j}(w)$ denote the set of factors of w of length k with at least j
occurrences.

A semigroup S is called l-threshold k-testable if there is an alphabet Σ
[and a surjective morphism $\phi : \Sigma^+ \to S$] such that for all u, $v \in \Sigma^+$, if
$h_{k-1}(u) = h_{k-1}(v)$, $t_{k-1}(u) = t_{k-1}(v)$ and $F_{k,j}(u) = F_{k,j}(v)$ for all $j \leq l$,
then $u\phi = v\phi$.

An automaton is *l-threshold k-testable* if the syntactic semigroup of the automaton is *l-threshold k-testable*.

A semigroup S is *locally threshold testable* if it is *l*-threshold *k*-testable for some k and l.

x, y, z, t (sometimes with indices) denote variables from Σ.

u, v, w, s (sometimes with indices) denote words over Σ.

u^ρ denotes the class of a completely invariant congruence ρ on Σ^+ containing the element $u \in \Sigma^+$.

$oc_u(w)$ denotes the number of occurrences of the subword u in the word w.

$h_n(u)$ $[t_n(u)]$ denotes left [right] subword of the word u of length n.

$u \succeq v$ - the word v is a subword of u.

Let $F_k(w)$ denote the set of subwords of w of length k.

To a word u and integer k we associate a graph $\Gamma_k(u)$ with $F_k(u)$ as vertex set and suppose $s_1 \to s_2$ if $s_1 = h_k(s)$, $s_2 = t_k(s)$ for some $s \in F_{k+1}(s)$ (de Bruine graph [8], see [24] as well).

A word u will be called *indecomposable* iff every two vertices of the graph $\Gamma_k(u)$ belong to an oriented cycle.

The equality $u = v$ is called *balanced* if for any variable x $oc_x(u) = oc_x(v)$.

var S denotes the variety of semigroups generated by a semigroup S.

Let us denote $X_n = x_1...x_n$ and let T_n be the variety of *n*-testable semigroups. A finite basis of identities of the variety T_n is the following [24]:

$$\alpha_r : X_r^{m+1} X_p = X_r^{m+2} X_p \tag{1}$$

where $r \subset \{1,...n\}$, $p = n - 1 (mod\ r)$, $m - (n - p - 1)/r$,

$$\beta : X_{n-1} y X_{n-1} z X_{n-1} = X_{n-1} z X_{n-1} y X_{n-1}. \tag{2}$$

The number of identities α_r is equal to n.

Let us introduce the following identities for fixed k and l:

$$\delta : w_1 X_r^{m+1} X_p w_2 = w_1 X_r^{m+2} X_p w_2, \tag{3}$$

where for $w = w_1 X_r^{m+1} X_p w_2 = w(x_1,...x_r, y_1,...,y_s)$ we have $oc_{y_i} w = 1$, $oc_{y_i y_j} w = 0$, $mr + p = k - 1$, $r > p$ and every x_i belongs to some subword u of $X_r^{m+1} X_p$ such that $|u| = k$, $2l \geq oc_u(w) \geq l$.

The identity α_r is obtained from the identity δ in the case $l = 1$ and empty words w_1, w_2.

$$\epsilon_r : X_r^{m+1} X_p y X_r^m X_p = X_r^m X_p y X_r^{m+1} X_p, \tag{4}$$

where $r \subset \{1,...n\}$, $mr + p = k - 1$, $r > p$.

$$\gamma : t_{k-1}(u) y u z h_{k-1}(u) = t_{k-1}(u) z u y h_{k-1}(u), \tag{5}$$

$$\zeta : uzh_{k-1}(u)yt_{k-1}(u) = h_{k-1}(u)yt_{k-1}(u)zu, \qquad (6)$$

where $2k - 1 \geq |u| \geq k - 1$, y, z may be empty. The identity β is obtained from the identity γ in case $|u| = k - 1$.

$$\eta : X_i u Y_j v X_i us(Y_j) v X_i = X_i us(Y_j) v X_i u Y_j v X_i, \qquad (7)$$

where $|Y_j| = j, |X_i| = i$, $|s(Y_j)| < 2j+2$, $h(s(Y_j)) = t(s(Y_j)) = Y_j$, $i+j+|u| = i + j + |v| = k$.

The identities ϵ_r and the identities γ, ζ, η are balanced.

Let ρ be the completely invariant congruence on Σ^+ determined by the identities $\alpha_1, ... \alpha_k$, β.

Let τ be the completely invariant congruence on Σ^+ determined by the identities ϵ_r and all identities δ, γ, ζ, η.

Let λ be the completely invariant congruence on Σ^+ determined by the variety of l-threshold k-testable semigroups.

Our aim is to prove that the set of all identities δ, γ and ϵ_r form a finite basis of identities of the variety of l-threshold k-testable semigroups ($\lambda = \tau$).

3   The Identities and the Congruences

We can formulate the above mentioned result of Beauquier and Pin [3] in the following form:

Theorem 1 *[3] A language L is locally threshold testable if and only if the syntactic semigroup S of L is aperiodic and for any two idempotents e, f and elements a, u, b of S we have:*

$$eafuebf = ebfueaf. \qquad (8)$$

Lemma 1 *Every ρ-class of the completely invariant congruence ρ is a union of some λ-classes of the completely invariant congruence λ and every λ-class is a union of some τ-classes of the completely invariant congruence τ.*

Proof. The k-testability can be considered as 1-threshold k-testability and so k-testability implies l-threshold k-testability for any l. Hence $u^\lambda = v^\lambda$ implies $u^\rho = v^\rho$ and so $\lambda \subseteq \rho$. Identities ϵ_r, β, ζ, η and γ are balanced and therefore they are valid in i-threshold k-testable semigroups for any integer i. They are valid in k-testable semigroups too. The identities δ (3) are a consequence of the identities α_r (1) for k-testability, whence l-threshold k-testable semigroups satisfy the identities δ.

Hence $u^\tau = v^\tau$ implies $u^\lambda = v^\lambda$ and so $\tau \subseteq \lambda$.   □

Lemma 2 *All words from a given ρ-class (λ-class, τ-class) have the same graph Γ_{k-1}.*

The proof follows from Lemma 1 and from the definition of the graph Γ_{k-1} (see [23] too).

Lemma 3 *Let X be non-trivial τ-class of a completely invariant congruence τ. Then every word in X contains a subword $u(vu)^m$, where $|u(vu)^{m-1}| \geq k - 1$. Every word s such that $|s| < k$ and $oc_s(w)$ for distinct w from X is not constant is a subword of some such $u(vu)^m$, and in case of existence of such s the class X is infinite.*

Proof. Every non-trivial τ-class X of a completely invariant congruence τ is included in some non-trivial ρ-class of the congruence ρ (lemma 1).

Identities $\alpha_1, ...\alpha_k, \beta$ from the basis of identities of the variety T_k define non-trivial ρ-classes of completely invariant congruence ρ. Each right and left part W of these identities contains a subword of the kind $u(vu)^m$, such that $|u(vu)^{m-1}| = k - 1$. Moreover, every word s such that $|s| = k - 1$ and $oc_s(w)$ is not the same for distinct w from X, belongs, as a node of the graph $\Gamma_{k-1}(w)$, to a cycle of length not greater than $k - 1$, and therefore s is a subword of some such $u(vu)^m$. Any substitution could not change this situation, whence every word from X contains a subword of the kind $u(vu)^m$.

If $oc_s(w)$ is not the same for distinct w from X, then $oc_s(w) \geq l$ and we can use the identities δ (3) and obtain in this way an infinite set of words from X. □

Lemma 4 *Every H-class of a locally threshold testable semigroup contains only one element.*

Proof. A locally threshold testable semigroup contains only trivial subgroups. In view of Theorem 2.22 [8], the Schützenberger group G of an H-class is a homomorphic image of some semigroup with trivial subgroups. Therefore, G is trivial and any H-class is trivial too. □

Lemma 5 *Suppose that $u^\lambda = v^\lambda$, where the words u, v are indecomposable. Then $u^\tau = v^\tau$.*

Proof. Let us note that all words in $u^\rho = v^\rho$ are indecomposable, because all these words have the same graph $\Gamma_{k-1}(u)$. (Lemma 2). By Lemma 1, $\lambda \subseteq \rho$, whence all words from $u^\lambda = v^\lambda$ are indecomposable.

Let us consider the τ-class u^τ. Without loss of generality let us assume that $oc_s(u) \geq oc_s(v)$ for any s such that $|s| = k$. It is obvious when $oc_s(u) < l$ because in this case $oc_s(u) = oc_s(v)$.

In the opposite case, by Lemma 3, the τ-class u^τ is infinite, and the word s is a subword of some word $w = w_1 X_r^{m+1} X_p w_2$, where $oc_s(w) \geq l$. So we can take another suitable indecomposable word u from u^τ.

Our aim is to prove that for any integer $n \leq |v|$ there exists a word $w \in u^\tau$ such that $h_n(v) = h_n(w)$ and $oc_s(v) \leq oc_s(w)$ for any s. We proceed by induction on n. Since $h_1(v) = h_1(u)$, we can take $w = u$. By the definition of the local threshold testability and the congruences τ and λ, the induction assumption holds for $n = 1$ and, moreover, for any $n \leq k - 1$.

We can conclude that for any word $w \in u^\tau$ in view of $\Gamma_{k-1}(v) = \Gamma_{k-1}(u) = \Gamma_{k-1}(w)$ each subword of length k (and $k - 1$ too) from v belongs to w, and the subword $t_k(h_n(v))$ also belongs to w.

For $s = t_k(h_n(v))$ we have $oc_s(w) \geq oc_s(v)$ by choice w, and $oc_s(h_n(v)) > oc_s(h_n(w))$ because s is the last subword in $h_n(v)$, but not in $h_n(w)$. Therefore, some inclusion of s in v does not belong to $h_n(v)$.

According to the induction assumption, $h_{n-1}(v) = h_{n-1}(w)$. So $w = h_{n-1}(v)xw_1$ and $v = h_{n-1}(v)yv_1$. The case which may be considered is $x \neq y$. It is clear that we can assume $n \geq k-1$. Let us denote $T_{k-1} = t_{k-1}(h_{n-1}(v))$. Since v and w belong to one λ-class of the congruence λ, we have $w \succeq T_{k-1}y$. In view of $oc_s(w) \geq oc_s(v)$ for s such that $|s| = k - 1$, some inclusion of the word $s = T_{k-1}y$, does not belong to the subword $h_{n-1}(v)x$.

Suppose first that this inclusion of the word $T_{k-1}y$ has non-empty intersection with the subword $h_{n-1}(v)x$. So the word $T_{k-1}y$ only has its beginning in the subword $h_{n-1}(v)x$ and $w = h_{n-1}(v)xw_1 = w_3 T_{k-1}xw_1 = w_3 h_r(T_{k-1})T_{k-1}yw_2$ for some $h_r(T_{k-1})$, and some w_1, w_2, w_3. Let us take the minimal such r. Therefore, for some X_r, $h_{n-1}(v) = w_3 X_r^{m+1} X_p$, where X_p (and $X_p x$ as well) is some left subword of X_r, $m \geq 0$ is some integer. We have $T_{k-1} = X_r^{m+1} X_p$, whence $k - 1 = r(m+1) + p$. The subword $X_r^{m+2} X_p y$ belongs to w and its beginning $X_r^{m+1} X_p$ ends the subword $h_{n-1}(v)$ of w.

In the case $oc_{X_r}(w) > l$, let us use one of the identities δ (1), and by reducing X_r obtain a word with beginning $h_{n-1}(v)y$. In the case $oc_{X_r}(w) \leq l$ we have $oc_{X_r}(v) = oc_{X_r}(w)$. The word $s = t_k(X_r^{m+1} X_p x)$ can be found at the end of the word $h_n(w)$ and so $oc_s(w) < oc_s(v)$. Therefore, in view of the choice of X_r the word s belongs to the subword w_2. Hence we can use the identity ϵ_r (4), move X_r to s and replace x by y in this case too.

So let us suppose that the left subword $h_{n-1}(v)x$ of w has no intersection with some inclusion of the subword $T_{k-1}y$ of w.

Let us now consider the word $v = h_{n-1}(v)yv_1$. If the subword $T_{k-1}x$ of v belongs only to $h_n(v)$ then $oc_{T_{k-1}x}(v) \neq oc_{T_{k-1}x}(w)$. Therefore, $oc_{T_{k-1}x}(v) \geq l$. In view of Lemma 3, we can apply the identity δ (3) to the word ww and change some subword T_{k-1}. Then we use the identity ϵ_r and replace x by y at the end of the subword $h_n(w)$.

So we can suppose that the subword $T_{k-1}x$ of v does not belong only to the subword $h_n(v)$, and one can find it after this subword in v. If this inclusion of $T_{k-1}x$ in v has non-empty intersection with the subword $h_n(v)$ then its left subword T_{k-1} and the subword $T_{k-1}y = t_k(h_n(v))$ form a word $s = X_r T_{k-1}$, where $X_r = h_r(T_{k-1})$. Therefore, in the graph $\Gamma_{k-1}(v) = \Gamma_{k-1}(w)$ there exists a cycle with node T_{k-1} and edge $T_{k-1}y$ of length less than k. The word s belongs to w too. If s does not belong to $h_n(w)$ then s intersects with the complement of $h_n(v)$. So we can use the identity ϵ_r (4) for corresponding subword of w and replace x by y at the end of the subword $h_n(w)$. If s belongs to $h_n(w)$ then all subwords of s of length k lie outside $h_n(v)$ in view of $oc_s(w) \geq oc_s(v)$. Therefore, we first use the identity ϵ_r, then the identity η (7) and then ϵ_r. In this way we obtain a word beginning with $h_n(v)$ from w.

Let us suppose now that the left subword $h_{n-1}(v)y$ of v has no intersection with some subword $T_{k-1}x$ of v. Some path from the edge $T_{k-1}y$ in $\Gamma_{k-1}(v)$ to the edge $T_{k-1}x$ contains not less than k edges. Corresponding subwords of v of length k do not belong to $h_{n-1}(v)y$. Therefore, in view of $oc_s(w) \geq oc_s(v)$, for any s such that $|s| = k$ they belong to w but do not belong to $h_{n-1}(w)$. If the path is a cycle then its length is not less than k.

We have a word $w = ...T_{k-1}x...T_{k-1}yw_1....$ In case $w_1 \succeq T_{k-1}$ let us use the identity β (2) (a kind of identity γ) and obtain a word $...T_{k-1}y...T_{k-1}x....$ In a case where T_{k-1} begins in $T_{k-1}y$ and ends in w_1, we can use the identity γ and replace x by y in this case too.

Therefore, we can assume that the subword $T_{k-1}y$ is the last subword of w containing T_{k-1}. This implies that on the considered path there exists a node such that the corresponding word s of length $k - 1$ can be found in the word w twice: after the subword $T_{k-1}y$, and before this subword, but after $h_{n-1}(w)$. In view of the length of the path both inclusions of s have no intersection. Let us now go on to consider only a minimal subword of w containing both these inclusions of s, $T_{k-1}y$ and $h_n(w) = T_{k-1}x$.

In a case where the second inclusion of s has non-empty intersection with the subword $T_{k-1}y$ and the first inclusion of s has non-empty intersection with $h_n(w) = T_{k-1}x$, these intersections are distinct. Therefore, we can apply the identity ϵ_r (4) and replace x by y. So let us suppose that at least one of these intersections is empty.

When the considered inclusions of the subword s have intersections with the subword $T_{k-1}y$ of w, let us use the identity η (7), where the intersection of T_{k-1} with the first inclusion of s is denoted by X_i, and the intersection of T_{k-1} with the second inclusion of s is denoted by Y_j. Therefore, T_{k-1} is presented by $X_i u Y_j$ and s is presented by $Y_j v X_i$. We use the same identity η when the first inclusion of s intersects with $h_{n-1}(v)$ and $T_{k-1}y$.

Where the first inclusion of s intersects with $h_{n-1}(v)$, let us use the identity ζ (6), where a subword of w which begins in $h_{n-1}(v)$ by $T_{k-1}y$ and ends by s is denoted by the letter u from the identity ζ.

If only the second inclusion of s has intersection with $T_{k-1}y$, let us use the identity γ (5), where a subword of w_2 which begins with s and ends with T_{k-1} is denoted by the letter u from the identity γ. We replace the letter x by y at the end of the subword $h_n(w)$ in all considered cases.

This finishes the induction.

For $n = |v|$ we have $vv_1 \in u^\tau$ for some v_1. Analogously, $v_2 v \in u^\tau$ for some v_2 and $uu_1, u_2 u \in v^\tau$ for some u_2, u_1. Therefore, the words u, v belong to one H-class of the congruence τ, whence by Lemma 4 $u = v$. □

Theorem 2 *The set of all identities δ, γ, ζ, η, ϵ_r for given k and l form a finite basis of identities of the variety of l-threshold k-testable semigroups.*

Proof. We have to prove that $\tau = \lambda$. By Lemma 1, for any pair of words u, v we have $u^\tau = v^\tau \to u^\lambda = v^\lambda$. Therefore, it is sufficient to prove that $u^\lambda = v^\lambda \to u^\tau = v^\tau$. Let us assume that $u^\lambda = v^\lambda$ (and $u^\rho = v^\rho$ as well).

Suppose first that the word u is indecomposable. Then, by Lemma 2, v is also indecomposable, whence by Lemma 5 we have $u^\tau = v^\tau$.

Let us now go to the general case. Suppose $u = u_1 u_2 ... u_s$ and $v = v_1 v_2 ... v_t$ where u_i and v_i are maximal indecomposable subwords of u or v. By lemma 2, $\Gamma_{k-1}(v) = \Gamma_{k-1}(u)$. Maximal strongly connected components of these graphs correspond to some indecomposable words, and the graph is a chain of maximal strongly connected components. Therefore, $\Gamma_{k-1}(v_i) = \Gamma_{k-1}(u_i)$. From $u^\lambda = v^\lambda$ it follows that $u_i^\lambda = v_i^\lambda$. So $u_i^\tau = v_i^\tau$. Therefore, $u^\tau = v^\tau$.

The assertion of the theorem holds in this case too. □

Acknowledgments

I would like to express my gratitude to Stuart Margolis for posing problems of local threshold testability and to Shalom Feigelstock for improving the style of the paper.

References

[1] J. Almeida, Implicit operations on finite J-trivial semigroups and a conjecture of I. Simon, *Journal of Pure and Applied Algebra*, 69 (1990), 205–208.

[2] M.P. Beal and J. Senellart, On the bound of synchronization delay of local automata, *Theoretical Computer Science*, 205.1-2 (1998), 297–306.

[3] D. Beauquier and J.E. Pin, Factors of words. In Lecture Notes in Computer Science, 372. Springer, Berlin, 1989, 63–79.

[4] D. Beauquier and J.E. Pin, Languages and scanners, *Theoretical Computer Science*, 198.4 (1991), 3–21.

[5] J.-C. Birget, Strict local testability of the finite control of two-way automata and of regular picture description languages, *Journal of Alg. Comp.*, 1.2 (1991), 161–175.

[6] J.A. Brzozowski and I. Simon, Characterizations of locally testable events, *Discrete Mathematics*, 4 (1973), 243–271.

[7] P. Caron, LANGAGE: a Maple package for automaton characterization of regular languages. In Lecture Notes in Computer Science, 1436. Springer, Berlin, 1998, 46–55.

[8] A. Clifford and G. Preston, The algebraic theory of the semigroups, *Mathematical Surveys*, 7, American Mathematical Society, 1964.

[9] Z. Esik and I. Simon, Modelling literal morphisms by shuffle, *Semigroup Forum*, 56.2 (1998), 225–227.

[10] A. Ginzburg, About some properties of definite, reverse-definite and related automata, *IEEE Transactions on Electron. Comput.*, EC-15 (1966), 806–810.

[11] T. Head, Formal language theory and DNA: an analysis of the generative capacity of specific recombinant behaviors, *Bulletin of Mathematical Biology*, 49 (1987), 739–757.

[12] F. Hinz, Classes of picture languages that cannot be distinguished in the chain code concept and deletion of redundant retreats. In Lecture Notes in Computer Science, 349. Springer, Berlin, 1990, 132–143.

[13] S. Kim and R. McNaughton, Computing the order of a locally testable automaton. In Lecture Notes in Computer Science, 560. Springer, Berlin, 1991, 186–211.

[14] S. Kim, R. McNaughton and R. McCloskey, A polynomial time algorithm for the local testability problem of deterministic finite automata, *IEEE Transactions on Comput.*, 40 (1991), 1087–1093.

[15] G. Lallement, *Semigroups and Combinatorial Applications*. John Wiley, New York, 1979.

[16] S.W. Margolis and J.E. Pin, Languages and inverse semigroups. In Lecture Notes in Computer Science, 199. Springer, Berlin, 1985, 285–299.

[17] R. McNaughton and S. Papert, *Counter-Free Automata*. MIT Press, Cambridge, Mass., 1971.

[18] M. Minsky and S. Papert, *Perceptrons*. MIT Press, Cambridge, Mass., 1969.

[19] M. Perles, M.O. Rabin and E. Shamir, The theory of definite automata, *IEEE Transactions on Electron. Comput.*, EC-12 (1963), 233–243.

[20] J. Pin, Finite semigroups and recognizable languages. An introduction: semigroups and formal languages, *Math. and Ph. Sci.*, 466 (1995), 1–32.

[21] J. Reiterman, The Birkhoff theorem for finite algebras, *Algebra Universalis*, 14 (1982), 1–10.

[22] W. Thomas, Classifying regular events in symbolic logic, *Journal of Computer and System Sciences*, 25 (1982), 360–376.

[23] A.N. Trakhtman, The varieties of n-testable semigroups, *Semigroup Forum*, 27 (1983), 309–318.

[24] A.N. Trakhtman, Identities of locally testable semigroups, *Comm. in Algebra*, to appear.

[25] Th. Wilke, An algebraic theory for regular languages of finite and infinite words, *International Journal of Algebra and Computation*, 3.4 (1993), 447–489.

[26] Th. Wilke, Locally threshold testable languages of infinite words. In Lecture Notes in Computer Science, 665. Springer, Berlin, 1993, 63–79.

[27] Y. Zalcstein, Locally testable semigroups, *Semigroup Forum*, 5 (1973), 216–227.

IV

MODELS OF MOLECULAR COMPUTING

Soft Computing Modeling of Microbial Metabolism

Ruxandra Chiurtu

Alexandru Agapie

Manuela Buzoianu

Florin Oltean

Marius Giuclea

Roxana Vasilco

National Institute for Research and Development in Microtechnology
Bucharest, Romania
{`ruxandra,roxana`}`@oblio.imt.pub.ro`

Abstract. The preliminary work described here is part of a project that aims to develop a microbial environmental microsensor. The primary aim of this project is to research and establish the degree of correlation between some basic bacterial metabolic sequences and the corresponding mathematical model. The results are encouraging; the mathematical model proves to have a considerable capacity to generalize the set of biochemical reactions, with only very slight errors.

1 Introduction

In recent years, numerous approaches have been used to integrate soft computing with medicine and biosciences. Medical image processing by soft computing is a widely used approach, for example the detection of nuclei in the cell cluster in noisy medical images [7] or the storage of medical information about *Diabetes mellitus* in an information system, performed by J. Stepanink [6].

The use of soft computing methods to interpret microorganism behavior has many areas that have yet to be explored. In 1994, for example, Craven et al. [2] initiated an interesting project on detecting and classifying bacteria for medical applications.

In this paper, we describe a mathematical model based on a few bacterial metabolic sequences, and determine the correlation degree between this model and experimental reality.

2   Metabolic Preview

The metabolism can be defined as a complex process in which living organisms use polysaccharides, lipids and proteins to decompose into CO_2 and H_2O, with energy release. There are two types of reactions involved in cell metabolism:
- reactions which generate the energy necessary for cell functions,
- reactions which synthesize the new cell organites required by the biological processes.

These processes are closely related since a considerable portion of a cell's energy budget is expended on byosynthetic reactions. Protein biosynthesis accounts for about 60% and nutrient transport for about 18%. If the nutrients need to be reduced before they are assimilated into cell constituents, energy must be expended on the reductive process.

Only a few examples of each reaction are discussed below and they may be considered a "catalogue" of the principal energy yielding chemotrophic reactions found in bacteria. All these reactions may be termed *dissimilative* since neither of the reactants is assimilated [4].

a. *H_2 oxidations.*

Molecular hydrogen participation in biological reactions is mediated by hydrogenases. Hydrogen is highly reactive and rarely accumulates in any environment. It is produced in significant quantities only in anaerobic situations. H_2 / CO_2: The process:

$$H_2 + \frac{1}{4} CO_2 \rightleftharpoons \frac{1}{4} CH_4 + \frac{1}{2} H_2O$$

is carried out by all methanogenic bacteria. In *Clostridium aceticum* the following process may take place, and it occurs preferentially in sewage sludge:

$$H_2 + \frac{1}{4}CO_2 \rightleftharpoons \frac{1}{4}CH_3COO^- + \frac{1}{2}H_2O$$

H_2/SO_4^{2-} : Oxidation with sulfate is carried out by *Desulfovibrio*:

$$H_2 + \frac{1}{4}SO_4^{2-} + \frac{1}{4}H^+ \rightleftharpoons \frac{1}{4}HS^- + H_2O.$$

H_2/NO_3^-: The process:

$$H_2 + \frac{2}{5}NO_3^- + \frac{2}{5}H^+ \rightleftharpoons \frac{1}{5}N_2 + \frac{6}{5}H_2O$$

has been described for *Paracoccus (Micrococcus) denitrificans* and *Alcaligenes (Hydrogenomonas) eutrophus* -[4]. It yields considerable free energy.
H_2/O_2: The two above-mentioned bacteria may also carry out the reaction:

$$H_2 + \frac{1}{2}O_2 \rightleftharpoons H_2O$$

In living cells, the energy necessary for ADP phosphorylation is released by the glucose oxidation reaction:

$$Glucose + 6\,O_2 \rightleftharpoons 6\,CO_2 + 6\,H_2O.$$

b. *CHO oxidations*:

These oxidations are performed by *heterotrophic bacteria*; the reactions are very diverse. Possibly some aerobic bacteria can grow on carbon monoxide as an energy source; some methanogenic bacteria can grow on CO:

$$4\,CO + 2\,H_2O \rightleftharpoons 3\,CO_2 + CH_4.$$

c. *CH$_4$ oxidations*:

Methane is oxidized principally by molecular oxygen, and an oxygenase is necessary:
CH_4/SO_4^{2-} :

$$CH_4 + SO_4^{2-} \rightleftharpoons CO_2 + 2\,H_2O + HS^-.$$

CH_4/O_2: Species of *Pseudomonas, Methylomonas, Methylobacter* and *Methylococcus* oxidize methane according to:

$$CH_4 + 2\,O_2 \rightleftharpoons CO_2 + 2\,H_2O,$$

where the necessary methane is obtained according to:

$$CH_3COOH \leftrightarrow CH_4 + CO_2.$$

All these selected biochemical reactions were used in the mathematical modeling that is presented in the section bellow.

3   Mathematical Modeling

The above biochemical reactions were processed by abductive networks (ANs) and genetic algorithms (GAs).

The term "abduction" can be defined as reasoning under uncertainty. Abductive systems [1] are the fruit of almost thirty years of research and development in advanced statistics, neural networks and artificial intelligence, and they are based on two key facts:
- mathematical functions are an extremely powerful representation for numerical data,
- a network structure greatly simplifies the task of learning functional models.

The result is a network of powerful functional elements, called an *abductive network*, that is a network of functional nodes. Each node contains a mathematical function, which computes an output if it is provided with a number of inputs.

The power of abductive networks is derived from its ability to deal with complex problems by subdiving them into smaller, much simpler ones. Currently an AN has three types of nodes: singles, doubles and triples, depending on the number of inputs. These elements are at most third degree polynomial equations with one, two or three variables. The system automatically synthesizes a network from a database of numerical examples, making hypotheses about the network structure and selecting those that fit the data. Thus, it uses the advanced techniques of regression models as well as statistical modeling criterion, that is "the predicted squared error".

Abductive networks are more efficient than neural networks because they can automatically discover the optimal structure of the network for a given problem and they need less time.

Genetic algorithms are probabilistic search algorithms which start with an initial population of likely problem solutions and then evolve towards better solutions. From a theoretical point of view, GAs differ in some fundamental ways from other optimization algorithms, namely -see [5]:

- GAs work with coding parameters, not with parameters themselves;

- GAs search from a population of points, not a single point;

- GAs use payoff (objective function) information, not derivatives or other auxiliary knowledge;

- GAs use probabilistic transition rules, not deterministic rules.

3.1   Experimental Results

We chose to work with genetic algorithms and also abductive networks in order to build mathematical models for a certain group of biochemical reactions. Therefore, we selected the following set of oxidation reactions:

$$R_1: \quad H_2 + \tfrac{1}{4} CO_2 \leftrightarrow \tfrac{1}{4} CH_4 + \tfrac{1}{4} H_2O$$
$$R_2: \quad H_2 + \tfrac{1}{2} CO_2 \leftrightarrow \tfrac{1}{4} CH_3COO^- + \tfrac{1}{2} H_2O$$
$$R_3: \quad H_2 + \tfrac{1}{4} SO_4^{2-} \leftrightarrow \tfrac{1}{4} HS^- + H_2O$$
$$R_4: \quad H_2 + \tfrac{2}{5} NO_3^- + \tfrac{2}{5} H^+ \leftrightarrow \tfrac{1}{5} N_2 + \tfrac{6}{5} H_2O$$
$$R_5: \quad H_2 + \tfrac{1}{2} O_2 \leftrightarrow H_2O$$
$$R_6: \quad CH_3COOH \leftrightarrow CH_4 + CO_2$$
$$R_7: \quad C_6H_{12}O_6 + 6\,O_2 \leftrightarrow 6\,CO_2 + 6\,H_2O$$
$$R_8: \quad 4\,CO^- + 2\,H_2O \leftrightarrow 3\,CO_2 + CH_4$$
$$R_9: \quad CH_4 + SO_4^{2-} \leftrightarrow CO_2 + 2\,H_2O + HS^-$$
$$R_{10}: \quad CH_4 + 2\,O_2 \leftrightarrow CO_2 + 2\,H_2O$$

We tested these reactions with two models: one based on polynomials and genetic algorithms and the other one based on abduction networks. The difference between the two models is that for GA training we used all ten reactions, whereas for AN training we separated them into two sets of nine reactions and one reaction. In the case of the second model, we used the first nine reactions for training and the last one for testing (in a different order). Thus, we can expect the AN model to be more "stabile", with a better generalization capacity.

4 Results Obtained with the GA Model

A mathematical representation of the chemical reactions from above can be formally written as follows:

$$
\begin{pmatrix} H_2 & CO_2 & SO_4 & NO_3 & H & O_2 & CH_4 & H_2O \\ X_1 & X_2 & X_3 & X_4 & X_5 & X_6 & X_7 & X_8 \end{pmatrix}
\begin{pmatrix}
1 & 1 & 1 & 1 & 1 & 0 & 0 & 0 & 0 & 0 \\
\tfrac{1}{4} & \tfrac{1}{2} & 0 & 0 & 0 & 1 & 1 & \tfrac{3}{4} & 0 & 0 \\
0 & 0 & \tfrac{1}{4} & 0 & 0 & 0 & 0 & 0 & \tfrac{1}{2} & 0 \\
0 & 0 & 0 & \tfrac{2}{5} & 0 & 0 & 0 & 0 & 0 & 0 \\
0 & 0 & 0 & \tfrac{2}{5} & 0 & 0 & 0 & 0 & 0 & 0 \\
0 & 0 & 0 & 0 & \tfrac{1}{2} & 0 & 0 & 0 & 0 & 1 \\
0 & 0 & 0 & 0 & 0 & 1 & 0 & \tfrac{1}{4} & \tfrac{1}{2} & \tfrac{1}{2} \\
0 & 0 & 0 & 0 & 0 & 0 & 1 & 0 & 0 & 0
\end{pmatrix} \leftrightarrow
$$

$$
\begin{pmatrix} CH_4 & H_2O & CH_3COOH & HS & N_2 & C_6H_{12}O_6 & O_2 & CO & CO_2 \\ Y_1 & Y_2 & Y_3 & Y_4 & Y_5 & Y_6; & Y_7 & Y_8 & Y_9 \end{pmatrix}
\begin{pmatrix}
\tfrac{1}{4} & 0 & 0 & 0 & 0 & 0 & 0 & 0 & 0 \\
\tfrac{1}{2} & \tfrac{1}{2} & 1 & \tfrac{6}{5} & 1 & 0 & 0 & \tfrac{1}{2} & 1 & 1 \\
0 & \tfrac{1}{4} & 0 & 0 & 0 & 1 & 0 & 0 & 0 & 0 \\
0 & 0 & \tfrac{1}{4} & 0 & 0 & 0 & 0 & 0 & \tfrac{1}{2} & 0 \\
0 & 0 & 0 & \tfrac{1}{5} & 0 & 0 & 0 & 0 & 0 & 0 \\
0 & 0 & 0 & 0 & 0 & 0 & \tfrac{1}{6} & 0 & 0 & 0 \\
0 & 0 & 0 & 0 & 0 & 0 & 1 & 0 & 0 & 0 \\
0 & 0 & 0 & 0 & 0 & 0 & 0 & 1 & 0 & 0 \\
0 & 0 & 0 & 0 & 0 & 0 & 0 & 0 & \tfrac{1}{2} & \tfrac{1}{2}
\end{pmatrix}
$$

Based on this representation, we constructed a polynomial P= P $(X_1, X_2, X_3,$ $X_4, X_5, X_6, X_7, X_8)$, which, in the points: $\left(1, \frac{1}{4}, 0, 0, 0, 0, 0, 0\right)$ and $\left(1, \frac{1}{2}, 0, 0, 0, 0, 0, 0\right)$, take the values that correspond to the variables $Y_1, Y_2,$ ..., Y_9, which are the column vectors of the second number matrix. Consequently, the polynomial P has dimension nine so that P=$(P_1,\ P_2,\ P_3,\ P_4,$ $P_5,\ P_6,\ P_7,\ P_8,\ P_9)$.

The results obtained by applying the GA program endowed with adaptive mutation (implemented in SOFT COMPUTING GENETIC TOOL, within the framework of IMT -[3]) are as follows:

$$P_1 = -0.0819\,X_1 - 0.1789\,X_2 - 0.3335\,X_3 + 0.1399\,X_4 - 0.0861\,X_5$$
$$-0.2512\,X_6 - 0.0049\,X_7 - 0.1322\,X_8 + 0.1953$$

$$P_2 = -0.0251\,X_1 - 0.9432\,X_2 + 0.0962\,X_3 + 0.4481\,X_4 + 0.1911\,X_5$$
$$+0.0984\,X_6 - 0.031\,X_7 - 0.0236\,X_8 + 0.9612$$

$$P_3 = 0.2293\,X_1 + 0.0589\,X_2 - 0.8848\,X_3 + 0.0269\,X_4 - 0.5736\,X_5$$
$$-0.4503\,X_6 + 0.9092\,X_7 - 0.0405\,X_8 - 0.0074$$

$$P_4 = -0.0863\,X_1 - 0.099\,X_2 + 0.3375\,X_3 - 0.5526\,X_4 - 0.4263\,X_5$$
$$-0.1179\,X_6 - 0.0379\,X_7 - 0.0379\,X_8 + 0.1369$$

$$P_5 = 0.1555\,X_1 - 0.2408\,X_2 - 0.3287\,X_3 - 0.3095\,X_4 + 0.6022\,X_5$$
$$-0.1644\,X_6 + 0.4753\,X_7 + 0.3141\,X_8 - 0.0733$$

$$P_6 = 0.021\,X_1 - 0.1929\,X_2 - 0.261\,X_3 - 0.0779\,X_4 - 0.0854\,X_5$$
$$-0.1305\,X_6 - 0.1724\,X_7 + 0.3153\,X_8 + 0.0443$$

$$P_7 = -0.0808\,X_1 + 0.1628\,X_2 + 0.1107\,X_3 + 0.0836\,X_4 - 0.0124\,X_5$$
$$+0.0553\,X_6 - 0.2154\,X_7 + 0.7848\,X_8 + 0.0524$$

$$P_8 = -0.1156\,X_1 - 0.041\,X_2 - 0.1946\,X_3 + 0.0033\,X_4 - 0.0983\,X_5$$
$$-0.0973\,X_6 - 0.1126\,X_7 - 0.1126\,X_8 + 0.1536$$

$$P_9 = 0.021\,X_1 - 0.1929\,X_2 - 0.261\,X_3 - 0.0779\,X_4 - 0.0854\,X_5$$
$$-0.1305\,X_6 - 0.1724\,X_7 + 0.3153\,X_8 + 0.0443$$

We can verify these results by comparing the reaction R_{10} of methane oxidation with the polynomial value: $P\left(0, 0, 0, 0, 0, 1, \frac{1}{2}, 0\right)$ =(-0.058, 1.044, -0.003, 5e-05, -5e-05, 1.11e-16, 1.67e-16, 0, 0.4743).

In a future chemical reaction approach within the framework of evolutionary algorithms, we will try to generalize this result to obtain a polynomial model for many more reactions that describe a complete biochemical process and to find the inverse of the polynomial function in order to suggest the reversible character of the reactions.

4.1 Results Obtained with the AN Model

Reaction number	variable name	expected value	network estimate	error
1	CH_4	0.25000	0.25000	0.0000
1	H_2O	0.50000	0.50000	1.3323e-15
1	CH_3COOH	0.0000	8.3267e-17	8.3267e-17
1	HS	0.0000	0.0000	0.0000
1	N_2	0.0000	0.0000	0.0000
1	$C_6H_{12}O_6$	0.0000	2.0817e-17	2.0817e-17
1	O_2	0.0000	1.1102e-16	1.1102e-16
1	CO	0.0000	0.0000	0.0000
1	CO_2	0.0000	1.4294e-15	1.4294e-15
2	CH_4	0.0000	4.7531e-16	4.7531e-16
2	H_2O	0.50000	0.50000	2.2204e-16
2	CH_3COOH	0.25000	0.25000	-2.7756e-16
2	HS	0.0000	0.0000	0.0000
2	N_2	0.0000	0.0000	0.0000
2	$C_6H_{12}O_6$	0.0000	2.0817e-17	2.0817e-17
2	O_2	0.0000	1.1102e-16	1.1102e-16
2	CO	0.0000	0.0000	0.0000
2	CO_2	0.0000	1.4294e-15	1.4294e-15
3	CH_4	0.0000	7.5981e-16	7.5981e-16
3	H_2O	1.2000	1.2000	-6.6613e-16
3	CH_3COOH	0.0000	1.2212e-15	1.2212e-15
3	HS	0.0000	0.0000	0.0000
3	N_2	0.20000	0.20000	0.0000
3	$C_6H_{12}O_6$	0.0000	2.0817e-17	2.0817e-17
3	O_2	0.0000	1.1102e-16	1.1102e-16
3	CO	0.0000	0.0000	0.0000
3	CO_2	0.0000	1.4294e-15	1.4294e-15
4	CH_4	0.0000	7.5981e-16	7.5981e-16
4	H_2O	1.0000	1.0000	-1.3323e-15
4	CH_3COOH	0.0000	8.8818e-16	8.8818e-16
4	HS	0.0000	0.0000	0.0000
4	N_2	0.0000	0.0000	0.0000
4	$C_6H_{12}O_6$	0.0000	2.0817e-17	2.0817e-17
4	O_2	0.0000	1.1102e-16	1.1102e-16
4	CO	0.0000	0.0000	0.0000
4	CO_2	0.0000	1.4294e-15	1.4294e-15
5	CH_4	0.0000	5.7593e-16	5.7593e-16
5	H_2O	0.0000	5.1070e-15	5.1070e-15
5	CH_3COOH	1.0000	1.0000	-2.2204e-15
5	HS	0.0000	0.0000	0.0000
5	N_2	0.0000	0.0000	0.0000
5	$C_6H_{12}O_6$	0.0000	2.0817e-17	2.0817e-17
5	O_2	0.0000	1.1102e-16	1.1102e-16
5	CO	0.0000	1.4599e-14	1.4599e-14
5	CO_2	0.0000	7.6744e-15	7.6744e-15

6	CH_4	0.0000	5.7593e-16	5.7593e-16
6	H_2O	0.0000	5.1070e-15	5.1070e-15
6	CH_3COOH	0.0000	1.1102e-16	1.1102e-16
6	HS	0.0000	0.0000	0.0000
6	N_2	0.0000	0.0000	0.0000
6	$C_6H_{12}O_6$	0.17000	0.17000	0.0000
6	O_2	1.0000	1.0000	-8.8818e-16
6	CO	0.0000	0.0000	0.0000
6	CO_2	0.0000	1.4294e-15	1.4294e-15
7	CH_4	0.0000	0.0000	0.0000
7	H_2O	0.50000	0.50000	-5.5511e-17
7	CH_3COOH	0.0000	0.0000	0.0000
7	HS	0.0000	0.0000	0.0000
7	N_2	0.0000	0.0000	0.0000
7	$C_6H_{12}O_6$	0.0000	2.0817e-17	2.0817e-17
7	O_2	0.0000	1.1102e-16	1.1102e-16
7	CO	1.0000	1.0000	0.0000
7	CO_2	0.0000	0.0000	0.0000
8	CH_4	0.0000	4.2674e-16	4.2674-16
8	H_2O	1.0000	1.0000	-1.3323e-15
8	CH_3COOH	0.0000	1.0547e-15	1.0547e-15
8	HS	0.50000	0.50000	0.0000
8	N_2	0.0000	0.0000	0.0000
8	$C_6H_{12}O_6$	0.0000	2.0817e-17	2.0817e-17
8	O_2	0.0000	1.1102e-16	1.1102e-16
8	CO	0.0000	0.0000	0.0000
8	CO_2	0.50000	0.50000	0.0000
9	CH_4	0.0000	4.2674e-16	4.2674e-16
9	H_2O	1.0000	1.0000	-1.3323e-15
9	CH_3COOH	0.0000	1.2768e-15	1.2768e-15
9	HS	0.0000	0.0000	0.0000
9	N_2	0.0000	0.0000	0.0000
9	$C_6H_{12}O_6$	0.0000	2.0817e-17	2.0817e-17
9	O_2	0.0000	1.1102e-16	1.1102e-16
9	CO	0.0000	0.0000	0.0000
9	CO_2	0.50000	0.50000	0.0000

Individual Error Statistics

For AN training (working with the *AIM* software produced by *AbTech Corporation, SUA,* -[1]) we used only nine reactions, namely: R_1, R_2, R_4, R_5, R_6, R_7, R_8, R_9, R_{10}. Reaction R_3 was used for testing.

We should point out that every equation corresponds to a sub-table (the equation index being represented in the first column of the table). In the second column we put the chemical substances -the potential participants in the reactions. In the third column we put the relative coefficient (the number of molecules, with norm 1) of the substance in the reaction. The fourth column shows the same coefficient computed with the AN and, finally, the last column shows the error -that is to say, the difference between the real and the estimated coefficient.

Afterwards, the AN trained on the nine equations was tested on the following reaction:

$$H_2 + \frac{1}{4} SO_4^{2-} \rightarrow \frac{1}{4} HS^- + H_2O$$

Reaction number	variable name	expected value	network estimate	error
1	CH_4	0.0000	7.5981e-16	7.5981e-16
1	H_2O	1.0000	1.0000	-1.3323e-15
1	CH_3COOH	0.0000	6.9389e-16	6.9389e-16
1	HS	0.25000	0.25000	4.4409e-16
1	N_2	0.0000	0.0000	0.0000
1	$C_6H_{12}O_6$	0.0000	2.0817e-17	2.0817e-17
1	O_2	0.0000	1.1102e-16	1.1102e-16
1	CO	0.0000	0.0000	0.0000
1	CO_2	0.0000	1.4294e-15	1.4294e-15

As can be seen from the last column, the errors are very small and this shows that the model based on AN has a very good generalization capacity. Furthermore, the model can be extended by introducing more reactions in the training and testing sets.

5 Conclusions

We consider that the experimental results are extremely important and original, since the mathematical model we obtained "knows" how to yield the end-products of a given metabolic sequence, after they have been developed from another group of metabolic reactions.

Our results -and in particular the fact that we have confirmed that the basic original energetic process can be overreached and developed by means of highly modern methods of mathematical modeling- have finally pointed out the strong connections between life sciences and positive sciences.

References

[1] AIM, *User's Manual*. AbTech Corporation, Experts in Advanced Computing Technologies, Charlottesville, USA.

[2] M.A. Craven et al., Application of an artificial neural network based electronic nose to the classification of bacteria. In *Proceedings of EUFIT'94, 2nd European Congress on Intelligent Techniques and Soft Computing,* Aachen, 1994, 768–774.

[3] A.H. Dediu, Soft computing genetic tool. In *Proceedings of EUFIT'96, 4th European Congress on Intelligent Techniques and Soft Computing,* Aachen, 1996, 415–419.

[4] T. Fenchel and T.H. Blackburn, *Bacteria and Mineral Cycling.* Institute of Ecology and Genetics, Aarhus, 1997.

[5] D.E. Goldberg, *Genetic Algorithms in Search, Optimization and Machine Learning.* Addison-Wesley, Reading, Mass., 1989.

[6] J. Stepanink, Rough set based data mining in diabetes mellitus data table. In *Proceedings of EUFIT'98, 6th European Congress on Intelligent Techniques and Soft Computing,* Aachen, 1998, 980–985.

[7] E. Uchino et al., Detection of nuclei in the cell cluster in noisy medical image by combinatorial fuzzy Hough transform. In *Proceedings of EUFIT'95, 3rd European Congress on Intelligent Techniques and Soft Computing,* Aachen, 1995, 1680–1685.

DNA Hybridization, Shifts of Finite Type, and Tiling of the Integers

Ethan M. Coven

Department of Mathematics
Wesleyan University
Middletown, CT, U.S.A.
`ecoven@wesleyan.edu`

Nataša Jonoska

Department of Mathematics
University of South Florida
Tampa, FL, U.S.A.
`jonoska@math.usf.edu`

Abstract. We show a relationship between shifts of finite type, two models for DNA based computation and tiling of the integers. Using the Frobenius-Perron Theorem we show that the informational entropy of the Adleman model for generating paths in a graph and the language of a simple H-system can be easily computed. For both these models we associate tiling shifts that tile the integers. We show the relationship between the entropies of the DNA based computing models and the tiling shifts.

1 Introduction

Simple H-systems were introduced in [5]. There, the reaction of a set of endonucleases with a ligase on a set of DNA molecules was modelled by formal language theory. More general H-systems have also been studied and it has been shown that they are capable of universal computation [13, 6]. Recent preliminary experimental results have shown that it is possible for different enzymes to react in the same buffer [7]. In his seminal article [1], L. Adleman described an experiment using DNA molecules to solve a special case of the directed Hamiltonian path problem. In both this and other experiments, a mix of molecules is allowed in a reaction in which longer molecules are produced. The mix can produce potentially highly complex sets of molecules. Even though there have been extensive studies of the theoretical computational power of this and other models of biomolecular compution, the complexity of the reactions involved has been largely ignored. An initial attempt to define computational complexity for biomolecular computation was made in [4]. There, the physico-chemical properties within the tube and among the molecules was used as a measure of complexity.

In this article, we consider another measure of the complexity of a mix of DNA molecules, the *Shannon entropy* associated with a formal language. For certain models, the entropy is easy to compute using the Frobenius-Perron Theorem on nonnegative matrices. We illustrate this with two models. The first is Adleman's model for generating paths in a directed graph [1], Section 3. The entropy of the mix of molecules created within the tube in this experiment is $\frac{1}{20} \log 2$. In Section 4, we describe a simple way of associating a finite directed labelled graph with a simple H-system. Using this graph, we compute the entropy of the language associated with the H-system and the entropy of the language made from the DNA strands encoded by the simple H-system. In Section 5, we show that there is a set of subsets of the integers, called *prototiles* that encode the DNA bases A, C, G, T and that the integers can be tiled by these prototiles; equivalently, arbitrarily long double-stranded DNA molecules can be modelled by strings of these prototiles. We associate sets of prototiles with the models from Sections 3 and 4, and compute the entropies of the corresponding tiling shifts. Notions from symbolic dynamics, in particular that of *shift of finite type*, appear throughout this paper. In Section 2, we list the relevant definitions and properties.

2 Basic Definitions: Shifts of Finite Type, Entropy

In this section we define *shift of finite type*, one of the basic notions of symbolic dynamics, and we recall some basic properties. For more results in symbolic dynamics, see [8].

Let $\mathcal{A}$ be a finite alphabet. Finite sequences of symbols in $\mathcal{A}$ are called *words* or *blocks*, and words of length k are called *k-words* or *k-blocks*. The set of all words (including the empty word) over $\mathcal{A}$ is $\mathcal{A}^* = \bigcup_{k \geq 0} \mathcal{A}^k$. With the operation concatenation, it is the free monoid generated by $\mathcal{A}$. With the discrete topology on $\mathcal{A}$ and the product topology on $\mathcal{A}^\mathbb{Z}$, the set of all doubly infinite sequences over $\mathcal{A}$, $\mathcal{A}^Z$ is a compact metric space, homeomorphic to the Cantor set. The *shift map* $\sigma : \mathcal{A}^Z \to \mathcal{A}^Z$, defined by $(\sigma(x))_i = x_{i+1}$, $i \in Z$, is a homeomorphism of $\mathcal{A}^Z$ onto itself. $\mathcal{A}^Z$ is called the *full shift*. A nonempty subset S of $\mathcal{A}^Z$ is called a *subshift* of $\mathcal{A}^Z$ (or sometimes just a *shift*) if and only if S is closed (in the topology of $\mathcal{A}^Z$) and shift-invariant, i.e. $\sigma(S) = S$.

Associated with a subshift S is a set of words $F(S)$, defined by $w \in F(S)$ if and only if there are left-infinite and right-infinite sequences v' and v'' such that $v'wv'' \in S$. $F(S)$ is known by various names: the set of *finite words which appear in S*, the *factor set* of S, and the *factor language* of S. Similarly, for a language $L \subseteq \mathcal{A}^Z$ we have $F(L)$, the *factor language* of L, and for a point $x \in \mathcal{A}^Z$ we have $F(x)$, the set of *words which appear in x*. It is straightforward to show that a subshift is determined by its factor language.

A nonempty subset Γ of $\mathcal{A}^Z$ is called a *shift of finite type* if and only if there exist $k \geq 1$ and a set $\mathcal{B}$ of k-blocks such that $x \in \Gamma$ if and only if every k-block which appears in x is in $\mathcal{B}$. The full shift $\mathcal{A}^Z$ is a shift of finite type: $k = 1$ and $\mathcal{B} = \mathcal{A}$. The factor language of a shift of finite type is strictly locally testable, hence regular.

A shift of finite type Γ can be represented as the set of all doubly infinite paths in a finite directed graph G as follows. Suppose that k and $\mathcal{B}$ are as above. The set of vertices of G is $\mathcal{A}^{k-1}$, and the set of edges is identified with $\mathcal{B}$, the set of k-blocks appearing in Γ. For an edge $e = a_1 \cdots a_k$, the starting vertex is $a_1 \cdots a_{k-1}$ and the terminal vertex is $a_2 \cdots a_k$. Labelling this edge with the symbol a_k gives a one-to-one correspondence between the doubly infinite labelled walks in G and the points in Γ.

The directed graph G can be regarded as a finite state automaton where each vertex is both an initial and a terminal state. The language recognized by this automaton is $F(\Gamma)$. Moreover, the labelling is deterministic.

One of the measures of the complexity of a shift of finite type Γ is its *Shannon entropy*, $h(\Gamma)$, defined by:

$$h(\Gamma) = \lim_{n \to \infty} \frac{1}{n} \log |B_n(F(\Gamma))|,$$

where $B_n(F(\Gamma)) = A^n \cap F(\Gamma)$ is the set of n-blocks appearing in Γ and $|\cdot|$ denotes cardinality. The same formula defines the entropy of any language.

Suppose that a shift of finite type Γ is represented by a directed graph G. Let A_G be its adjaceny matrix, i.e. A_G is indexed by the vertices of G, and the (v, w)-th entry of A_G is the number of edges from vertex v to vertex w. By the Frobenius-Perron Theorem on nonnegative matrices [8], A_G has an eigenvalue λ_G, called its *Perron value*, such that $\lambda_G \geq 0$ and $\lambda_G \geq |\lambda|$ for every eigenvalue λ. The entropy of Γ is $h(\Gamma) = \log \lambda_G$. If $\Gamma \neq \emptyset$, then $\lambda_G \geq 1$. In particular, the full shift over a k-letter alphabet is represented by a graph with one vertex and k edges. The adjacency matrix of this graph is $[k]$ and its Perron value is k. Hence, the entropy of the full shift is $\log k$.

3   DNA Hybridization

In this section, we briefly describe some of the structure of DNA and how the hybridization process is used in some models of biomolecular computation.

The information in a DNA molecule is stored in a sequence of nucleotides, also called bases, A,G,C,T (adenine, guanine, cytosine, thymine). Nucleotides are joined together by phosphodiester bonds. Each nucleotide has two distinguished ends, 3' and 5', such that the phosphodiester bond binds a 3' end of one nucleotide with a 5' end of another. Hence, a single strand of DNA is oriented: it has a "beginning" (usually denoted 5') and an "end" (denoted 3'). Nucleotides come in complementary pairs: A and T, C and G. This is the so called Watson-Crick complementarity. Two single-stranded DNA molecules with complementary sequences of nucleotides and opposite orientation are joined by hydrogen bonds to form a double-stranded molecule that appears as a double helix in space. When double-stranded molecules are heated to $95^{o}C$, they disentangle (denature) into single-stranded molecules. When single-stranded molecules are cooled, the nucleotides seek their complements and reanneal into double-stranded form. These properties are used in many models of DNA computers.

We will compute the (informational) complexity of Adleman's model [1] for constructing paths in a graph. Adleman solved a special case of the Hamiltonian path problem for a directed graph G by generating paths using DNA molecules. This graph is depicted in Fig. 2. Similar approaches have been used by several researchers [12, 9, 11]. As a measure of the (informational) complexity of such experiments, we calculate the entropy of the experiment using the Frobenius-Perron Theorem. First we describe the connection between Adleman's model for constructing paths in a directed graph and shifts of finite type.

Define a directed graph G in which the vertices are represented by single-stranded DNA oligos twenty (randomly chosen) nucleotides long. If an edge starts at vertex v_1 and ends at vertex v_2, represent it by a 20-nucleotide sequence of single-stranded DNA as follows. The first ten nucleotides of the edge oligo are complementary to the last ten nucleotides of the vertex oligo representing v_1, and the second ten nucleotides of the edge oligo are complementary to the first ten nucleotides of the vertex oligo representing v_2 (see Fig. 1).

A path $e_1 \cdots e_k$ of length k in G is represented by a double-stranded DNA molecule of length $20k$ base pairs with ten nucleotides (single-stranded) overhanging from each end. If we consider G as a finite state automaton with alphabet the set of edges and such that every vertex of G is both initial and

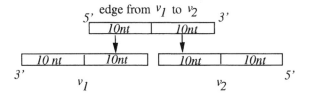

Figure 1: Joining two vertex oligos with an edge oligo.

terminal, then G accepts the strictly locally testable language $L(G)$. The set of biinfinite paths in G represents a shift of finite type Γ whose factor language is contained in $L(G)$.

Denote the set of DNA molecules that represent paths in G by $L(H) \subseteq \{A, G, C, T\}^*$. Let $\mathcal{B}_k(G)$ denote the set of paths (words in $L(G)$) of length k and $\mathcal{B}_k(H)$ the set of DNA molecules in $L(H)$ of length k. Then, recalling that $|\cdot|$ denotes cardinality, we have $|\mathcal{B}_k(G)| = |\mathcal{B}_{20k}(H)|$. So:

$$
\begin{aligned}
h(L(G)) &= \lim_{k \to \infty} \tfrac{1}{k} \log |\mathcal{B}_k(G)| \\
&= \lim_{k \to \infty} \tfrac{1}{k} \log |\mathcal{B}_{20k}(H)| \\
&= \lim_{k \to \infty} \tfrac{20}{20k} \log |\mathcal{B}_{20k}(H)| \\
&= 20 \lim_{k \to \infty} \tfrac{1}{k} |\mathcal{B}_k(H)| \\
&= 20 h(L(H)).
\end{aligned}
$$

More generally, we have:

Proposition 1 *Let G be a directed graph and $L(H)$ be the set of DNA molecules that represent paths in G. If the vertices of G are represented by DNA oligos of length n, then:*

$$
h(L(H)) = \frac{1}{n} \log \lambda_{A_G},
$$

where λ_{A_G} is the Perron value of the adjacency matrix A_G.

For example, consider the graph G that was used in Adleman's experiment. This graph is presented in Fig. 2.

The adjacency matrix A_G of G is:

$$
\begin{bmatrix}
0 & 1 & 0 & 1 & 0 & 0 & 1 \\
0 & 0 & 1 & 0 & 1 & 0 & 0 \\
0 & 0 & 0 & 1 & 0 & 1 & 0 \\
0 & 1 & 0 & 0 & 1 & 0 & 0 \\
0 & 1 & 0 & 1 & 0 & 0 & 0 \\
0 & 0 & 0 & 1 & 1 & 0 & 1 \\
0 & 0 & 0 & 0 & 0 & 0 & 0
\end{bmatrix}
$$

and the Perron value of A_G is $\lambda_{A_G} = 2$. Hence, the entropy of the set of molecules generated is $h(H) = \frac{1}{20} \log 2$.

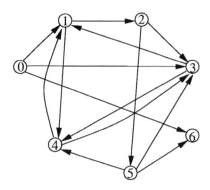

Figure 2: Adleman's graph.

4   Simple H-systems and Shifts of Finite Type

Restriction enzymes (endonucleases) of type II recognize specific sites (recognition sites) of a double-stranded DNA molecule and cut the molecule (by destroying the phosphodiester bonds), often in a way that leaves a small single-stranded overhang. For example, the enzyme *EcoR*I recognizes the site $5' - GAATTC - 3'$ and cuts a double stranded molecule as in Fig. 3. For a more detailed description of the action of these restriction enzymes, see [13, 6].

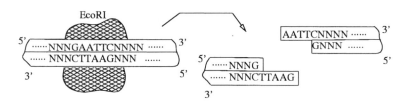

Figure 3: Action of the *EcoR*I endonuclease.

Once cut, molecules with complementary overhangs can join together and the phosphodiester bonds can be reestablished with the aid of a ligase enzyme. What words (strings of DNA molecules) can be obtained in a single tube by allowing a set of restriction enzymes and a ligase to react? This question was first considered by T. Head [5] and then followed by extensive studies of H-systems by other researchers, primarily G. Paun. Here we will concentrate on simple H-systems. For more general H-systems, see [13, 6].

Let $\mathcal{A}$ be an alphabet, $\mathcal{M} \subseteq \mathcal{A}$, and $L \subseteq \mathcal{A}^*$, and let $\sigma_{\mathcal{M}}(L)$ be defined by $z \in \sigma_{\mathcal{M}}(L)$ if and only if there exist $x, y \subset L$ such that for some $c \subset \mathcal{M}$, $x = x'cx''$, $y = y'cy''$, and $z = x'cy''$. A *simple H-system* is a triple $H = (\mathcal{A}, S, \mathcal{M})$, where $\mathcal{A}$ is a finite alphabet, $S \subseteq \mathcal{A}^*$ is a finite set of initial words, called *axioms*, and $\mathcal{M} \subseteq \mathcal{A}$. The language generated by H is $L(H) = \bigcup_{i=0}^{\infty} L^{(i)}$, where the sets $L^{(i)}$ are defined recursively by $L^{(0)} = S$ and $L^{(i+1)} = L^{(i)} \cup \sigma_{\mathcal{M}}(L^{(i)}), i \geq 0$. It is not difficult to prove that the language generated by a simple H-system is strictly locally testable [5, 13, 6]. We will reprove this fact by constructing a labelled directed graph such that the set of labels of the paths is the language generated by a simple H-system.

Let H be a simple H-system. Associate with H a graph G_H in the following way. First let G'_H be the graph with vertices PREF∪CENT∪SUFF, where:

$$\begin{aligned}
\text{PREF} &= \{xc \mid F(x) \cap \mathcal{M} = \emptyset, c \in \mathcal{M}, xc \in F(S)\}, \\
\text{CENT} &= \{cyc' \mid F(y) \cap \mathcal{M} = \emptyset, c, c' \in \mathcal{M}, cyc' \in F(S)\}, \\
\text{SUFF} &= \{cz \mid F(z) \cap \mathcal{M} = \emptyset, c \in \mathcal{M}, cz \in F(S)\}.
\end{aligned}$$

Note that some of the words x, y, z in the defintions of PREF, CENT, and SUFF may be empty. There is an edge from vertex v to vertex v' if and only if there is a symbol $c \in \mathcal{M}$ such that $v = v_1 c$ and $v' = cv_2$ for some v_1 and v_2. The edge from $v_1 c$ to cv_2 is labelled c.

The graph G_H is obtained from G'_H by replacing each vertex $v \in \{uc, cuc, cu\}$ with a path p labelled u. If an edge in G'_H ends at vertex v, then in G_H this edge ends at the first vertex of path p, and if an edge in G'_H starts at vertex v, then in G_H this edge starts at the final vertex of p. The graph G'_H, and hence G_H, is not necessarily deterministic.

We consider G_H as a finite state automaton. If xc is in PREF, then the first vertex of the path labelled x is an initial state of G_H. Similarly, if cz is in SUFF, then the last vertex of the path labelled z is a terminal state of G_H. We leave it to the reader to show that the language recognized by the automaton G_H is $L(H)$.

Example 1 *Let H be the simple H-system given with $H = (\mathcal{A}, S, \mathcal{M})$, where $\mathcal{A} = \{a, b, c\}, S = \{abaca, acaba\}$ and $\mathcal{M} = \{b, c\}$. The graph G'_H has six vertices: ab, bac, ca, ac, cab, ba — see Fig. 4 (a). The graph G_H is obtained from G'_H by substituting a path for each a in a string which represents a vertex of G'_H — see Fig. 4 (b). In this case the path is an edge, since the only factor of S that does not contain a symbol from $\mathcal{M}$ is a. The double circled vertices represent the initial states of G_H and the filled circles represent the terminal states of G_H.*

Proposition 2 *Let $H = (\mathcal{A}, S, \mathcal{M})$ be a simple H-system and $k = \max\{|u| : u \in F(S), F(u) \cap \mathcal{M} = \emptyset\}$. If $p_1 \cdots p_{k+2}$ and $q_1 \cdots q_{k+2}$ are paths that start at the same vertex v and have the same label $x_1 \cdots x_{k+2}$, then $p_1 = q_1$.*

Proof. First note that if $x_1 \notin \mathcal{M}$, then x_1 is a label of a path in G_H and this edge is not in G'_H. Hence there is only one edge coming out of v, and we have $p_1 = q_1$.

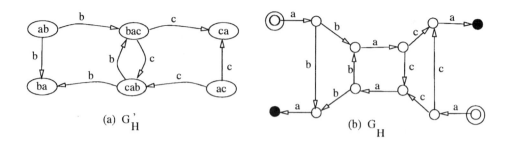

(a) G'_H (b) G_H

Figure 4: Graphs G'_H and G_H associated with the H-system $(\mathcal{A}, S, \mathcal{M})$.

Assume that $x_1 \in \mathcal{M}$. If $x_2 \in \mathcal{M}$, then $x_1 x_2$ is a vertex of G'_H, and so of G_H. Thus, $p_1 = q_1$ is an edge of G'_H from v to $x_1 x_2$ and so of G_H. Assume that $x_2 \notin \mathcal{M}$. Then x_2 is the first symbol of a label of a path in G_H that does not appear in G'_H. So, for some $s \leq k + 2$, the word $w = x_1 \cdots x_s$ is of the form $w = cuc'$ and is in CENT. Hence, w is a vertex of G'_H. This means that $p_1 = q_1$ is an edge of G'_H from v to w, and so $p_1 = q_1$ in G_H.   □

The labelling described in the proposition above is known in symbolic dynamics as *right closing* (with *delay* $k+1$). Right closing is a generalization of deterministic — deterministic graphs are right closing with delay 0. Right closing graphs preserve entropy [8], so for a simple H-system H we have $h(L(H)) = \lambda$, where λ is the Perron value of the adjacency matrix A_{G_H} of the graph G_H.

Following [14], we recall the definition of constants for a language. A word c is a *constant* for a language L if and only if $xcy' \in L$ whenever $xcy, x'cy' \in L$. For a given simple H-system $H = (A, S, \mathcal{M})$, it is clear that if $c \in \mathcal{M}$, then c is a constant for $L(H)$. But even more is true: if $u \in F(L(H))$ is such that $F(u) \cap \mathcal{M} \neq \emptyset$, then u is a constant for $L(H)$. For let $u = u_1 c u_2$, where $c \in \mathcal{M}$, and suppose that $xuy, x'uy' \in L(H)$. Since $c \in \mathcal{M}$, the word $xu_1 cu_2 y'$ is in $\sigma_{\mathcal{M}}(L(H)) = L(H)$, (consider $xu_1 cu_2 y$ and $x'u_1 cu_2 y'$), so $xuy' \in L(H)$. It follows that if $k = \max\{|u| : u \in F(S), F(u) \cap \mathcal{M} = \emptyset\}$, then every word of length greater than k is a constant for $L(H)$. This is another characterization of strictly locally testable languages [10]. (See also Theorem 4.6 in [6].) Thus, the shift space represented by graph G_H is a shift of finite type.

We now return to the action of endonucleases on DNA segments. Assume that we have a finite number of distinct double-stranded DNA segments, and an infinite supply of each segment. Assume further that we have a finite number of restriction endonucleases and a ligase. Denote by E the set of restriction endonucleases. Consider the alphabet $\mathcal{A} = \{A, C, G, T\}$. Write the sequences of double-stranded DNA segments in direction 5' to 3', and consider them as elements of $\mathcal{A}^*$. For example, if we have a segment

$^{5'}A \quad A \quad G \quad C \quad C \quad T^{3'}$
$_{3'}T \quad T \quad C \quad G \quad G \quad A_{5'}$, then we will write $AAGCCT$ and $AGGCTT$.

For each restriction endonuclease ϵ, consider its restriction site ρ_ϵ (again in direction $5'$ to $3'$) and denote the overhang produced by the action of the enzyme by α_ϵ. For example, for $EcoRI$, the restriction site is $\rho = GAATTC$ and the overhang is $\alpha = AATT$. Denote by S' the set of all initial DNA segments, written as words in $\mathcal{A}^*$, as described above. Let $\{\alpha_1, \ldots, \alpha_k\}$ be the set of all distinct overhangs produced by the restriction enzymes in E. For $i = 1, 2, \ldots, k$, let $\bar{\alpha}_i$ be a new symbol encoding the sequence α_i, and let $\Sigma = \{\bar{\alpha}_1, \ldots, \bar{\alpha}_k\}$. Extend $\mathcal{A}$ to $\bar{\mathcal{A}} = \mathcal{A} \cup \Sigma$ and define a simple H-system $\bar{H} = (\bar{\mathcal{A}}, S, \Sigma)$ where:

$$S = \{xu\bar{\alpha}_i u'y \mid xu\alpha_i u'y \in S' \text{and there exists } \epsilon \text{ such that } \rho_\epsilon = u\alpha_i u'$$
$$\text{and } \alpha_\epsilon = \alpha_i\}.$$

It is easy to see that the language $L(\bar{H})$ generated by the simple H-system $\bar{H}$ is the set of distinct molecules produced by the action of the set of endonucleases and a ligase on the set of molecules S'.

In order to calculate the entropy of the language $L(\bar{H})$, first we form the graph $G_{\bar{H}}$ as above. We change the graph $G_{\bar{H}}$ into a new graph G such that every edge in $G_{\bar{H}}$ labelled $\bar{\alpha} \in \Sigma$ is replaced by a path labelled α. For example, if an overhang $\alpha = AATT$ is produced by an enzyme, we replace the edge labelled $\bar{\alpha}$ in $G_{\bar{H}}$ by a path labelled $AATT$.

Proposition 3 *Let L be the language produced by a set of DNA segments, a finite set of restriction enzymes, and a ligase. Then $h(L) = \log \lambda$, where λ is the Perron value of the adjacency matrix of the graph G described above.*

5   DNA Hybridization and Tiling of the Integers

The connection between tiling of the integers and symbolic dynamics was examined by E. Coven, W. Geller, S. Silberger and W. Thurston [2]. As in tilings of the plane, let $\mathcal{P} = \{P_1, \ldots, P_k\}$ be a finite collection of finite subsets of the integers, i.e. *prototiles*. A translate of a prototile P by an integer t is called a *tile* and is denoted $t + P$. A set of prototiles $\mathcal{P}$ *tiles the integers* if and only if the integers are a union of mutually disjoint tiles, i.e. $Z = \bigcup(t_j + P_{i_j})$, where $(t_j + P_{i_j}) \cap (t_{j'} + P_{i_{j'}}) = \emptyset$ if $j \neq j'$. This expression is called a *tiling of the integers*. The tiling of the integers, $\bigcup(t_j + P_{i_j})$, corresponds to the point $x \in \{1, \ldots, k\}^Z$ defined by $x_i = j$ if and only if $i \in t_j + P_j$. It may be useful to think of the set $\{1, \ldots, k\}$ as a set of colors, and a tiling of the integers as a doubly infinite sequence of colors. The set of all doubly infinite sequences that are produced by tilings of the integers with a given set of prototiles will be called a *tiling shift*. (In [2], this set is called a tiling system.)

For example, let $\mathcal{P} = \{\{0\}, \{0, 2\}\}$. Associate distinct colors, say a and b, with the prototiles and think of the prototiles as a and b_b. Prototile a can be placed in the middle of prototile b_b to form bab, and prototile b_b can be shifted one space and placed within itself to form $bbbb$. Therefore, the resulting tiling shift is the set of all doubly infinite concatenations of the blocks a, bab and $bbbb$. There are (uncountably) many ways to tile the integers with these prototiles.

In [2] it was proved that the set of topological entropies of tiling shifts is the same as that of the set of shifts of finite type. We construct a set of prototiles that correspond to the nucleotides $\{A, G, C, T\}$ that make up DNA.

Consider the following prototiles:

$$
\begin{array}{ll}
A : & a\_a\_\_\_\_\_aa\_\_aa\_a\_a \\
T : & t\_t_____tt\_\_\_t\_t \\
C : & c\_c_____c\_\_\_c\_c \\
G : & g\_g\_\_\_\_\_gg\_ggg\_g\_g
\end{array}
$$

These prototiles encode the nucleotides in direction $5'$ to $3'$. In double-stranded DNA molecules, the second strand contains these prototiles written in reverse order, $3'$ to $5'$. We include four additional prototiles, $\bar{A}, \bar{G}, \bar{C}, \bar{T}$, written in the reverse orders of A, G, C, T. The complementary prototile of a prototile $X \in \{A, G, C, T\}$ is the prototile $\bar{Y}$, where Y is the Watson-Crick complement of X. The first four and the last four places of a prototile allows joining with other nucleotides in the same direction (say $5'$ to $3'$), representing the phosphodiester bonds. The second four places are empty. The first four positions of the complementary prototile occupy these four places. This allows the complementary prototiles to be joined in a sequence, representing the phosphodiester bonds of the second strand of a double stranded DNA molecule. The nucleotide itself is encoded by the remaining six places.

Proposition 4 *The prototiles $\{A, G, C, T, \bar{A}, \bar{G}, \bar{C}, \bar{T}\}$ tile the integers. Moreover, every tiling of the integers corresponds to a doubly infinite double stranded DNA molecule.*

We leave the proof of this proposition to the reader. It is not difficult, but there are many cases to be checked.

Since the prototiles are rather lengthy, we illustrate stacking of the tiles for only a small DNA segment. Consider $\begin{smallmatrix} 5' A & G & T^{3'} \\ 3' T & C & A_{5'} \end{smallmatrix}$. The corresponding stacking of tiles is:

$$a\_a\_t\_t\_aa\overline{tt}aa g a g u \overline{c}\overline{c}\overline{c} yy \overline{c} yyy tg tg a \overline{c}a\overline{c}a\overline{a}att\overline{aa}\_t\;t\;\bar{a}\;\bar{a},$$

where bars denote the complementary tiles that appear in the second strand.

It is easy to see from the construction that if a double-stranded DNA molecule is n base pairs long, then the corresponding stacked tile will have a $(14(n-1)+6)$-long body with no empty spaces in between the eight postions in the begining and the end, where the empty spaces alternate. Since the first strand determines the second, and every sequence of nucleotides within a DNA molecule is possible, the set of double-stranded DNA molecules can be thought of as $\{A, C, G, T\}^*$, the factor language of the full shift over a four letter alphabet. Hence the entropy of the set of all possible double-stranded molecules is $\log 4$, and the entropy of the tiling shift defined by the prototiles $\{A, G, C, T, \bar{A}, \bar{G}, \bar{C}, \bar{T}\}$ is $\frac{1}{14} \log 4$.

Recall the model of making longer DNA segments from shorter ones described in Section 3 (see Fig. 1). With each single-stranded DNA oligo we can associate a prototile (stacked tiles of A, G, C, T tiles). If the oligo has length n, then the corresponding prototile will have length $14n + 4$. From the proposition above, a tiling of the integers with these prototiles is possible if the oligos can form arbitrarily long double-standed DNA segments.

Let G be a graph and, as before, let $L(G)$ be the set of double-stranded DNA molecules that represent paths in the graph. The single-stranded DNA molecules that represent vertices and edges of G define prototiles made from the basic prototiles A, G, C, T and their complements. Let S be the tiling subshift obtained from these prototiles.

Proposition 5 *If the vertices of G are represented by DNA oligos of length n then the entropy of the tiling shift S is:*

$$h(S) = \frac{1}{14n} \log \lambda,$$

where λ is the Perron value of the matrix A_G.

References

[1] L. Adleman, Molecular computation of solutions of combinatorial problems, *Science*, 266 (1994), 1021–1024.

[2] E. Coven et al., The Symbolic Dynamics of Tiling the Integers, preprint.

[3] E. Coven and M. Paul, Sofic systems, *Israel Journal of Mathematics*, 20 (1975), 165–177.

[4] M. Garzon, N. Jonoska and S. Karl, Bounded complexity of DNA computing. To appear in *BioSystems*.

[5] T. Head, Formal language theory and DNA: an analysis of the generative capacity of specific recombinant behaviors, *Bulletin of Mathematical Biology*, 49 (1987), 737–759.

[6] T. Head, Gh. Păun and D. Pixton, Language theory and molecular genetics. In G. Rozenberg and A. Salomaa (eds.) *Handbook of Formal Languages*. Springer, Berlin, 1997, vol. 2, 295–358.

[7] E. Laun and K.J. Reddy, Wet splicing systems. In H. Rubin and D.H. Wood (eds.), *DNA Based Computers III*, DIMACS series AMS 48 (1999), 73–83.

[8] D. Lind and B. Marcus, *An Introduction to Symbolic Dynamics*. Cambridge University Press, New York, 1995.

[9] R. Lipton, DNA solution of hard computational problems, *Science*, 268 (1995), 542–545.

[10] A. de Luca and A. Restivo, A characterization of strictly locally testable languages and its applications to subsemigroups of a free semigroup, *Information and Control*, 44 (1980), 300–319.

[11] N. Morimoto, M. Arita and A. Suyama, Solid phase DNA solution to the Hamiltonian path problem. In H. Rubin and D.H. Wood (eds.), *DNA Based Computers III*, DIMACS series AMS 48 (1999), 193-206.

[12] Q. Ouyang et al., DNA solution to the maximal clique problem, *Science*, 278 (1997), 446–449.

[13] Gh. Păun, G. Rozenberg and A. Salomaa, *DNA Computing: New Computing Paradigms*. Springer, Berlin, 1998.

[14] M.P. Schützenberger, Sur certaines opérations de fermeture dans les langages rationnels, *Symposia Mathematica*, 15 (1975), 245–253. (Also in *RAIRO Information Theory*, 1 (1974), 55-61.)

Generalized Homogeneous P-Systems

Rudolf Freund

Institute of Computer Languages
Technical University of Vienna
Austria
rudi@logic.at

Franziska Freund

De La Salle Schools
Vienna, Austria
ffreund@logic.at

Abstract. Recently, Gheorghe Păun introduced P-systems as a new model for computations based on membrane structures. Many variants of P-systems have been shown to have universal computational power. Using the membranes as a kind of filter when transferring specific objects into an inner compartment or out into the surrounding compartment turned out to be a very powerful mechanism if combined with suitable rules that are applied within the membranes in the model of generalized P-systems, GP-systems for short. GP-systems can simulate graph controlled grammars of arbitrary type based on productions working on single objects; moreover, several variants of GP-systems that use splicing or cutting and recombination of strings were shown to have universal computational power, too. In this paper, we consider GP-systems with homogeneous membrane structures, GhP-systems for short, that cut and recombine string objects with specific markers at the ends of the strings which can be interpreted as electrical charges; these electrical charges determine the permeability of the membranes to the string objects. We show that GhP-systems have universal computational power; moreover, a very restricted variant of GhP-systems characterizes minimal linear languages.

1   Introduction

The most important feature of the P-systems model –as introduced by Gheorghe Păun in [8]– is the membrane structure (for a chemical variant of this idea see [1]) consisting of membranes hierarchically embedded in the outermost *skin* membrane. Every membrane encloses a *region* that may contain other membranes; the part delimited by the membrane labelled by k and its inner membranes is called *compartment* k. A region delimited by a membrane may enclose not only other membranes but also specific objects and operators, which in general are considered as multisets, as well as evolution rules, which in *generalized P-systems (GP-systems)* as introduced in [3] and [4] are evolution rules for the operators. GP-systems take into account ground operators as well as transfer operators (simple rules of this kind are called travelling rules in [11]); these transfer operators transfer objects or operators (or even rules) either to the outer compartment or to an inner compartment delimited by a specific membrane, and also check for permitting and/or forbidding conditions on the objects to be transferred. In this way, the membranes act like a test-tube filter system (e.g. see [9] and [6]). In [4] it was shown how GP-systems with splicing or cutting and recombination rules can simulate test tube systems using the corresponding type of molecular rules.

In the specific model of generalized P-systems with special homogeneous membrane structures considered in this paper, the objects (which are assumed to be available in an unbounded number) can pass the membranes at a specific depth of the membrane structure depending on their electrical charges only. In contrast to the original definition of P-systems, in G(h)P-systems no priority relations on the rules are used, and we do not enforce parallelism guarded by a universal clock, because these features are not motivated biologically and their implementation, at least "in vitro", seems to be very difficult ("in silicio" these features might be implemented more easily). On the other hand, both the membrane structure and the operations used therein are motivated by nature; yet despite this biological background, real implementations in the lab ("in vitro") or in electronic media ("in silicio") remain topics for future interdisciplinary research.

In the section below, we start with some preliminary notions from formal language theory and then give a general definition of a molecular system that also captures the notion of cutting/recombination systems of strings. In section three, we introduce the model of GhP-systems considered in this paper and, as a first result, we show that a restricted variant of GhP-systems characterizes minimal linear languages. Our main result which shows that GhP-systems have universal computational power, is discussed in section four. A short summary and some suggestions for future research topics conclude the paper.

2 Preliminary Definitions

We denote by V^* the free monoid generated by an alphabet V under the operation of concatenation; the *empty string* is denoted by λ, and $V^* \setminus \{\lambda\}$ is denoted by V^+. Any subset of V^+ is called a λ-*free (string) language*. $\mathbf{N}$ denotes the set of non-negative integers.

A minimal linear grammar G is a quadruple $(\{S\}, V_T, P, S)$, where S is the only non-terminal symbol and the start symbol of the grammar, V_T is the terminal alphabet, and P is the set of linear productions of the forms $S \to w$ with $w \in V_T^+$ or $S \to uSw$ with $uw \in V_T^+$.

A *molecular system* is a quadruple $\sigma = (B, B_T, P, A)$, where B and B_T are sets of *objects* and *terminal objects*, respectively, with $B_T \subseteq B$, P is a set of *productions*, and A is a set of axioms from B. A production p in P in general is a partial recursive relation $\subseteq B^k \times B^m$, for some $k, m \geq 1$, where we also demand that the domain of p is recursive (i.e. given $w \in B^k$, it is decidable if there exists some $v \in B^m$ with $(w, v) \in p$) and, moreover, that the range for every w is finite, i.e. for any $w \in B^k$, $card(\{v \in B \mid (w, v) \in p\}) < \infty$. For any two sets L and L' over B, we say that L' is computable from L by a production p if and only if $\{w_1, ..., w_k\} \subseteq L$ and $L' = L \cup \{v_1, ..., v_m\}$, for some $(w_1, ..., w_k) \in B^k$ and $(v_1, ..., v_m) \in B^m$ with $(w_1, ..., w_k, v_1, ..., v_m) \in p$; we also write $L \Longrightarrow_p L'$ and $L \Longrightarrow_\sigma L'$. A computation in σ is a sequence $L_0, ..., L_n$ such that $L_i \subseteq B$, $0 \leq i \leq n$, $n \geq 0$, as well as $L_i \Longrightarrow_\sigma L_{i+1}$, $1 \leq i \leq n$; in this case we write $L_0 \Longrightarrow_\sigma^n L_n$, and we also write $L_0 \Longrightarrow_\sigma^* L_n$ if $L_0 \Longrightarrow_\sigma^n L_n$ for some $n \geq 0$. The *language generated by* σ is:

$$L(\sigma) = \{w \mid A \Longrightarrow_\sigma^* L, \ w \in L \cap B_T\}.$$

The special productions on string objects that we shall consider below are the cutting and recombination operations.

A *cutting/recombination scheme* (a *CR-scheme* for short) is a quadruple (V, M, C, R), where V is a finite alphabet; M is a finite set of *markers*; V and M are disjoint sets; C is a set of *cutting rules* of the form $u\#l\$m\#v$, where $u \in V^* \cup MV^*$, $v \in V^* \cup V^*M$, and $m, l \in M$, and $\#, \$$ are special symbols not in $V \cup M$; $R \subseteq M \times M$ is the recombination relation representing the *recombination rules*. Cutting and recombination rules are applied to objects from $O(V, M)$, where we define:

$$O(V, M) = V^+ \cup MV^* \cup V^*M \cup MV^*M.$$

For $x, y, z \in O(V, M)$ and a cutting rule $c = u\#l\$m\#v$, we define $x \Longrightarrow_c$ (y, z) if and only if for some $\alpha \in V^* \cup MV^*$ and $\beta \in V^* \cup V^*M$ we have $x = \alpha uv\beta$ and $y = \alpha ul$, $z = mv\beta$. For $x, y, z \in O(V, M)$ and a recombination rule $r = (l, m)$ from R, we define $(x, y) \Longrightarrow_r z$ if and only if for some $\alpha \in V^* \cup MV^*$ and $\beta \in V^* \cup V^*M$ we have $x = \alpha l$, $y = m\beta$, and $z = \alpha\beta$. For a CR-scheme $\sigma = (V, M, C, R)$ and any language $L \subseteq O(V, M)$, $\sigma(L)$ then denotes the set of all objects obtained by applying one cutting or one recombination rule to objects from L. We also define $\sigma^0(L) = L$ and $\sigma^{i+1}(L) = \sigma(\sigma^i(L))$ for all $i \geq 0$, as well as $\sigma^{(0)}(L) = L$ and $\sigma^{(i+1)}(L) = \sigma^{(i)}(L) \cup \sigma(\sigma^{(i)}(L))$ for all $i \geq 0$; moreover, we denote $\sigma^*(L) = \bigcup_{i=0}^{\infty} \sigma^{(i)}(L)$. An *extended CR-system* is a molecular system of type CR σ, $\sigma = (O(V, M), O(V_T, M_T), P, A)$, where $V_T \subseteq V$ is the set of terminal symbols, $M_T \subseteq M$ is the set of terminal markers, A is the set of axioms, P is the union of the relations (productions) defined by the cutting rules from C ($\subseteq O(V, M) \times O(V, M)^2$) and the recombination rules from R ($\subseteq O(V, M) \times O(V, M)$), and (V, M, C, R) is the underlying CR-scheme.

Throughout this paper we shall restrict ourselves to markers that can be interpreted as electrical charges of ions, i.e. we shall write $[+k]$ and $[-k]$, $k \in \mathbf{N}$, for these special markers. In this sense, the recombination rules we use will be of the simple forms $([+k], [-k])$ and $([-k], [+k])$, $k \in \mathbf{N}$.

In [6] we showed that test tube systems with only two test tubes that use cutting and recombination rules in the tubes and filters of a special type between the tubes have the computational power of arbitrary grammars and Turing machines, respectively (which also holds true for test tube systems with splicing rules).

3   Generalized Homogeneous P-Systems

In this section, we quite informally describe the model of GhP-systems discussed in this paper. Only the features not captured by the original model of P-systems as described in [2], [7], and [8] will be defined in more detail. In these papers, only the number of symbols are counted in the multiset sense, whereas in [10] at least the outputs are strings. In generalized P-systems (for the basic definition of GP-systems the reader is referred to [3] and [4]) the objects are usually strings or graphs, etc.

The basic ingredient of a $(G(h))$P-system is a *membrane structure* that consists of several membranes within one unique surrounding membrane, the so-called skin membrane. All the membranes can be labelled (in a one-to-one manner) by natural numbers; the outermost membrane (skin membrane) is always labelled by 0. In this way, a membrane structure can be uniquely described by a string of correctly matching parentheses, in which each pair corresponds to a membrane. For example, the membrane structure depicted in Figure 1 shows a skin membrane that contains two inner membranes, 1 and 2, which in turn contain membranes 3 and 4, respectively. This structure is described by $[_0[_1[_3]_3]_1[_2[_4]_4]_2]_0$. Figure 1 also shows that a membrane structure can be graphically represented by a Venn diagram, where two sets can be disjoint or one set can be the subset of the other. In this representation, every membrane encloses a *region* that may contain other membranes; the part delimited by the membrane labelled by k and its inner membranes is called *compartment k* below. The space outside the skin membrane is called the *outer region*.

Informally, in [7] and [8] *P-systems* were defined as membrane structures containing multisets of objects in the compartments k as well as evolution rules for the objects. A priority relation on the evolution rules guarded the application of the evolution rules to the objects, which had to be affected in parallel (if possible according to the priority relation). The output was obtained in a designated compartment from a halting configuration (i.e. a configuration of the system in which rules can no longer be applied).

A *generalized homogeneous P-system (GhP-system) of molecular type X* is a construct γ, $\gamma = (B, B_T, P, A, \mu, I, in, out)$, where:

- (B, B_T, P, A) is a molecular system of type X (we shall restrict ourselves to molecular systems of type CR in the following);

- μ is a membrane structure (with the membranes labelled by natural numbers $0, ..., n$);

- $I = (I_0, ..., I_n)$, where I_k is the initial content of compartment k containing a set of objects from A as well as a set of rules from P;

- *in* for each $j \in \{0, 1, ..., n\}$ specifies a condition an object must fulfill in order to be able to pass into the inner compartment of a membrane k, $k \in \{1, ..., n\}$, in the region enclosed by membrane j;

- *out* specifies a condition an object must fulfill in order to be able to pass a membrane k, $k \in \{1, ..., n\}$, into its surrounding compartment.

According to the definition given above, we only allow one unique out-condition, whereas the in-conditions specified by *in* may depend on the region the membrane to be passed lies in; yet in the following we will also demand that the in-conditions of regions at the same level of the membrane structure coincide.

A *computation* in γ starts with an initial configuration in which I_k are the contents of compartment k. A transition from one configuration to another is performed by applying a rule (from P) in I_k to an object in compartment

k or by moving an element out of a compartment k or into a compartment k according to the conditions given by *out* and *in*. The language generated by γ is the set of all terminal objects $w \in B_T$ obtained in the terminal compartment 0 by some computation in γ.

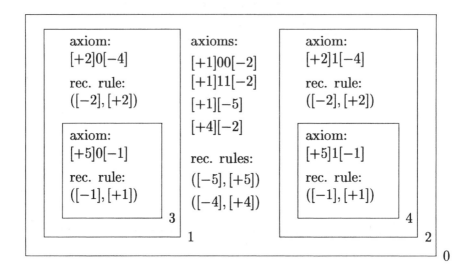

Figure 1: GhP-system generating $\{ww^r \mid w \in \{0,1\}^+\}$.

Example 1 *The main ingredients of a GhP-system (of type CR) generating the language* $\{ww^r \mid w \in \{0,1\}^+\}$ *(where w^r denotes the mirror image of w) are depicted in Figure 1. Moreover, the permeability of the membranes is specified in such a way that objects with a sum of electrical charges equal to $+1$ can leave every compartment, whereas objects with a sum of electrical charges equal to -1 and -3 can enter compartments $1, 2$ and $3, 4$, respectively. In the skin membrane, we start with (terminal) objects of the form $[+1]\,ww^r\,[-2]$ for some $w \in \{0,1\}^+$; these can enter compartment 1 (or 2), where the application of the recombination rule yields $[+1]ww^r0[-4]$ ($[+1]ww^r1[-4]$), which can pass into compartment 3 (4), where the application of the corresponding recombination rule yields $[+5]0ww^r0[-4]$ ($[+5]1ww^r 1[-4]$). These objects can pass the membranes back to compartment 0, where the objects are reduced to the terminal objects $[+1]0ww^r0[-2]$ ($[+1]1ww^r 1[-2]$).* □

It is easy to see that due to the different electrical charges of the axioms and the intermediate objects evolving in compartments 1 and 3 (2 and 4) we could even take the union of these compartments and work in the simpler membrane structure $[_0[_1]_1[_2]_2]_0$ (compare with Figure 2).

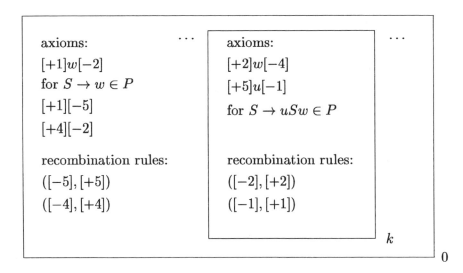

Figure 2: GhP-system generating $L(G)$, where $G = (\{S\}, V_T, P, S)$ is a minimal linear grammar.

Lemma 1 *Any minimal linear language generated by a minimal linear grammar G, $G = (\{S\}, V_T, P, S)$, can be generated by a GhP-system with membrane structure $[_0[_1]_1...[_n]_n]_0$, where n is the number of linear productions of the form $S \to uSw$, with $uw \in V_T^+$ in P.*

Proof. As permeability conditions for the membranes, we specify that objects with a sum of electrical charges equal to $+1$ can leave every compartment k, $1 \leq k \leq n$, whereas objects with a sum of electrical charges equal to -1 can enter any compartment k. The main parts of the GhP-system are depicted in Figure 2. The terminal objects $w \in L(G)$ can be extracted from compartment 0 in the form $[+1]w[-2]$. $\square$

Without proof, we mention that GhP-systems of the form described in the preceding lemma exactly characterize minimal linear languages.

4 The Computational Power of GhP-Systems

The main result we prove below establishes the universal computational power of GhP-systems.

Theorem 1 *Any recursively enumerable language L can be generated by a GhP-system γ with membrane structure:*

$$[_0[_1[_{n+1}]_{n+1}]_1...[_n[_{2n}]_{2n}]_n]_0.$$

Proof. The idea of the proof –to simulate the productions of a grammar that generates L as in Post systems normal form– has already been used in many papers on splicing systems and cutting/recombination systems (see [9] and [6]). Instead of a grammar G' generating L, we consider a grammar G generating the language $L\{d\}$, where the end marker d in any derivation of a word $w'd$, for $w' \in L$, is generated exactly in the last step of this derivation in G. A string w appearing in a derivation of such a grammar G, $G = (V_N, V_T, P, S)$, generating $L\{d\}$ is represented by its rotated versions $[+1]\, Xw_2Bw_1Y\,[-1]$, where $w = w_1w_2$, B is a special symbol indicating the beginning of the string within the rotated versions and X, Y are special symbols marking the ends of the strings. Final strings first appear in the form $[+1]\, XBw'dY\,[-1]$, where w' is the final result from L which we want to get, and finally appear as $[+1]\, w'\,[-1]$ in compartment 0.

In compartment 0 we start with the axiom $[+1]XBSY[-1]$; the objects passed back to compartment 0 are objects of the form $[+1]\, Xw_2Bw_1Y\,[-1]$, $w_2w_1 \in (V_N \cup V_T \cup \{d\})^+$, or terminal objects of the form $[+1]w'[-1]$, $w' \in V_T^+$. Whereas the last region $[_n[_{2n}]_{2n}]_n$ is used to obtain the terminal objects $[+1]\, w'\,[-1]$, the other regions $[_k[_{n+k}]_{n+k}]_k$, $1 \le k < n$, are used to simulate a production $\alpha \to \beta$ or the rotation of a symbol b (thus simulating a production $b \to b$) in the following way:

In compartment k, $1 \le k < n$, we use the cutting rules $u\#[-2]\$[+2]$ $\#\alpha Y[-1]$ and $[+1]X\#[-4]\$[+4]\#v$ for suitable strings u, v in order to obtain $[+4]w[-2]$ from $[+1]Xw\alpha Y[-1]$. This object $[+4]w[-2]$ with the sum of electrical charges equal to $+2$ can pass through membrane $n + k$. In compartment $n + k$ we then obtain the string object $[+1]X\beta wY[-1]$ by using the axioms $[+1]X\beta[-4]$ and $[+2]Y[-1]$ as well as the corresponding recombination rules $([-4], [+4])$ and $([-2], [+2])$.

The "neutral" objects with the sum of electrical charges equal to 0 can leave every compartment j, $1 \le j \le 2n$, and re-enter every compartment k, $1 \le k \le n$, from compartment 0.

Due to lack of space, the remaining technical details of the proof have to be left to the reader.   □

Obviously, we can encode arbitrary additional data u in the axiom, i.e. we can take $[+1]XBSuY[-1]$ instead of $[+1]XBSY[-1]$ in compartment 0; hence, from the preceding theorem we also obtain the following result:

Corollary 1 *Any partial recursive function can be computed by a GhP-system.*

5   Conclusion

The idea of membrane structures offers a nearly unlimited variety of variants as has already been pointed out in [8]. In this paper, we have considered homogeneous (static) membranic structures that contain cutting and recombination rules and which are only permeable to those objects that depend on the electrical charges of the string objects. The formal definition of molecular systems also allows us to consider objects other than strings, e.g. graphs and the corresponding cutting and recombination rules (see [5]). A thorough investigation of the generative power of such variants of GhP-systems and their complexity for solving special problems remains for future research.

Acknowledgements

We gratefully acknowledge all the interesting and fruitful discussions with Gheorghe Păun concerning his brilliant ideas for various models of P-systems.

References

[1] G. Berry and G. Boudol, The chemical abstract machine, *Theoretical Computer Science*, 96 (1992), 217–248.

[2] J. Dassow and Gh. Păun, On the Power of Membrane Computing, TUCS Research Report 217, December 1998.

[3] R. Freund, Generalized P-systems. In *Proceedings of Fundamentals of Computation Theory, FCT'99*, Ia i, 1999.

[4] R. Freund, Generalized P-systems with splicing and cutting/recombination. In *Proceedings of the Workshop on Formal Languages and Automata, WFLA'99*, Ia i, 1999.

[5] R. Freund and F. Freund, Cutting and recombination of graphs. In *Proceedings of Automata and Formal Languages, AFL'99*, Vasszeczeny, 1999.

[6] R. Freund and F. Freund, Test tube systems: when two tubes are enough. In *Proceedings of Developments in Language Theory, DLT'99*, Aachen, 1999.

[7] Gh. Păun, Computing with Membranes, TUCS Research Report 208, November 1998.

[8] Gh. Păun, Computing with membranes: an introduction, *Bulletin of the European Association for Theoretical Computer Science*, 67 (February 1999), 139–152.

[9] Gh. Păun, G. Rozenberg and A. Salomaa, *DNA Computing: New Computing Paradigms*. Springer, Berlin, 1998.

[10] Gh. Păun, G. Rozenberg and A. Salomaa, Membrane Computing with External Output, TUCS Research Report 218, December 1998.

[11] I. Petre, A normal form for P-systems, *Bulletin of the European Association for Theoretical Computer Science*, 67 (February 1999), 165–172.

Crossing-Over on Languages: A Formal Representation of Chromosomes Recombination

Lucian Ilie

Victor Mitrana

Faculty of Mathematics
University of Bucharest
Romania
{ilie,mitrana}@funinf.math.unibuc.ro

Abstract. We propose that the crossing-over operation on languages is a formal model of the linkage and recombination of genes in chromosomes, and we present results concerning the power of this operation on the languages in the Chomsky's hierarchy.

1 Introduction

One was observed that the linkage between genes were never complete because of the exchange events between homologous chromosomes. This recombination process by exchanging of segments between homologous chromosomes was called crossing-over.

Each gene occupies a well-defined site or locus in its chromosome, and has corresponding locations in the pair of homologous chromosomes. Crossing-over has the following features:

1. The exchange of segments occurs after the chromosomes have replicated.

2. The exchange process involves a breaking and rejoining of the two chromatids, which results in the exchange of equal and corresponding segments between them.

3. Crossing-over occurs more or less at random throughout the length of a chromosome pair.

We introduce an operation, which has the same above features, on strings and languages. The operation is applicable to a pair of strings of the same length as the crossing-over between homologous chromosomes.

Each string is cut into several fragments, in the same sites and the fragments are crossed by ligases. A new string, of the same length, is formed by starting at the left end of one parent, copying a segment, crossing over to the next site in the other parent, copying a substring, crossing back to the first parent and so on until the right end of one parent is reached. Obviously, a new string can be obtained by starting with the other parent.

We should point out that the crossing over on words is similar to the crossing over on chromosomes:

- the words are of the same length,

- the segments which exchange with each other are of the same length,

- the number of sites in strings where the crossing-over occurs is arbitrarily large,

- the sites in strings where crossing-over ocurs are spread randomly along the strings.

In [1] T. Head introduced an operation called splicing with motivations related to DNA recombinant behaviours; we refer to this paper and also to [2] for more motivations and for further references.

The papers [4], [7], [6], [8], [9] deal with different types of splicing systems. A generalization of the splicing systems is proposed in [5] namely, crossover systems.

Our approach differs from the others in the following aspects:

- all the other operations can be applied to pairs of strings of different length,

- the number of sites where the strings are cut is limited to one in [4], [7], [6], [8], [9] or to a predefined integer in [5],

- the corresponding fragments are of different lengths,

- the rules for splitting are given as finite or infinite sets of pairs, quadruples, or n-tuples.

We consider that our operation more accurately reflects the crossing-over between chromosomes (the real situation).

In this paper, we consider the crossing-over operation and we study the relation between this operation and other operations in formal language theory.

In the next section we shall give the basic definitions. Section 3 contains some examples and counterexamples.

We then investigate the power of the non-iterated case of the crossing over operation and establish a strong conection between crossing-over and other operations in formal language theory and parallel programming theory.

The last section deals with the iterated case, where some open problems are formulated.

2 Basic Definitions

Let V be an alphabet; we denote by V^* the set of all strings over V, by ϵ the empty word, $V^+ = V^* - \{\epsilon\}$, and by $|w|$ the length of $w \in V^*$. The families of finite, regular, linear, context-free, context-sensitive, and recursively enumerable languages are denoted by FIN, REG, LIN, CF, CS, RE, respectively. For further elements of formal language theory, the reader is referred to [3], [10].

Let M be a finite subset of V^+ whose elements are called *markers* and w_1, w_2 be two strings of equal length in V^*. (Observe that ϵ cannot be a marker.)

We define the relation:

$$(w_1, w_2) \Longrightarrow_M w$$

iff the following conditions hold:

(i) $w_1 = x_1 x_2 \ldots x_n, w_2 = y_1 y_2 \ldots y_n$, for some $n \geq 2$, $x_i, y_i \in V^*, 1 \leq i \leq n$,

(ii) $|x_i| = |y_i|$, for any $2 \leq i \leq n$,

(iii) $Pref(x_i) \cap M \neq \emptyset, Pref(y_i) \cap M \neq \emptyset$, for any $2 \leq i \leq n$ (for a string x, $Pref(x)$ denotes the set of all prefixes of x),

(iv) $w = \begin{cases} x_1 y_2 x_3 y_4 \ldots x_{n-2} y_{n-1} x_n, & \text{if } n \text{ is odd}, \\ x_1 y_2 x_3 y_4 \ldots y_{n-2} x_{n-1} y_n, & \text{if } n \text{ is even}. \end{cases}$

Remark. Obviously $(w_2, w_1) \Longrightarrow_M y_1 x_2 y_3 x_4 \ldots$.

For a pair of strings (w_1, w_2) we denote:

$$CO_M(w_1, w_2) = \begin{cases} \{w \mid (w_1, w_2) \Longrightarrow_M w\}, & \text{if } |w_1| = |w_2|, \\ \emptyset, & \text{otherwise}. \end{cases}$$

For a language $L \subseteq V^*$ we define the *non-iterated crossing-over* operation as follows:

$$CO_M(L) = L \cup \bigcup_{w_1, w_2 \in L} CO_M(w_1, w_2).$$

If we denote:

$$CO_M^1(L) = CO_M(L),$$
$$CO_M^{i+1}(L) = CO_M(CO_M^i(L)), \text{ for any } i \geq 1,$$

then the *iterated crossing-over* operation is:

$$CO_M^*(L) = \bigcup_{n \geq 1} CO_M^n(L).$$

For a family F of languages, we denote:

$$CO(F) = \{CO_M(L) \mid L \in F, M \in FIN, \epsilon \notin M\},$$
$$CO^*(F) = \{CO_M^*(L) \mid L \in F, M \in FIN, \epsilon \notin M\}.$$

We say that the family F is closed under the CO or CO^* operation if and only if $CO(F) = F$ or $CO^*(F) = F$, respectively.

We also give the definitions of two well known operations in formal language theory and parallel programming theory which are connected with our operation, as will be seen in the following sections. The first operation is defined, for strings $x, y \in V^*$, as follows:

$$Shuf(x,y) = \{x_1 y_1 x_2 y_2 \ldots x_p y_p \mid x = x_1 \ldots x_p, y = y_1 \ldots y_p,$$
$$p \geq 1, x_i, y_i \in V^*, 1 \leq i \leq p\}.$$

The second operation is defined only for strings $x, y \in V^*$ of equal length in the following way.

$$SShuf(x,y) = a_1 b_1 a_2 b_2 \ldots a_p b_p, a_i, b_i \in V, 1 \leq i \leq p, x = a_1 \ldots a_p,$$
$$y = b_1 \ldots b_p.$$

For two languages $L_1, L_2 \subseteq V^*$, we define:

$$Shuf(L_1, L_2) = \bigcup_{x \in L_1, y \in L_2} Shuf(x, y),$$

$$SShuf(L_1, L_2) = \{SShuf(x,y) \mid x \in L_1, y \in L_2, |x| = |y|\}.$$

3   Examples and Counterexamples

Example 1 *For every language $L \subseteq V^*$ we can find a (finite) set of markers M such that $CO_M(L) = CO_M^*(L) = L$. Indeed, if we take a new symbol $c \notin V$ and put $M = \{c\}$, then clearly we have the equality above.*

Example 2 *Consider the alphabet $V = \{a, b, c, d, e, f\}$, the set of markers $M = \{f\}$, and the language $L = \{a^n f d^m f c^m \mid m, n \geq 1\} \cup \{a^m f b^n f e^n \mid m, n \geq 1\}$. We obtain:*

$$CO_M(L) = CO_M^*(L) = L \cup \{a^n f b^n f c^n \mid n \geq 1\} \cup \{a^n f d^n f e^n \mid n \geq 1\},$$

because the only way to obtain new strings is to take $w_1 = a^n f d^m f c^n$, $w_2 = a^n f b^n f e^n$, $n \geq 1$, the derivations being $(w_1, w_2) \Longrightarrow_M a^n f b^n f c^n$, $(w_2, w_1) \Longrightarrow_M a^n f d^n f e^n$.

Example 3 *For an alphabet* $V = \{a_1, a_2, \ldots, a_n\}$, *consider* $M = V$ *and* $L = \cup_{i=1}^n \{a_i^m \mid m \geq 0\}$. *It follows that:*

$$CO^*(L) = V^*.$$

One of the inclusions is obvious. For the other one, consider a string $w = a_{i_1} a_{i_2} \ldots a_{i_p} \in V^*$. *The string* w *can be obtained from strings in* L, *using markers in* M *as follows:*

$$
\begin{aligned}
(a_{i_1}^p, a_{i_2}^p) &\Longrightarrow_M a_{i_1} a_{i_2}^{p-1}, \\
(a_{i_1} a_{i_2}^{p-1}, a_{i_3}^p) &\Longrightarrow_M a_{i_1} a_{i_2} a_{i_3}^{p-2}, \\
(a_{i_1} a_{i_2} a_{i_3}^{p-2}, a_{i_4}^p) &\Longrightarrow_M a_{i_1} a_{i_2} a_{i_3} a_{i_4}^{p-3}, \\
&\ldots, \\
(a_{i_1} a_{i_2} \ldots a_{i_{p-2}} a_{i_{p-1}}^2, a_{i_p}^p) &\Longrightarrow_M a_{i_1} a_{i_2} \ldots a_{i_p} = w.
\end{aligned}
$$

The following observation is obvious but important, not only here but also in sections below. For any strings $w_1, w_2, w \in V^*$ *with* $(w_1, w_2) \Longrightarrow_M w$ *(for arbitrary* V *and* M*), the characters in* w_1 *and* w_2 *preserve their places, i.e. the* k-th *character of* w, $1 \leq k \leq |w|$, *is the* k-th *character of* w_1 *or* w_2. *It follows that, for any language* $L \subseteq V^*$ *with* $CO_M^*(L) = V^*$ *for some set of markers* $M \subseteq V^*$, *the number of strings of the same length* p *in* L, $p \geq 1$, *is bigger or equal to the cardinality of* V.

In Example 3, the language L has this property (L has exactly $n = |V|$ strings of the same length p, for any $p \geq 1$) and M is minimal in the sense that there is no language $L' \subseteq V^*$ such that:

(i) for any $p \geq 1$, the number of strings of length p in L' is $n = |V|$,
(ii) there is a finite set $M' \subseteq V^+$ such that $CO_{M'}^*(L) = V^*$ and $|M'| < |M|$.

Suppose that there are some L', M' satisfying the conditions above.

For any fixed $p \geq 2$ and $1 \leq i \leq p - 1$, the number of combinations of two characters on the positions i and $i + 1$ in strings of length p in L is at most n (the number of all strings of length p in L) while the same number for V^* is, obviously, n^2. So, at least $n^2 - n$ combinations as above have to be obtained by derivations using markers in M'. Take a marker $m \in M'$ and denote by m_1 the first character of m. Because any new combination as above obtained by using m has m_1 as its second character (on the position $i+1$) and because we must have a combination with the second character m_1 from the beginning, in L, it follows that, using any marker in M', we cannot produce more than $n - 1$ such new combinations. We get that $(n - 1)|M'| \geq n^2 - n$ so $|M'| \geq n = |M|$, a contradiction which proves the minimality (in the mentioned sense) of M.

Example 4 *The language* $L = \{a^n b^n \mid n \geq 0\}$ *can not be obtained from some* $L' \subseteq \{a, b\}^*$ *and some finite* $M \subseteq \{a, b\}^+$ *in a proper way, i.e.* $CO_M(L') = L$ *and* $L' \subset L$. *If we have* $w_1, w_2 \in L', w \notin \{w_1, w_2\}$ *with* $(w_1, w_2) \Longrightarrow_M w$, *then there will be some* $n \geq 0$ *such that* $w_1 = w_2 = a^n b^n$ *(because* $|w_1| = |w_2|$*) and, as* $|w| = |w_1|$*, it follows that* $w \notin L$*; the affirmation is proved.*

In fact, any language which contains at most one string of any given length can not be obtained in a proper way in the sense above.

4   Non-Iterated Crossing-Over

In this section we investigate the power of the non-iterated crossing-over operation. Remember that a trio is a family of languages closed under ϵ-free homomorphisms, inverse homomorphisms, and intersection with regular sets. A full trio is a trio closed under arbitrary homomorphisms.

Lemma 1 *For any family* F *of languages,* $F \subseteq CO(F)$.

Proof. See Example 1.   □

Theorem 1 *The families FIN and RE are closed under the CO operation.*

Proof. Obviously, $CO_M(L)$ is finite for any set of markers M and any finite language L. Using Lemma 1, the closure of the family FIN is proved. For the family RE, we use Lemma 1 and Church's thesis.   □

Lemma 2 *If a family* F *is closed under limited erasing, inverse homomorphisms, intersection with regular sets, and SShuf operation, then* $CO(F) \subseteq F$.

Proof. Consider an alphabet V, a language L over V, and a (finite) set of markers $M \subseteq V^+, M = \{x_1, x_2, \ldots, x_m\}$.

For any character $a \in V$, we consider a new symbol $a' \notin V$ and denote $V' = \{a' \mid a \in V\}$. Consider also $2m$ new symbols $c_i, c_i', 1 \leq i \leq m$.

In the following, when we put a prime to a word (language) we shall understand that in this word (language) all the symbols are primed. For instance, if $w = a_1 a_2 \ldots a_n \in V^*$, then $w' = a_1' a_2' \ldots a_n'$ and if $A \subseteq V^*$, then $A' = \{w' \mid w \in A\}$.

Take the homomorphisms:

$$h_1 : (V \cup \{c_i \mid 1 \leq i \leq m\})^* \longrightarrow V^*, \ h_1(a) = a, \text{ for any } a \in V,$$
$$h_1(c_i) = x_i, \text{ for any } 1 \leq i \leq m,$$
$$h_2 : (V \cup \{c_i \mid 1 \leq i \leq m\})^* \longrightarrow (V \cup \{c_i \mid 1 \leq i \leq m\})^*,$$
$$h_2(a) = a, \text{ for any } a \in V,$$
$$h_2(c_i) = c_i x_i, \text{ for any } 1 \leq i \leq m,$$
$$h_3 : (V' \cup \{c_i' \mid 1 \leq i \leq m\})^* \longrightarrow V'^*, \ h_3(a') = a', \text{ for any } a' \in V',$$
$$h_3(c_i') = x_i', \text{ for any } 1 \leq i \leq m,$$
$$h_4 : (V' \cup \{c_i' \mid 1 \leq i \leq m\})^* \longrightarrow (V' \cup \{c_i' \mid 1 \leq i \leq m\})^*,$$
$$h_4(a') = a', \text{ for any } a' \in V',$$
$$h_4(c_i') = c_i' x_i', \text{ for any } 1 \leq i \leq m.$$

The language:

$$L_1 = SShuf(h_2(h_1^{-1}(L)), h_4(h_3^{-1}(L'))) \cap \left(\bigcup_{i,j=1}^{n} (VV')^* \{c_i c_j'\} \right)^* (VV')^*$$

is in F. A string in L_1 has the form:

$$a_{p_{1,1}} a_{q_{1,1}}' a_{p_{1,2}} a_{q_{1,2}}' \cdots a_{p_{1,n_1}} a_{q_{1,n_1}}' c_{i_1} c_{j_1}' a_{p_{2,1}} a_{q_{2,1}}' \cdots a_{p_{2,n_2}} a_{q_{2,n_2}}' c_{i_2} c_{j_2}' \cdots$$
$$c_{i_k} c_{j_k}' a_{p_{k+1,1}} a_{q_{k+1,1}}' \cdots a_{p_{k+1,n_{k+1}}} a_{q_{k+1,n_{k+1}}}',$$

where, for any $1 \leq r \leq k+1, 1 \leq s \leq n_r, a_{p_{r,s}} \in V, a_{q_{r,s}}' \in V'$, and, for any $2 \leq r \leq k+1$, there are some $m_r, l_r, 1 \leq m_r, l_r \leq n_r$ such that:

$$a_{p_{r,1}} a_{p_{r,2}} \cdots a_{p_{r,m_r}} = x_{i_{r-1}},$$
$$a_{q_{r,1}}' a_{q_{r,2}}' \cdots a_{q_{r,l_r}}' = x_{j_{r-1}}'.$$

Construct a ϵ-free gsm g which leads a string in L_1 as above to the string:

$$a_{p_{1,1}} a_{q_{1,1}}' a_{p_{1,2}} a_{q_{1,2}}' \cdots a_{p_{1,n_1}} a_{q_{1,n_1}}' c_{i_1} c_{j_1}' a_{p_{2,1}}' a_{q_{2,1}} a_{p_{2,2}}' a_{q_{2,2}} \cdots$$
$$a_{p_{2,n_2}}' a_{q_{2,n_2}} c_{i_2} c_{j_2}' a_{p_{3,1}} a_{q_{3,1}}' \cdots a_{p_{3,n_3}} a_{q_{3,n_3}}' c_{i_3} c_{j_3}' a_{p_{4,1}} a_{q_{4,1}}' \cdots.$$

In order to do that task, g works as follows:

- Start scanning the string in the state s_0.

- Leave all symbols $a_{p_{r,s}}, a_{q_{r,s}}'$ unchanged until either the right end of the string is detected (in which case **stop**) or the next c_i is reached.

- Leave c_i and c_j' unchanged but change the state into s_1.

- Change any $a_{p_{r,s}}$ into $a_{p_{r,s}}'$ and any $a_{q_{r,s}}'$ into $a_{q_{r,s}}$ until either the right end of the string is detected (in which case **stop**) or the next c_i is reached.

- Leave c_i and c_j' unchanged but change the state into s_0.

- Restart the work from the second step for the remaining part of the initial string.

Notice that both states of g are final.

Consider now the homomorphism:

$$
\begin{aligned}
&h_5 : (V \cup V' \cup \{c_i \mid 1 \le i \le m\} \cup \{c'_i \mid 1 \le i \le m\})^* \longrightarrow V^*, \\
&h_5(a) = a, \text{ for any } a \in V, \\
&h_5(a') = \epsilon, \text{ for any } a' \in V', \\
&h_5(c_i) = h_5(c'_i) = \epsilon, \text{ for any } 1 \le i \le m.
\end{aligned}
$$

Obviously, we have:

$$
CO_M(L) = h_5(g(L_1)).
$$

Because any family closed under ϵ-free homomorphisms, inverse homomorphisms, and intersection with regular sets is closed under ϵ-free gsm mappings, it follows that the language $g(L_1)$ is in F. As can be easily seen, h_5 is 4-limited erasing on $g(L_1)$. Consequently, $CO_M(L) \in F$ and the proof is over.   □

Theorem 2 *The families REG and CS are closed under CO operation.*

Proof. From Lemma 1, we obtain the inclusions $REG \subseteq CO(REG), CS \subseteq CO(CS)$. Because the families REG and CS are closed under the $SShuf$ operation and they are trios (any trio being closed under limited erasing and ϵ-free gsm mappings), the reverse inclusions follow from Lemma 2.   □

Lemma 3 *Any trio closed under union and the CO operation is closed under the SShuf operation.*

Proof. Let $L_1 \subseteq V_1^*, L_2 \subseteq V_2^*$ be two languages in F. Consider the alphabet $V_2' = \{a' \mid a \in V_2\}$ and the following homomorphisms:

$$
\begin{aligned}
&h_1 : V_1^* \longrightarrow V_1^*, &&h_1(a) = aa, \text{ for any } a \in V_1, \\
&h_2 : V_2'^* \longrightarrow V_2'^*, &&h_2(a') = a'a', \text{ for any } a' \in V_2', \\
&h_3 : V_2^* \longrightarrow V_2'^*, &&h_3(a) = a', \text{ for any } a \in V_2, \\
&h_4 : (V_1 \cup V_2')^* \longrightarrow (V_1 \cup V_2)^*, &&h_4(a) = a, \text{ for any } a \in V_1, \\
& &&h_4(a') = a, \text{ for any } a' \in V_2'.
\end{aligned}
$$

Consider also the set of markers $M = V_1 \cup V_2'$. In these conditions, it is easy to check that:

$$
SShuf(L_1, L_2) = h_4(CO_M(h_1(L_1) \cup h_2(h_3(L_2))) \cap (V_1 V_2')^*).
$$

Using the closure properties of the family F, we get $SShuf(L_1, L_2) \in F$ and the inclusion of the family F under the $SShuf$ operation is proved.   □

Theorem 3 *The family CF is not closed under the operation CO.*

Proof. The family CF is a trio closed under union, but it is not closed under $SShuf$. By Lemma 3, the conclusion follows.   □

Looking at Example 2, we notice that the language L there is linear while $CO_M(L)$ is not even context-free. Consequently, we obtain the following result, which is stronger than Theorem 3.

Theorem 4 *There exists a language $L \in LIN$ such that, for a suitable set of markers M, $CO_M(L) \notin CF$.*

The following result is obvious.

Corollary 1 *The family LIN is not closed under the CO operation.*

The next result is a connection between the CO operation and the $crossover_k, k \geq 1$, operation ([5]). Briefly, for some fixed $k \geq 1$, the $crossover_k$ operation works in a similar way to CO but the words, as well as the interchanged segments, are not necesarily of the same length and the number of sites is fixed. For more details, [5] can be consulted.

Theorem 5 *If F is a full trio closed under union and the CO operation, then F is closed under $crossover_k$, for any $k \geq 1$.*

Proof. By Lemma 2 in [5], any full trio closed under the $Shuf$ operation is closed under $crossover_k$. Consequently, it is enough to prove that F is closed under $Shuf$.

From Lemma 3, we get that F is closed under the $SShuf$ operation. Using this, we shall prove that F is closed under $Shuf$ using only the closure of F under limited erasing, intersection with regular sets, and substitution with ϵ-free regular sets. (It is well known that any trio is closed under these three operations.)

Take $L_1 \subseteq V_1^*, L_2 \subseteq V_2^*$ two languages in F and a symbol $c \notin V_1 \cup V_2$. Consider the substitution (with ϵ-free regular sets) $s : (V_1 \cup V_2)^* \longrightarrow 2^{(V_1 \cup V_2 \cup \{c\})^*}, s(a) = c^*ac^*$, for any $a \in V_1 \cup V_2$, and the homomorphism $h : (V_1 \cup V_2 \cup \{c\})^* \longrightarrow (V_1 \cup V_2)^*, h(a) = a$, for any $a \in V_1 \cup V_2, h(c) = \epsilon$. It is easy to check that:

$$Shuf(L_1, L_2) = h(SShuf(s(L_1), s(L_2)) \cap \{aa, ac, ca \mid a \in V_1 \cup V_2\}^*).$$

(Observe that h is 2-limited erasing on the language it is applied to.) Consequently, F is closed under $Shuf$ operation and the proof is ready.   □

5   The Iterated Crossing-Over

Lemma 4 *For any family of languages $F, F \subseteq CO^*(F)$.*

Proof. See Example 1.   □

Theorem 6 *The families FIN and RE are closed under the CO^* operation.*

Proof. Because $CO_M^*(L)$ is finite for any set of markers M and any finite language L, it follows that $CO^*(FIN) \subseteq FIN$. By Church's thesis, we get $CO^*(RE) \subseteq RE$. The reverse inclusions follows from Lemma 5.1.   □

Theorem 7 *There exists a language $L \in LIN$ such that, for a suitable set of markers $M, CO_M^*(L) \notin CF$.*

Proof. The language L in Example 2 is linear while $CO_M^*(L)$ is not context-free.   □

The following result is obvious.

Corollary 2 *The families LIN and CF are not closed under the CO^* operation.*

The closure of the families REG and CS under the CO^* operation is left as an open problem. For the family REG we belive that the answer to this problem is affirmative.

As far as the problem for the family CS is concerned, the following reasoning could be used. For a context-sensitive language L, there is a Turing machine which accepts $CO^*(L)$ and, because any string in $CO^*(L)$ is either in L or is obtained from some strings of the same length in L, the workspace could be bounded by a linear function of the length of the input, thus obtaining a linear bounded automaton. Consequently, we strongly conjecture that the answer to the problem for the family CS is affirmative.

References

[1] T. Head, Formal language theory and DNA: an analysis of the generative capacity of specific recombinant behaviours, *Bulletin of Mathematical Biology*, 49 (1987), 737–759.

[2] T. Head, Splicing schemes and DNA. In G. Rozenberg and A. Salomaa (eds.), *Lindenmayer Systems: Impacts on Theoretical Computer Science and Developmental Biology*. Springer, Berlin, 1992, 371–383.

[3] J.E. Hopcroft and J.D. Ullman, *Introduction to Automata Theory, Languages, and Computation*. Addison-Wesley, Reading, Mass., 1979.

[4] A. Mateescu, Gh. Păun, G. Rozenberg and A. Salomaa, Simple splicing systems, *Discrete Applied Mathematics*, 84 (1998), 145–163.

[5] V. Mitrana, Crossover systems: a generalization of splicing systems, *Journal of Automata, Languages and Combinatorics*, 2.3 (1997), 151–160.

[6] Gh. Păun, On the power of the splicing operation, *International Journal of Computer Mathematics*, 59 (1995), 27–35.

[7] Gh. Păun, On the splicing operation, *Discrete Applied Mathematics*, 70 (1996), 57–79.

[8] D. Pixton, Regularity of splicing languages. In *Proceedings of the First IEEE Symposium on Intelligence in Neural and Biological Systems*, Washington, 1995.

[9] D. Pixton, Context-Free Splicing Systems, manuscript, 1996.

[10] A. Salomaa, *Formal Languages*. Academic Press, New York, 1973.

Restricted Concatenation Inspired by DNA Strand Assembly

Carlos Martín-Vide

Research Group on Mathematical Linguistics
Rovira i Virgili University
Tarragona, Spain
cmv@astor.urv.es

Alfonso Rodríguez-Patón

Department of Artificial Intelligence
Faculty of Informatics
Polytechnical University of Madrid
Spain
arpaton@fi.upm.es

Abstract. Taking our inspiration from biochemistry and DNA computing, we introduce two variants of controlled concatenation of strings and languages: we give a finite set of pairs of strings and we concatenate two arbitrary strings only when a pair belonging to the control set can be found among their substrings. We consider two types of non-iterated and iterated conditional concatenations and we settle the closure properties of abstract families of languages. Then, we use the new concatenation operations as basic operations in Chomsky grammars: rewriting a nonterminal means concatenating a new string with the strings to the left and to the right of that nonterminal, so restricted concatenations can be used. Context-free grammars working with prefix-suffix conditions can generate recursively enumerable or context-sensitive languages, depending on whether or not they use erasing rules.

1   Introduction

The operation of concatenation is fundamental for formal language theory, automata theory, and combinatorics on words. There are also "practical" circumstances in which operations are similar but not identical to concatenation. This is the case of the heads-to-tails selective assembly of collagen molecules. A similar selective concatenation of strings (single stranded DNA molecules) is essentially used in a proposed evolutionary algorithm for solving the Hamiltonian path problem [1]: codes of vertices are concatenated by means of given "splints" which anneal selectively to their ends (blocks $\bar{u}\bar{v}$ are given, and strings of the form xu, vy are bound together because of the complementarity of $\bar{u}\bar{v}$ with the suffix u of the first string and the prefix v of the second one; after ligation, the double stranded portion is melted, $\bar{u}\bar{v}$ is washed out, and we get the string $xuvy$).

In the present paper, we consider two variants of such a conditional concatenation. Specifically, we use a given set of control pairs of strings. Two arbitrary strings can be concatenated only when the strings in a control pair appear in the two strings to be concatenated (1) as compact substrings, and (2) as a suffix and a prefix, respectively. We define the iterated version of each of these operations in the same way as we define the Kleene closure starting from the usual concatenation. Clearly, in the two cases we have restricted forms of the usual concatenation and Kleene closure operations (hence the family of linear languages is closed under none of these operations). All other families in the Chomsky hierarchy are closed under both non-iterated and iterated conditional concatenations.

These results are quite similar to those obtained with usual concatenation and Kleene closure. Rather different are the results obtained in the following framework: rewriting the symbol A in the string uAv means replacing it with some string x, which is concatenated to u and to v, leading to uxv. If we use any of our conditional operations instead of the non-restricted concatenation, then the power of context-free grammars is strictly increased. In the case of suffix-prefix conditions we obtain characterizations of recursively enumerable languages and of context-sensitive languages, depending on whether or not we use erasing rules. These results show that in certain circumstances the conditional concatenation can be very useful.

2   Formal Language Theory Prerequisites

In this section we introduce many intuitively simple notations that we shall subsequently use; for further details of formal language theory we refer to [4].

For an alphabet V we denote by V^* the free monoid generated by V under the operation of concatenation; the empty string is denoted by λ and $V^* - \{\lambda\}$

is denoted by V^+. (The elements of V^* are equivalently called *strings* and *words*.) The length of $x \in V^*$ is denoted by $|x|$ and the number of occurrences of $a \in V$ in x is denoted by $|x|_a$. If $x = x_1 x_2 x_3$, then we say that x_1 is a prefix, x_2 is a substring, and x_3 is a suffix of x. The sets of substrings, prefixes, and suffixes of a string $x \in V^*$ are denoted by $Sub(x), Pref(x), Suf(x)$, respectively. When we consider only substrings, prefixes, and suffixes of length at most t, then we write $Sub_{\leq t}(x), Pref_{\leq t}(x), Suf_{\leq t}(x)$, respectively. Moreover, for $x \in V^*$, we define:

$$Pref_{=t}(x) = \begin{cases} x, & \text{if } |x| \leq t, \\ u, & \text{for } x = uv, |u| = t, v \in V^*. \end{cases}$$

$Suf_{=t}(x)$ is defined in a similar way.

A morphism $h : V^* \longrightarrow U^*$ is said to be *restricted on* a language $L \subseteq V^*$ if there is a constant k such that $h(x) \neq \lambda$ for all $x \in Sub(L)$ with $|x| \geq k$. The morphism h is a *coding* if $h(a) \in U$ for all $a \in V$, a *weak coding* if $h(a) \in U \cup \{\lambda\}$ for all $a \in V$, and a *projection* if $h(a) \in \{a, \lambda\}$ for all $a \in V$.

The *left derivative* of a language $L \subseteq V^*$ with respect to a string $x \in V^*$ is defined by $\partial_x^l(L) = \{w \in V^* \mid xw \in L\}$; the *right derivative* of L with respect to x is defined by $\partial_x^r(L) = \{w \in V^* \mid wx \in L\}$.

A finite automaton is given in the form $A = (K, V, s_0, F, P)$, where K is the set of states, V is the alphabet, s_0 is the initial state, F is the set of final states, and P is the set of transitions, presented as rewriting rules of the form $sa \rightarrow s'$ (in the state s, the automaton reads the symbol a and passes to the state s').

A Chomsky grammar is denoted by $G = (N, T, S, P)$, where N is the nonterminal alphabet, T is the terminal alphabet, $S \in N$ is the axiom, and P is the finite set of rewriting rules, given in the form $x \rightarrow y, x, y \in (N \cup T)^*$, with x containing at least one nonterminal.

Finally, by *REG, LIN, CF, CS, RE* we denote the families of regular, linear, context-free, context-sensitive, recursively enumerable languages, respectively.

3   Restricted Variants of Concatenation

We define now the conditional variants of the concatenation operation as suggested in the Introduction.

Let V be an alphabet and $M \subseteq V^* \times V^*$ a finite set of pairs of strings. For $x, y \in V^*$ we define:

1. $BC_M(x, y) = \begin{cases} xy, & \text{if } (Sub(x) \times Sub(y)) \cap M \neq \emptyset, \\ \text{undefined}, & \text{otherwise}, \end{cases}$
   (*block subwords* are used as conditions on concatenated strings)

2. $HCM(x, y) = \begin{cases} xy, & \text{if } (Suf(x) \times Pref(y)) \cap M \neq \emptyset, \\ \text{undefined}, & \text{otherwise} \end{cases}$

   (*head-tail* conditions are imposed on the concatenated strings).

We denote by D the set $\{BC, HC\}$. The operations π_M, for $\pi \in D$, are extended in the natural way to languages $L_1, L_2 \subseteq V^*$:

$$\pi_M(L_1, L_2) = \{\pi_M(x, y) \mid x \in L_1, y \in L_2\}.$$

Then, we can iterate these operations: for $\pi \in D, M \subseteq V^* \times V^*$, and $L \subseteq V^*$, we define:

$$\pi_M^1(L) = L,$$
$$\pi_M^{i+1}(L) = \pi_M(\pi_M^i(L), L), \text{ for } i \geq 1,$$
$$\pi_M^+(L) = \bigcup_{i \geq 1} \pi_M^i(L).$$

Note that if $M = \{(\lambda, \lambda)\}$, then the two non-iterated operations π_M coincide with the usual concatenation and the two iterated operations π_M^+ coincide with the Kleene closure $+$.

We say that a family FL of languages is closed under the operation π (π^+) if $\pi_M(L_1, L_2) \in FL$ ($\pi_M^+(L) \in FL$, respectively), for all $L_1, L_2 \in FL$ ($L \in FL$, respectively), and for all finite sets $M \subseteq V^* \times V^*$.

4 Closure Properties

The aim of this section is to settle the closure properties of the families in the Chomsky hierarchy under the four operations defined in the previous section. We address this question in a more general framework.

Lemma 1 *If FL is a family of languages which is closed under concatenation with symbols, concatenation, intersection with regular languages, and restricted morphisms, then FL is closed under the operations $\pi \in D$.*

Proof. Let $L_1, L_2 \subseteq V^*$ be two languages in FL, $M \subseteq V^* \times V^*$ be a finite set, and d be a new symbol. We define the morphism (in fact, it is a projection) $h : (V \cup \{d\})^* \to V^*$ by $h(a) = a$ for $a \in V$ and $h(d) = \lambda$. For all $\pi \in D$, we obtain:

$$\pi_M(L_1, L_2) = h((L_1\{d\}L_2) \cap R_\pi(M)),$$

where:

$$R_{BC}(M) = \bigcup_{(u,v) \in M} V^*\{u\}V^*\{d\}V^*\{v\}V^*,$$
$$R_{HC}(M) = \bigcup_{(u,v) \in M} V^*\{u\}\{d\}\{v\}V^*.$$

It is easy to see that the languages $R_\pi(M)$ check the existence of control pairs from M, in the form requested by π (block, suffix-prefix, etc.), in the strings from $L_1\{d\}L_2$ proposed as candidates for concatenation. Thus, the equality follows.

All the languages $R_\pi(M)$ are regular (the operations used to define them preserve their regularity, and M is a finite set, so we always get finite unions of regular languages). Taking into account the closure properties of the family FL, we find that $\pi_M(L_1, L_2) \in FL$. $\qquad\qquad\qquad\qquad\square$

Theorem 1 *All families REG, CF, CS, RE are closed under the operations $\pi \in D$; the family LIN is closed under none of these operations.*

We now go on to investigate the closure under the iterated operations π^+, for $\pi \in D$. A general result like that in Lemma 1 is again true. In fact, the closure under concatenation in the premise in Lemma 1 can be replaced by closure under Kleene + and we get closure under π^+ instead of closure under π. The proof is the same for the two operations, although the construction of the regular language used is more elaborate than in the proof in Lemma 1.

Lemma 2 *If FL is a family of languages which is closed under concatenation with symbols, Kleene +, intersection with regular languages, and restricted morphisms, then FL is closed under the operations π^+, $\pi \in D$.*

Proof. Let $L \subseteq V^*$ be a language in FL, $M \subseteq V^* \times V^*$ be a finite set, and d be a new symbol. We define the morphism $h : (V \cup \{d\})^* \to V^*$ by $h(a) = a$ for $a \in V$ and $h(d) = \lambda$. For all $\pi \in D$, we obtain:

$$\pi_M^+(L) = h((L\{d\})^+ \cap R_\pi),$$

where the regular languages R_π are given by means of finite automata, $R_\pi = L(A_\pi)$, as constructed below.

We denote $s = \max\{|u| \mid (u, v) \in M\}$ and $t = \max\{|v| \mid (u, v) \in M\}$. For any constant r, we denote $V^{\leq r} = \{x \in V^* \mid |x| \leq r\}$.

A. $\pi = BC$.
We construct $A_{BC} = (K, V \cup \{d\}, s_0, F, P)$, where:

$$
\begin{aligned}
K \;=\;& \{[H; x] \mid H \subseteq V^{\leq s}, x \in V^{\leq s}\} \\
\cup\;& \{[H_1; H_2; x_2; H_3; x_3] \mid H_1, H_3 \subseteq V^{\leq s}, H_2 \subseteq V^{\leq t}, x_2 \in V^{\leq t}, \\
& x_3 \in V^{\leq s}\}, \\
s_0 \;=\;& [\emptyset; \lambda], \\
F \;=\;& \{[H; \emptyset; \lambda; \emptyset; x] \mid H \subseteq V^{\leq s}, x \in V^{\leq s}\},
\end{aligned}
$$

and the following transitions, for all possible $H, H_1, H_2, H_3, x, x_2, x_3$:

1. $[H; x]a \rightarrow [H \cup Sub_{\leq s}(xa); Suf_{=s}(xa)]$, for $a \in V$,

2. $[H; x]d \rightarrow [H; \emptyset; \lambda; \emptyset; x]$,

3. $[H_1; H_2; x_2; H_3; x_3]a \rightarrow$
   $[H_1; H_2 \cup Sub_{\leq t}(x_2 a); Suf_{=t}(x_2 a); H_3 \cup Sub_{\leq s}(x_3 a); Suf_{=s}(x_3 a)]$,
   for $a \in V$,

4. $[H_1; H_2; x_2; H_3; x_3]d \rightarrow [H_1 \cup H_3; \emptyset; \lambda; \emptyset; x_3]$
   if and only if $(H_1 \times H_2) \cap M \neq \emptyset$.

The idea behind this construction is the following. Our aim is to concatenate strings from L. We first consider a string of the form $w = z_1 d z_2 d \dots z_n d$, for $n \geq 1, z_i \in L, 1 \leq i \leq n$. We then check (by the intersection with $L(A_{BC})$) the condition imposed by M. We start from the left of w, symbol by symbol, with states of the form $[H; x]$; H memorizes the subwords of length at most s of the parsed string, while x memorizes the last s parsed symbols (or fewer, if in total there are less than s parsed symbols). The need of x appears when passing across the next symbol d: the suffix of the completed z_i and the prefix of the next string, z_{i+1}, can introduce substrings of the currently obtained string which can be used by M and do not appear as substrings of the separated strings z_i, z_{i+1}.

After crossing the first occurrence of d, we pass from a state of the form $[H; x]$ to a state of the form $[H_1; H_2; x_2; H_3; x_3]$, and we continue with states of this type till the end of the parsing. The components have the following meanings. During the parsing of z_2, H_1 is the set of substrings in z_1 of length at most s. At a later passing over some d, H_1 contains all the substrings of length at most s of the string placed to the left of that d, that is, the string already checked for correctness with respect to M. H_2 is the set of substrings of length at most t of the string which follows the last parsed occurrence of d (this string is to be checked whether or not its concatenation with the string constructed up to now is allowed with respect to M). To this end, we memorize in x_2 the last (at most) t scanned symbols. H_3 and x_3 are used to update the set H_1 at the moment that a further block z_i is accepted; H_3 stores the substrings of length at most s of the last scanned block z_i and x_3 consists of the last (at most) s scanned symbols.

When we pass over a symbol d in such a complex state, we check whether or not $H_1 \times H_2$ contains at least a pair (u, v) that is also an element of M. If this is the case, the parsing can continue; otherwise the automaton is blocked and no terminal state can be reached. When crossing over a symbol d, the set of subwords to the left is $H_1 \cup H_3$, the set H_2 becomes empty (ready for storing the subwords of a new string z_i, if any), and x_3 remains unchanged.

From these explanations, the equality $BC_M^+(L) = h((L\{d\})^+ \cap R_{BC})$ can easily be proved. Hence the lemma follows for this case.

B. $\pi = HC$.

We construct the automaton $A_{HC} = (K, V \cup \{d\}, s_0, F, P)$ with:

$$
\begin{aligned}
K &= \{[x] \mid x \in V^{\leq s}\} \cup \{[x_1; x_2; x_3] \mid x_1, x_3 \in V^{\leq s}, x_2 \in V^{\leq t}\}, \\
s_0 &= [\lambda], \\
F &= \{[x_1; \lambda; x_3] \mid x_1, x_3 \in V^{\leq s}\},
\end{aligned}
$$

and the following transitions, for all possible strings x, x_1, x_2, x_3:

1. $[x]a \rightarrow [Suf_{=s}(xa)]$, for $a \in V$,

2. $[x]d \rightarrow [x; \lambda; \lambda]$,

3. $[x_1; x_2; x_3]a \rightarrow [x_1; Pref_{=t}(x_2a); Suf_{=s}(x_3a)]$, for $a \in V$,

4. $[x_1; x_2; x_3]d \rightarrow [Suf_{=s}(x_1x_3); \lambda; x_3]$ if and only if $(Suf(x_1) \times Pref(x_2)) \cap M \neq \emptyset$.

The equality $HC_M^+(L) = h((L\{d\})^+ \cap R_{HC})$ can easily be checked in the same way as in the previous case. The two cases in the lemma are covered.
$\square$

Theorem 2 *All families REG, CF, CS, RE are closed under the operations π^+, $\pi \in D$; the family LIN is closed under none of these operations.*

5   Grammars with Restricted Concatenation

We have observed in the Introduction that the rewriting operation can be considered to involve a double concatenation. These concatenations can be restricted, either at the level of each rule (associating a condition set M to each rule) or at the level of the whole grammar (considering a unique set M for all rules). In this way we get two types of grammars with regulated rewriting.

A *grammar with a local π conditional rewriting*, $\pi \in D$, is a construct $G = (N, T, S, P)$, where N, T, S are as they are in a usual grammar (nonterminal alphabet, terminal alphabet and axiom, respectively), and P is a finite set of pairs $r = (x \rightarrow y, M_r)$, where $x \rightarrow y$ is a usual rewriting rule over $N \cup T$ and M_r is a finite subset of $(N \cup T)^* \times (N \cup T)^*$.

For $w, z \in (N \cup T)^*$ and $r \in P$ we write $w \Longrightarrow_r z$ if and only if $w = uxv$ and $z = \pi_M(\pi_M(u, y), v)$. (That is, the concatenation of u, y and of uy, v are correct in the sense of π_M. Of course, when $u = \lambda$ or $v = \lambda$, no concatenations, and therefore no conditions, appear in that side. When $y = \lambda$, we have $z = \pi_M(u, v)$, so the strings u, v must observe the condition.) We denote by $\Longrightarrow$ any of the relations $\Longrightarrow_r, r \in P$, and by $\Longrightarrow^*$ the reflexive and transitive closure of $\Longrightarrow$. The language generated by G is $L(G) = \{z \in T^* \mid S \Longrightarrow^* z\}$.

When all sets $M_r, r \in P$, are equal (to a set M), we say that G has a *global* π conditional rewriting and we write it in the form $G = (N, T, S, P_0, M)$, where P_0 are usual rewriting rules.

We consider here only context-free grammars and we denote by $K_l CF(\pi)$, the family of languages generated by λ-free context-free grammars with local π conditional rewriting, $\pi \in D$. When λ-rules are allowed, we write $K_l CF^\lambda(\pi)$. In the case of grammars with a global conditional rewriting we replace the subscript l with g.

The following relations are direct consequences of the definitions:

Lemma 3 *For all $\pi \in D$ we have: (i) $CF \subseteq K_g CF(\pi)$; (ii) $K_g CF(\pi) \subseteq K_l CF(\pi)$, $K_g CF^\lambda(\pi) \subseteq K_l CF^\lambda(\pi)$; (iii) $K_l CF(\pi) \subseteq K_l CF^\lambda(\pi)$, $K_g CF(\pi) \subseteq K_g CF^\lambda(\pi)$; (iv) $K_l CF(\pi) \subseteq CS$, $K_l CF^\lambda(\pi) \subseteq RE$.*

The result $K_g CF^\lambda(\pi) - CF \neq \emptyset$, for $\pi = BC$, is shown in [2]. But a much stronger version is settled for the prefix-suffix conditions, $\pi = HC$.

Theorem 3 $K_g CF(HC) = CS$, $K_g CF^\lambda(HC) = RE$.

Proof. Let $L \subseteq T^*$ be a context-sensitive language. We write:

$$L = \{x \in L \mid |x| \leq 2\} \cup \bigcup_{a,b \in T} \{a\} \partial_a^l (\partial_b^r(L))\{b\}.$$

The languages $L_{ab} = \partial_a^l(\partial_b^r(L))$ are context-sensitive for all $a, b \in T$. Let $G_{ab} = (N_{ab}, T, S_{ab}, P_{ab})$ be a grammar for the language L_{ab}, in the Kuroda normal form, that is with the rules in P_{ab} of the following types: $A \to c$, $A \to BC$, $AB \to CD$, for $A, B, C, D \in N, c \in T$. We assume that the sets $N_{ab}, a, b \in T$ are mutually disjoint and that all non-context-free rules in sets P_{ab} are labeled in a one-to-one manner, $r : AB \to CD$, with elements in a set L_{ab}. Let us denote by N the union of all sets N_{ab} and by P the union of all sets P_{ab}, for all $a, b \in T$.

We construct the grammar with HC global conditional rewriting $G = (N', T, S, P', M)$, where:

$$
\begin{aligned}
N' \;=\; & N \cup \{A', A'' \mid A \in N\} \cup \{S\} \\
& \cup\; \{[r, C], (r, D) \mid r : AB \to CD \in P\}, \\
P' \;=\; & \{S \to x \mid x \in L, |x| \leq 2\} \\
& \cup\; \{S \to a S_{ab} b \mid a, b \in T\} \\
& \cup\; \{A \to x \mid A \to x \in P\} \\
& \cup\; \{A \to [r, C],\; B \to (r, D), \\
& \qquad [r, C] \to C',\; (r, D) \to D'', \\
& \qquad C' \to C,\; D'' \to D \mid r : AB \to CD \in P\},
\end{aligned}
$$

and with the following condition set:

$$
\begin{aligned}
M \;=\; & \{(\alpha, \beta) \mid \alpha, \beta \in N \cup T\} \\
\cup \; & \{([r, C], (r, D)), \; (\alpha, [r, C]), \; ([r, C], \alpha), \\
& \quad ((r, D), \alpha), \; (\alpha', (r, D)) \mid r : AB \to CD \in P, \alpha \in N \cup T\} \\
\cup \; & \{(\alpha, \beta'), \; (\alpha', \beta''), (\alpha'', \beta), \; (\alpha, \beta'') \mid \alpha, \beta \in N \cup T\}.
\end{aligned}
$$

The strings of length at most 2 are introduced by initial rules. Each rule $S \to a S_{ab} b$ will determine the simulation in G of a derivation in the grammar G_{ab}, $a, b \in T$. The context-free rules in P_{ab} can be used without restriction, because we have in M all pairs (α, β) for $\alpha, \beta \in N \cup T$. Note that no such pair contains primed or double primed symbols, or symbols of the form $[r, C], (r, D)$. The key point of the proof is the simulation in G of non-context-free rules in P_{ab}. Consider such a rule, $r : AB \to CD$. Take a sentential form $a w_1 A B w_2 b$ (even for empty w_1, w_2 we have at least a and b to the left and right of AB, respectively). Among the rules associated with r in G only $A \to [r, C]$ can be applied. Again the continuation is uniquely determined. The derivation steps are: $a w_1 A B w_2 b \Longrightarrow a w_1 [r, C] B w_2 b \Longrightarrow a w_1 [r, C] (r, D) w_2 b \Longrightarrow a w_1 C' (r, D) w_2 b \Longrightarrow a w_1 C' D'' w_2 b \Longrightarrow a w_1 C D'' w_2 b \Longrightarrow a w_1 C D w_2 b$.

Therefore, all derivations in all G_{ab} can be simulated in G and all derivations in G correspond to a derivation in some G_{ab}. In conclusion, $L(G) = L$, which proves the theorem for the context-sensitive case.

In the recursively enumerable case the construction is the same; the only difference is that we also have rules of the form $A \to \lambda$ in the sets P_{ab}; such rules are context-free and raise no problem (the pairs (α, β) in M allow their use). Hence the same assertion follows. $\qquad\square$

6 Final Remarks

There is a natural extension to the conditional variants of the concatenation operation presented in this paper. Scattered, Parikh and reduced Parikh subwords may be used as conditions. These cases and also a representation of regular languages using finitely many operations π, π^+, for $\pi \in D$ and a coding, are discussed in [2].

References

[1] J.M. Barreiro, J. Rodrigo and A. Rodríguez-Patón, Evolutionary biomolecular computing, *Romanian Journal of Information Science and Technology*, 1.4 (1998), 287–382.

[2] J. Dassow, C. Martín-Vide, Gh. Păun and A. Rodríguez-Patón, Conditional concatenation, *Fundamenta Informaticae*, in press.

[3] J. Dassow and Gh. Păun, *Regulated Rewriting in Formal Language Theory*. Springer, Berlin, 1989.

[4] G. Rozenberg and A. Salomaa (eds.), *Handbook of Formal Languages*. Springer, Berlin, 1997.

DNA Tree Structures[1]

George Rahonis

30 Kyprou st.
Perea, Thessaloniki, Greece
grahonis@ccf.auth.gr

Abstract. We present a new model of a tree transducer with two inputs and one output, which is called the DNA tree transducer. We prove that this type of transducer implements the splicing operation on trees with recognizable sets of rules, and we investigate its properties.

1 Introduction

DNA computing is one of the most promising fields of theoretical computer science, which will hopefully lead to the construction of new types of computers. After the seminal work of L. Adleman [1] in 1994, a large number of researchers all over the world have dealt with this new topic, and a great number of papers have already been published in this area (see [8, 10]). From the formal language point of view, the basic model for DNA computing is the H scheme, and the crucial operation on words is the splicing operation. Recently, variants of DNA computing such as membrane computing have been defined, thus extending the power of computations and the culture of computer science [9, 11, 12, 13, 14]. Moreover, the splicing operation has been defined for non-linear objects such as trees. Of course, in this case the definition can be made in several ways. In the bibliography up to now, two types of tree-splicing have been used, [4] and [18]. By using the method defined in [4], we derived interesting results for the known families *REC* of

---

[1]Supported by the State Scholarship Foundation.

recognizable forests, OCF of one counter forests, ALG of algebraic forests, and GST of generalized synchronized forests. The case of the iterated splicing with finite and recognizable sets of rules has also been investigated [15].

In this paper, we introduce a model of a tree transducer, the so-called DNA tree transducer, for the implementation of the splicing operation on trees. This machine has two inputs and one output, the result of the splicing operation on the input trees. The set of the rules of the underlying H tree scheme is assumed to be recognizable. Properties of the DNA tree transducer model are also established.

2 Preliminaries

The reader is assumed to be familiar with basic notions from formal language theory [19], DNA computing [10], and tree languages [5, 6].

We denote by $P(V)$ the power set of a set V, and we write V^* for the free monoid generated by V. For a positive natural number n, $[n] = \{1, ..., n\}$.

If Σ is a *finite ranked alphabet*, Σ_n denotes the set of symbols of Σ whose rank equals n. A symbol $\sigma \in \Sigma$ might have more than one rank. The degree of Σ, $\deg(\Sigma)$ is the maximal number n such that $\Sigma_n \neq \emptyset$ and $\Sigma_k = \emptyset$ for each $k > n$.

Let $X = \{x_1, x_2, ...\}$ be a countably infinite set of variables and $X_m = \{x_1, ..., x_m\}$, for $m \geq 0$. The set of all trees over Σ, indexed by the variables $x_1, ..., x_m$ is denoted by $T_\Sigma(X_m)$. A *forest* F is a set of trees over an alphabet Σ, possibly with variables, that is $F \subseteq T_\Sigma(X_m)$, $m \geq 0$. We shall also need a second countable set of variables which will be denoted by $Y = \{y_1, y_2, ...\}$.

For $t \in T_\Sigma(X_n)$ and $t_1, ..., t_n \in T_\Sigma(X_m)$, we write $t(t_1, ..., t_n)$ for the result of substituting t_i for x_i in t.

Another way to define trees is by using tree domains. A *tree domain* D [6, 20] is a subset of N_+^* (N denotes the set of natural numbers) which satisfies the following conditions:

(i) If $u \in D$ then $v \in D$ for each prefix v of u.

(ii) For each $u \in D$, there exists $i \in N$, such that $uj \in D$ for each $j \in [i]$ (if u has no sons then $i = 0$).

For a set V, a $V-labeled\ tree$ is a mapping $t : D \to V$, where D is a tree domain. We call the elements of D the *nodes* of the tree and we denote D by $dom(t)$. A node $u \in D$ is labeled by $t(u) \in V$. If v is a proper prefix of $u \in dom(t)$, then u is called a *successor* of v.

In this way, each tree $t \in T_\Sigma(X)$ can be considered as a $\Sigma \cup X$-labeled tree $t : dom(t) \to \Sigma \cup X$, such that each node of t labeled by an element of rank $n \geq 0$, has exactly n immediate successors (sons) and variables do not have successors. The nodes of a tree t labeled by constants are called *leaves*, and the set of all leaves of a tree t is denoted by $leaf(t)$.

Now let Σ, Γ be ranked alphabets with $\deg(\Sigma) = n$. Assume that for each $k \in [n]$, there is a mapping $h_k : \Sigma_k \to T_\Gamma(X_k)$. Then the mappings h_k, $k \in [n]$ constitute a *tree homomorphism* $h : T_\Sigma \to T_\Gamma$, which is defined inductively by:

$h(\sigma(t_1, ..., t_k)) = h_k(\sigma)(h(t_1), ..., h(t_k))$, for $k \geq 0$, $\sigma \in \Sigma_k$ and $t_1, ..., t_k \in T_\Sigma$.

A tree homomorphism $h : T_\Sigma \to T_\Gamma$ will be called *linear*, if for each $\sigma \in \Sigma_k$, $k \geq 0$, $h_k(\sigma)$ is a linear tree, that is each variable from X_k appears at most once in $h_k(\sigma)$. Moreover, a linear tree homomorphism $h : T_\Sigma \to T_\Gamma$ is *alphabetic*, if for each $\sigma \in \Sigma_k$, $k \geq 0$, either:

$h_k(\sigma) = \gamma(x_{i_1}, ..., x_{i_m})$, $\gamma \in \Gamma_m$, or
$h_k(\sigma) = x_n$, $1 \leq n \leq k$.

Consider now a finite ranked alphabet Σ and let $\Gamma = \Sigma \cup \{\delta, \sigma_1, ..., \sigma_n\}$, with $rank(\delta) = n$ and $rank(\sigma_i) = m_i$, $i \in [n]$. The (n+1)-tuple $(\delta, \sigma_1, ..., \sigma_n)$ is called *a connected list in a forest* $F \subseteq T_\Gamma$, if whenever one of the above symbols appears in a tree $t \in F$, then it appears in the fork $\delta(\sigma_1, ..., \sigma_n)$, that is none of these symbols appear in trees of F out of the above fork.

Let Δ be the alphabet $\Delta = \Sigma \cup \{\sigma\}$, with $rank(\sigma) = \sum_{i=1}^n m_i = m$.

The *shift operation on the connected list* $(\delta, \sigma_1, ..., \sigma_n)$, denoted $sh : T_\Gamma \to T_\Delta$, is the inverse linear tree homomorphism $h : T_\Delta \to T_\Gamma$, which is defined by:

$-h_k(\gamma(x_1, ..., x_k)) = \gamma(x_1, ..., x_k)$, for each $\gamma \in \Sigma_k$, $k \geq 0$,
$-h_m(\sigma(x_1, ..., x_m)) =$
$\quad \delta(\sigma_1(x_1, ..., x_{m_1}), \sigma_2(x_{m_1+1}, ..., x_{m_1+m_2}), ..., \sigma_n(x_{m_1+...+m_{n-1}+1}, ..., x_m))$.

Thus, we write:

$sh(\delta(\sigma_1(x_1, ..., x_{m_1}), \sigma_2(x_{m_1+1}, ..., x_{m_1+m_2}), ..., \sigma_n(x_{m_1+...+m_{n-1}+1}, ..., x_m)))$
$= \sigma(x_1, ..., x_m)$

and:

$sh(\gamma(x_1, ..., x_k)) = \gamma(x_1, ..., x_k)$ for each $\gamma \in \Sigma_k$, $k \geq 0$.

The shift operation can obviously be extended to more than one connected list.

A *nondeterministic top-down finite tree automaton* is a four-tuple $\mathcal{A} = (\Sigma, Q, Q_0, \alpha)$, with Σ the finite ranked alphabet of input symbols, Q the finite set of states, $Q_0 \subseteq Q$ is the set of initial states and α is the family of state transitions which is defined by the mappings $\alpha_\sigma : Q \to P(Q^n)$, where $\sigma \in \Sigma_n$, $n \geq 0$. If $\sigma \in \Sigma_0$, then $\alpha_\sigma \subseteq Q$.

A *computation* of $\mathcal{A} = (\Sigma, Q, Q_0, \alpha)$ on an input tree $t \in T_\Sigma$ is a Q-labeled tree $r : dom(t) \to Q$ satisfying the following conditions:

(i) $r(\lambda) \in Q_0$.

(ii) Suppose that $u \in dom(t)$ has m successors $u_1, ..., u_m$ and $t(u) = \sigma \in \Sigma_m$. Then $(r(u_1), ..., r(u_m)) \in \alpha_\sigma(r(u))$.

(iii) If $u \in leaf(t)$ and $t(u) = \sigma$ $(\in \Sigma_0)$, then $r(u) \in \alpha_\sigma$.

The set of all computations of $\mathcal{A}$ on t is denoted by $com_\mathcal{A}(t)$ and the forest recognized by $\mathcal{A}$ is:

$$L(\mathcal{A}) = \{t \in T_\Sigma \ / \ com_\mathcal{A}(t) \neq \emptyset\}.$$

The family of forests recognized by nondeterministic top-down tree automata, is the family of recognizable forests and is denoted by REC [5, 6].

As usual, we denote by FIN the family of finite forests, by ALG the family of *algebraic forests* [7], by OCF the family of *one counter forests* [3] and by GST the family of *generalized synchronized forests* [17].

The notion of an alphabetic tree transduction was defined in [2], and used as the basis to build the AFL theory to the tree case [3, 16].

A tree transduction $\tau : T_\Sigma \to P(T_\Gamma)$ is called *alphabetic*, if there exists a ranked alphabet Δ, a recognizable forest $R \subseteq T_\Delta$, and two alphabetic homomorphisms $\varphi : T_\Delta \to T_\Sigma$, $\psi : T_\Delta \to T_\Gamma$, such that $\#\tau = \{(\varphi(t), \psi(t)) \ / \ t \in R\}$, where $\#\tau$ denotes the graph of τ.

An *alphabetic cone* is a family of forests closed under alphabetic transductions whereas *a sheaf of forests* is an alphabetic cone closed under the rational tree operations, i.e.: *union, top-catenation, α-product* and *a-star* [3].

Proposition 1 [3, 17] *The classes REC of recognizable, ALG of algebraic, OCF of one counter, and GST of generalized synchronized forests are sheaves.*

3   Splicing on Trees

We start from the definition of an *H tree scheme*.

Definition 1 *Let Σ be a finite ranked alphabet and $@, \#, \$$ be three new symbols not belonging to Σ, with $rank(@) = 3, 4, 5, 6, 7$ and $rank(\#) = rank(\$) = 0$. An H tree scheme is a pair $f = (\Sigma, L)$, with L the forest of rules, $L \subseteq @(\omega, \#, \omega, \$, \omega, \#, \omega)$, where ω stands either for T_Σ or for nothing.*

Thus, a rule $r \in L$ is a tree of the form $@(s_1, \#, s_2, \$, s_0, \#, s_4)$, $s_i \subset T_\Sigma$, $i = 1, 2, 3, 4$, where some s_i may be missed, for example $@(\#, s_2, \$, \#, s_4)$, $@(\#, \$, s_3, \#)$ can be elements of L.

L can be either finite or infinite.

An H tree scheme $f = (\Sigma, L)$ is called of $\mathcal{F}$ *type* if L belongs to the family of forests $\mathcal{F}$.

Let now $t, u, z \in T_\Sigma$. We say that z is obtained by *splicing* t, u, and we write $(t, u) \to_r z$, if $t = \sigma(t_1, ..., t_{i-1}, s_1, s_2, t_{i+2}, ..., t_n)$, $u = \sigma(u_1, ..., u_{i-1}, s_3, s_4, u_{i+2}, ..., u_n)$, $z = \sigma(t_1, ..., t_{i-1}, s_1, s_4, u_{i+2}, ..., u_n)$, with $\sigma \in \Sigma_n$, $n > 0$, $t_1, ..., t_{i-1}, t_{i+2}, ..., t_n, u_1, ..., u_{i-1}, u_{i+2}, ..., u_n \in T_\Sigma$ and $r = @(s_1, \#, s_2, \$, s_3, \#, s_4)$ is a rule in L.

Thus, splicing on trees means "vertically cutting" of two trees with the same root, at certain indices under the presence of a rule, and "connecting" the leftmost part of the first tree with the rightmost part of the second tree.

For $F \subseteq T_\Sigma$, and $f = (\Sigma, L)$ an H tree scheme, we write:

$$f(F) = \{z \in T_\Sigma \ / \ \exists \ t, u \in F, \exists \ r \in L, \text{ such that } (t, u) \to_r z\}$$

and:

$$f^{(1)}(F) = F \cup f(F).$$

If $\mathcal{F}_1, \mathcal{F}_2$ are two families of forests, we define the new families:

$$S(\mathcal{F}_1, \mathcal{F}_2) = \{f(F_1) \ / \ F_1 \in \mathcal{F}_1, f = (\Sigma, L) \text{ is an H tree scheme with } L \in \mathcal{F}_2\}$$

and:

$$S_1(\mathcal{F}_1, \mathcal{F}_2) = \{f^{(1)}(F_1) \ / \ F_1 \in \mathcal{F}_1, f = (\Sigma, L) \text{ is an H tree scheme with } L \in \mathcal{F}_2\}.$$

For the families REC, ALG, OCF and GST the following propostion holds:

Proposition 2 [4, 15] $S_1(REC, FIN) = REC$, $S_1(ALG, FIN) = ALG$, $S_1(OCF, FIN) = OCF$, $S_1(GST, FIN) = GST$, $S_1(REC, REC) = REC$, $S_1(ALG, REC) = ALG$, $S_1(OCF, REC) = OCF$, $S_1(GST, REC) = GST$.

More generally:

Theorem 1 [4] *If $\mathcal{F}$ is a sheaf of forests closed under the shift operation on connected lists, then $S_1(\mathcal{F}, FIN) = \mathcal{F}$ and $S_1(\mathcal{F}, REC) = \mathcal{F}$.*

Consider now an H tree scheme $f = (\Sigma, L)$. By passing to the yield of L, we obtain an H string scheme $\sigma = (\Sigma_0, yield(L))$ [8].

For $F \subseteq T_\Sigma$, the inclusion below is, in general, strict:

$$yield(f(F)) \subseteq \sigma(yield(F)).$$

In the case of recognizable forests, the next proposition holds.

Proposition 3 [4] $yield(S_1(REC, FIN)) = S_1(yield(REC), yield(FIN))$, *where the symbol S_1 has an analogous meaning in the word case.*

4   DNA Tree Transducers

In this section, we define the model of a DNA tree transducer and we prove that it implements the splicing operation on trees.

Let $f = (\Sigma, L)$ be an H tree scheme with $L \in REC$. There is a non-deterministic top-down finite tree automaton $\mathcal{A} = (\Sigma, Q, Q_0, \alpha)$ such that $L = L(\mathcal{A})$. Without loss of generality, we can assume that $\alpha_{\$} = q_{\$} \in Q$ and $\alpha_{\#} = q_{\#} \in Q$, and the states $q_{\$}, q_{\#}$ do not take part in any other computation of $\mathcal{A}$.

Definition 2 *The DNA tree transducer associated to the H scheme f is a five-tuple $\mathcal{M}_f = (P, \Sigma, \Sigma, P_0, R)$, with $P = (Q - \{ q_{\$}, q_{\#} \}) \cup \{p, p_1, p_2\}$ the set of states (p, p_1, p_2 are new states), Σ is the input and output alphabet, $P_0 = Q_0$ is the set of initial states, and R is the finite set of rules of the following types:*

(I) $q(\sigma(x_1, ..., x_n), \sigma(y_1, ..., y_n)) \rightarrow$
$\sigma(p(x_1), ..., p(x_{i-1}), p_1(q_1(x_i), q_2(x_{i+1})), p_2(q_3(y_i), q_4(y_{i+1})), p(y_{i+2}), ..., p(y_n))$
for each $q \in P_0$, $\sigma \in \Sigma_n$, $n \geq 1$, $i \in [n]$, and $(q_1, q_{\#}, q_2, q_{\$}, q_3, q_{\#}, q_4) \in \alpha_{@}(q)$.
(II) $p(\sigma(x_1, ..., x_n)) \rightarrow \sigma(p(x_1), ..., p(x_n))$, for each $\sigma \in \Sigma_n$, $n \geq 0$.
(III) $q(\sigma(x_1, ..., x_n)) \rightarrow \sigma(q_1(x_1), ..., q_n(x_n))$, for $q, q_1, ..., q_n \in P$, $\sigma \in \Sigma_n$, $n \geq 0$, and $(q_1, ..., q_n) \in \alpha_{\sigma}(q)$.
(IV) $< p_1(\sigma(x_1, ..., x_n), \gamma(y_1, ..., y_m)); T_{\Sigma} > \rightarrow \sigma(p(x_1), ..., p(x_n))$, for each $\sigma \in \Sigma_n$, $n \geq 0$, $\gamma \in \Sigma_m$, $m \geq 0$.
(V) $< p_2(\sigma(x_1, ..., x_n), \gamma(y_1, ..., y_m)); T_{\Sigma} > \rightarrow \gamma(p(y_1), ..., p(y_m))$, for each $\sigma \in \Sigma_n$, $n \geq 0$, $\gamma \in \Sigma_m$, $m \geq 0$.

We denote by $\Rightarrow_{\mathcal{M}_f}$ the one step computation of $\mathcal{M}_f$. Classically $\Rightarrow_{\mathcal{M}_f}$ is a binary relation on $T_{\Sigma \cup P}$, which is defined in the following manner: $t \Rightarrow_{\mathcal{M}_f} z$, if and only if:

(i) there is a rule $q(\sigma(x_1, ..., x_n), \sigma(y_1, ..., y_n)) \rightarrow$
$\sigma(p(x_1), ..., p(x_{i-1}), p_1(q_1(x_i), q_2(x_{i+1})), p_2(q_3(y_i), q_4(y_{i+1})), p(y_{i+2}), ..., p(y_n))$
of type (I), and
(ii) t has a terminal subtree $t' = q(\sigma(t_1, ..., t_n), \sigma(t'_1, ..., t'_n))$, with $t_i, t'_i \in T_{\Sigma}$, $i \in [n]$, and z is obtained by substituting:

$\sigma(p(t_1), ..., p(t_{i-1}), p_1(q_1(t_i), q_2(t_{i+1})), p_2(q_3(t'_i), q_4(t'_{i+1})), p(t'_{i+2}), ..., p(t'_n))$
for an occurrence of t' in t, or:
(i) there is a rule $p(\sigma(x_1, ..., x_n)) \rightarrow \sigma(p(x_1), ..., p(x_n))$ of type (II), and
(ii) t has a terminal subtree $t' = p(\sigma(t_1, ..., t_n))$ with $t_i, \in T_{\Sigma}$, $i \in [n]$, and z is obtained by substituting $\sigma(p(t_1), ..., p(t_n))$ for an occurrence of t' in t, or:
(i) there is a rule $q(\sigma(x_1, ..., x_n)) \rightarrow \sigma(q_1(x_1), ..., q_n(x_n))$ of type (III), and
(ii) t has a terminal subtree $t' = q(\sigma(t_1, ..., t_n))$ with $t_i, \in T_{\Sigma}$, $i \in [n]$, and z is obtained by substituting $\sigma(q_1(t_1), ..., q_n(t_n))$ for an occurrence of t' in t,

or:

(i) there is a rule $< p_1(\sigma(x_1, ..., x_n), \gamma(y_1, ..., y_m)); T_\Sigma > \rightarrow \sigma(p(x_1), ..., p(x_n))$ of type (IV), and

(ii) t has a terminal subtree $t' = p_1(\sigma(t_1, ..., t_n), \gamma(t'_1, ..., t'_m))$ with $t_1, ..., t_n$, $t'_1, ..., t'_m \in T_\Sigma$, and z is obtained by substituting $\sigma(p(t_1), ..., p(t_n))$ for an occurrence of t' in t, or:

(i) there is a rule $< p_2(\sigma(x_1, ..., x_n), \gamma(y_1, ..., y_m)); T_\Sigma > \rightarrow \gamma(p(y_1), ..., p(y_m))$ of type (V), and

(ii) t has terminal subtree $t' = p_2(\sigma(t_1, ..., t_n), \gamma(t'_1, ..., t'_m))$ and z is obtained by substituting $\gamma(p(t'_1), ..., p(t'_m))$ for an occurrence of t' in t.

Observe that the rules of type (IV) and (V) are applied in a look-ahead manner, that is they can be applied only if the trees that are substituted for the variables $x_1, ..., x_n, y_1, ..., y_m$ belong to T_Σ.

We denote by $\Rightarrow^*_{\mathcal{M}_f}$ the reflexive and transitive closure of $\Rightarrow_{\mathcal{M}_f}$. If there is no danger of confusion we just write $\Rightarrow$ and $\Rightarrow^*$ for $\Rightarrow_{\mathcal{M}_f}$ and $\Rightarrow^*_{\mathcal{M}_f}$, respectively.

The relation computed by $\mathcal{M}_f$ is a subset $\mid \mathcal{M}_f \mid \subseteq (T_\Sigma \times T_\Sigma) \times T_\Sigma$:

$$\mid \mathcal{M}_f \mid = \{((t, u), z) \in (T_\Sigma \times T_\Sigma) \times T_\Sigma \ / \text{ there is a } q \in P_0, \text{ such that:}$$
$$q(t, u) \Rightarrow^*_{\mathcal{M}_f} z\}.$$

If $(t, u) \in T_\Sigma \times T_\Sigma$, we denote by $\mathcal{M}_f((t, u))$ the set:

$$\mathcal{M}_f((t, u)) = \{z \in T_\Sigma \ / \ ((t, u), z) \in \mid \mathcal{M}_f \mid\}$$

whereas for $F \subseteq T_\Sigma$:

$$\mathcal{M}_f(F) = \bigcup_{t, u \in F} \mathcal{M}_f((t, u)).$$

Intuitively, the corresponding tree transducer $\mathcal{M}_f$ to an H tree scheme $f = (\Sigma, L)$, starts in a initial state with inputs t, u from T_Σ having the same root, and applies a rule of type (I). This rule "fixes" the indices $i, i+1$, and the transducer search for sites of a rule r from f. In order to check that a site of r appears in the inputs at the fixed indices, $\mathcal{M}_f$ uses rules of type (III), whereas it does not take concern itself with the other branches of the trees, by using rules of type (II). If it realizes that the checked branches belong to a site from r in L, then the rules of type (IV) and (V) compose a tree $z \in T_\Sigma$, which is in fact the tree obtained in f, from t, u, by the application of the rule r, that is $(t, u) \rightarrow_r z$.

Conversely, from the construction of $\mathcal{M}_f$, for each pair of trees $t, u \in T_\Sigma$ with the same root $\sigma \in \Sigma_n$, for each rule $r = @(s_1, \#, s_2, \$, s_3, \#, s_4)$, and for each $i \in [n]$, there is a rule in R that guesses the checking of the appearance of the sites (s_1, s_2) and (s_3, s_4) at the indices $i, i+1$ at t and u, respectively. The checking is implemented by the rules of type (III).

After this discussion, the reader will have no difficulties giving a formal proof for the next:

Theorem 2 *Let $f = (\Sigma, L)$ be an H tree scheme with $L \in REC$, and $\mathcal{M}_f$ be the corresponding DNA tree transducer. Then, for each forest $F \subseteq T_\Sigma$, it holds $f(F) = \mathcal{M}_f(F)$.*

We now investigate some properties of DNA tree transducers.

Since the application of different rules of an H tree scheme $f = (\Sigma, L)$, to a pair of trees $t, u \in T_\Sigma$, produce in general different trees, we have:

Proposition 4 *DNA tree transducers are not in general confluent.*

For a pair of trees $t, u \in T_\Sigma$, their *height difference* or simply *difference* is the nonegative integer $\mid hg(t) - hg(u) \mid$, where hg denotes the height of a tree.

A DNA tree transducer $\mathcal{M}_f$ has *bounded difference* if there is an integer n such that $\mid \max\{hg(t), hg(u)\} - hg(z) \mid \leq n$, for each $((t, u), z) \in \mid \mathcal{M}_f \mid$. By proposition 13 of [15], it is straightforward that:

Proposition 5 *DNA tree transducers have bounded difference.*

On the other hand, starting from each pair of trees (t, u), and each state $q \in P_0$, obviously there is no infinite sequence $q(t, u) \Rightarrow_{\mathcal{M}_f} t_1 \Rightarrow_{\mathcal{M}_f} \ldots t_k \Rightarrow_{\mathcal{M}_f} \ldots$. Thus:

Proposition 6 *DNA tree transducers are Noetherian.*

5   Conclusion

The model of the DNA tree transducer defined in the previous section can be used for the implementation of the splicing operation on trees, with recognizable sets of rules.

We can use similar models of DNA tree transducers, if the set of rules belongs to the families *OCF*, *GST*, *ALG* or either to the family *NST* of nondeterministic synchronized forests [20]. In these cases, we have to modify the rules of type (III), in order to simulate the operation of the underlying tree automaton. Such models of transducers will be studied in a forthcoming paper.

References

[1] L.M. Adleman, Molecular computation of solutions to combinatorial problems, *Science*, 226 (1994), 1021–1024.

[2] S. Bozapalidis, Alphabetic tree relations, *Theoretical Computer Science*, 99 (1992), 177–211.

[3] S. Bozapalidis and G. Rahonis, On two families of forests, *Acta Informatica*, 31 (1994), 235–260.

[4] S. Bozapalidis and G. Rahonis, H tree schemes with finite and recognizable sets of rules, *Romanian Journal of Information Science and Technology*, 4.1 (1998), 307–318.

[5] F. Gécseg and M. Steinby, *Tree Automata*. Akadémiai Kiadó, Budapest, 1984.

[6] F. Gécseg and M. Steinby, Tree languages. In G. Rozenberg and A. Salomaa (eds.), *Handbook of Formal Languages*. Springer, Berlin, 1997, vol. 3, 1–68.

[7] I. Guessarian, Pushdown tree automata, *Mathematical Systems Theory*, 16 (1983), 237–263.

[8] T. Head, Gh. Păun and D. Pixton, Language theory and molecular genetics: generative mechanisms suggested by DNA recombination. In G. Rozenberg and A. Salomaa (eds.), *Handbook of Formal Languages*. Springer, Berlin, 1997, vol. 2, 295–360.

[9] Gh. Păun, Computing with Membranes, TUCS Report 208, 1998.

[10] Gh. Păun, G. Rozenberg and A. Salomaa, *DNA Computing: New Computing Paradigms*. Springer, Berlin, 1998.

[11] Gh. Păun, G. Rozenberg and A. Salomaa, Membrane Computing with External Output, TUCS Report 218, 1998.

[12] Gh. Păun and T. Yokomori, Membrane computing based on splicing. In E. Winfree and D. Giffoed (eds.), *Preliminary Proceedings of the Fifth International Meeting on DNA Based Computers*. MIT Press, Cambridge, Mass., 1999, 213–227.

[13] Gh. Păun and T. Yokomori, Simulating H systems by P systems, *Journal of Universal Computer Science*, 5 (1999).

[14] Gh. Păun and S. Yu, On synchronization in P systems, *Fundamenta Informaticae*, 38.4 (1999), 397–410.

[15] G. Rahonis, Splicing on trees: the iterated case, *Journal of Universal Computer Science*, 5.9 (1999), 599–609.

[16] G. Rahonis, Alphabetic and synchronized tree transducers, *Theoretical Computer Science*, to appear.

[17] G. Rahonis and K. Salomaa, On the size of stack and synchronization alphabets of tree automata, *Fundamenta Informaticae*, 36 (1998), 57–69.

[18] Y. Sakakibara and C. Ferretti, Splicing on tree-like structures, *Theoretical Computer Science*, 210.2 (1999), 227–243.

[19] A. Salomaa, *Formal Languages*. Academic Press, New York, 1973.

[20] K. Salomaa, Synchronized tree automata, *Theoretical Computer Science*, 127 (1994), 25–51.